Post-processing Techniques for Additive Manufacturing

This text defines and covers different themes of post-processing techniques based on mechanical, chemical/electrochemical, and thermal energy. It will serve as an ideal reference text for senior undergraduate and graduate students in diverse engineering fields including manufacturing, industrial, aerospace, and mechanical.

This book:

- covers the fundamentals and advancements in the post-processing techniques for additive manufacturing;
- explores methods/techniques for post-processing different types of materials used in additive manufacturing processes;
- gives insight into the process selection criteria for post-processing of additive manufactured products made from different types of materials;
- discusses hybrid processes used for post-processing of additive manufacturing parts; and
- highlights post-processing techniques for properties enhancement.

The primary aim of the book is to give the readers a well-informed layout of the different post-processing techniques that range from employing mechanical energy to chemical, electrochemical, and thermal energy to perform the intended task.

Post-processing Techniques for Additive Manufacturing

Edited by
Zafar Alam
Faiz Iqbal
Dilshad Ahmad Khan

CRC Press
Taylor & Francis Group
Boca Raton London New York

CRC Press is an imprint of the
Taylor & Francis Group, an **informa** business

Front cover image: Halyna Khvedchik/Shutterstock

First edition published 2024
by CRC Press
2385 NW Executive Center Drive, Suite 320, Boca Raton, FL 33431

and by CRC Press
4 Park Square, Milton Park, Abingdon, Oxon, OX14 4RN

CRC Press is an imprint of Taylor & Francis Group, LLC

© 2024 selection and editorial matter, Zafar Alam, Faiz Iqbal, and Dilshad Ahmad Khan; individual chapters, the contributors

Library of Congress Cataloging-in-Publication Data
Names: Alam, Zafar, editor. | Iqbal, Faiz, editor. | Ahmad Khan, Dilshad, editor.
Title: Post- processing techniques for additive manufacturing / edited by Zafar Alam, Faiz Iqbal, and Dilshad Ahmad Khan.
Description: First edition. | Boca Raton : CRC Press, [2023] | Includes bibliographical references and index.
Identifiers: LCCN 2023000891 (print) | LCCN 2023000892 (ebook) | ISBN 9781032265100 (hbk) | ISBN 9781032265117 (pbk) | ISBN 9781003288619 (ebk)
Subjects: LCSH: Additive manufacturing. | Finishes and finishing.
Classification: LCC TS183.25 .P67 2023 (print) | LCC TS183.25 (ebook) | DDC 621.9/88--dc23/eng/20230302
LC record available at https://lccn.loc.gov/2023000891
LC ebook record available at https://lccn.loc.gov/2023000892

ISBN: 978-1-032-26510-0 (hbk)
ISBN: 978-1-032-26511-7 (pbk)
ISBN: 978-1-003-28861-9 (ebk)

DOI: 10.1201/9781003288619

Typeset in Sabon
by SPi Technologies India Pvt Ltd (Straive)

Contents

Preface

The book *Post-Processing Techniques for Additive Manufacturing* comprehensively covers traditional and advanced techniques for post-processing of additively manufactured components for a variety of industrial applications. Recent engineered products need high levels of precision and finish due to the requirements related to their enhanced surface characteristics and morphologies in high-tech applications such as aerospace, biomedical, laser, optical, and defence etc. Additive manufacturing has enabled highly complex features being easily manufactured for these applications. One key requirement that remains un-sufficed for these parts is the surface finish of the final manufactured part. Apart from this, additive manufacturing also requires other post-processing before a part can be finally employed in service. This book details a wide range of techniques which can be employed for finishing of such products that may include machine components, scientific instruments, and miniature parts, amongst others. The book elaborates on various technologies invented and employed for post-processing of products produced by additive manufacturing. The aim of this book is to first introduce the readers about the wide variety of additive manufacturing processes that are available for producing parts through this technology and then the primary aim is to describe the advancements in the processes that help achieve the post-processing of these parts. The book will include post-processing techniques ranging from mechanical techniques, vibration assisted techniques, laser-based techniques, and chemical based techniques for additive manufactured parts.

The book is well organized to impart in-depth knowledge to cater to the need of a variety of industries. Containing ten chapters in total, it starts with an introduction to additive manufacturing (Chapter 1), Mechanical post-processing techniques for metal additive manufacturing (Chapter 2 and 5), Vibratory surface finishing of additive components (Chapter 3), Ball burnishing of additively manufactured parts (Chapter 4), Chemical post-processing techniques (Chapter 6), Post processing for wire arc additive manufacturing (Chapter 7), post-processing techniques for improving fatigue strength of metal AM parts (Chapter 8), Ultrasonic and shot peening treatments of additively manufactured parts (Chapter 9), and Post-processing

Techniques of Additively Manufactured Ti-6Al-4V alloy (Chapter 10). The chapters are organized to cover all aspects of the post-processing including basic working principles, physics involved behind the processing, and detail of equipment involved in these followed by limitations where applicable.

Every chapter is supported by related diagrams, figures, and actual pictures, for a better realization. The content of this book is proposed to cater the needs of academicians, engineers, researchers, practitioners, and postgraduates. This book is an equal combined effort of Dr. Zafar Alam, Dr. Faiz Iqbal, and Dr. Dilshad Ahmad Khan. It includes novel and unique contributions from a wide range of authors from across the globe.

We will happily acknowledge the suggestions from the readers of this book to improve the content, and shall attempt to incorporate appropriate suggestions in future editions.

Zafar Alam
(Email: zafar@iitism.ac.in)

Faiz Iqbal
(Email: fiqbal@lincoln.ac.uk)

Dilshad Ahmad Khan
(Email: dilshad@nith.ac.in)

About the editors

Dr Zafar Alam is an Assistant Professor at the Department of Mechanical Engineering, Indian Institute of Technology (Indian School of Mines) Dhanbad, India. Prior to joining IIT (ISM) Dhanbad, he served as an Assistant Professor of Mechanical Engineering at Zakir Husain College of Engineering and Technology, Aligarh Muslim University, India for two years. He received his B.Tech in Mechanical Engineering from Jamia Millia Islamia, India in 2012. He has received his M.Tech and PhD in Production Engineering from Indian Institute of Technology Delhi, India in 2014 and 2019 respectively. As an academician and researcher, he has published an authored book, several book chapters, and more than two dozen research papers in peer reviewed international journals and conferences. He also has to his credit six Indian patents, and has received two international and three national awards including the critically acclaimed GYTI (Gandhian Young Technological Innovation) award for his contribution in the field of research and innovation. His research interests include, but are not limited to, advanced finishing/polishing processes, non-conventional machining, additive manufacturing, industrial automation, and motion control.

Dr Faiz Iqbal received his B.Tech in Mechanical and Automation Engineering from Maharshi Dayanand University, India in 2011. He received his M.Tech in Mechatronics Engineering from Amity University, India in 2013 and PhD in Manufacturing Automation from Indian Institute of Technology Delhi, India in 2019. He is currently working as a Lecturer in the School of Engineering, University of Lincoln, Lincoln, United Kingdom. As an academician and researcher, he has to his credit several research papers in peer reviewed international journals and conferences. He also has to his credit four Indian patents and has received two international and three national awards for his contribution in the field of research and innovation. He secured funding for COVID-19 project which he successfully delivered and was nominated for Scottish Knowledge Exchange award in COVID-19 collaborative response category and was awarded the COVID-19 Engineering medal by the School of Engineering, University of Edinburgh. His research interests include, but are not limited to, Mechatronics; Industrial

Automation; Smart Manufacturing; Fluidic logic/Soft Robotics; Advanced Machining Processes; Non-Conventional Machining; Industry 4.0 – Cyber Physical Systems; Advanced Metrology.

Dr Dilshad Ahmad Khan is an Assistant Professor at the Department of Mechanical Engineering, NIT Hamirpur, India. He is a prominent academician and researcher. He is a Chartered Engineer (CEng) from Institution of Mechanical Engineers, London. He has received his PhD in Manufacturing from IIT Delhi, India in 2018 and Master of Technology (M.Tech) from Aligarh Muslim University, India, in 2010. He received his Bachelor of Engineering (B.E) in Mechanical Engineering from Dr. B.R. Ambedkar University (formerly Agra University), India in 2006. He has published a good number of research articles in international journals and conferences. He got published an authored book entitled as Magnetic Field Assisted Finishing: Methods, Applications, and Process Automation. He is the author of several book chapters on various topics of engineering. He has filed several Indian patents for his innovations. He has received various National and International Awards for his research and innovations including Gandhian Young Technological Innovation Award (GYTI) hosted by President House (Rashtrapati Bhawan), New Delhi. His research interests include advanced finishing/polishing processes, non-conventional machining, additive manufacturing, mechatronic systems, and industrial automation.

Contributors

Ahmad Baharuddin Abdullah
Metal Forming Research Lab, School
of Mechanical Engineering, USM
Engineering Campus,
Nibong Tebal, Penang, Malaysia

Shadab Ahmad
School of Mechanical Engineering,
Shandong University of Technology,
Zibo 255000, China

Zafar Alam
Indian Institute of Technology
(Indian School of Mines)
Dhanbad, Jharkhand, India

Aref Azami
Centre for Precision Manufacturing,
DMEM, University of
Strathclyde, Glasgow G1 1XJ, UK

V. V. Dzhemelinskyi
National Technical University of
Ukraine "Igor Sikorsky Kyiv
Polytechnic Institute",
Kyiv, Ukraine

Severo R. Fernandez-Vidal
School of Engineering, University
of Cadiz, Av. Universidad
de Cádiz,
Puerto Real, Cádiz, Spain

Beyza Gavcar
Yildiz Technical University,
Istanbul, Turkey

Abdul Wahab Hashmi
Malaviya National Institute of
Technology,
Jaipur, India

Mir Irfan Ul Haq
Shri Mata Vaishno Devi
University,
Katra, Jammu and Kashmir, India

Zuhailawati Hussain
USM Engineering Campus,
Nibong Tebal, Penang,
Malaysia

Sridhar Idapalapati
Nanyang Technological
University,
Singapore

Faiz Iqbal
University of Lincoln,
Lincoln, United Kingdom

Pramod Kumar Jain
IIT Roorkee, Uttarakhand, India

Ramon Jerez-Mesa
Department of Material Science and
    Metallurgical Engineering,
    Barcelona East School of
    Engineering (EEBE), Universitat
    Politècnica de Catalunya (UPC),
Barcelona 08019, Spain

Dilshad Ahmad Khan
National Institute of Technology
Hamirpur, Himachal Pradesh, India

Hamaid Mahmood Khan
Fatih Sultan Mehmet Vakif University
Istanbul, Turkey

Ebubekir Koç
Fatih Sultan Mehmet Vakif University
Istanbul, Turkey

Santosh Kumar
IIT (BHU),
Varanasi, India

Maria Elizete Kunkel
Universidade Federal de São Paulo,
Brazil

A. Lamikiz
University of the Basque Country,
Bilbao, Spain
and
CFAA – University of the Basque
    Country
Zamudio, Spain

D.A. Lesyk
National Technical University of
    Ukraine, Igor Sikorsky Kyiv
    Polytechnic Institute
Kyiv, Ukraine
and
G.V. Kurdyumov Institute for Metal
    Physics of the NAS of Ukraine
Kyiv, Ukraine

and
West Pomeranian University of
    Technology
Szczecin, Poland

Jordi Lluma-Fuentes
Department of Material Science and
    Metallurgical Engineering,
    Barcelona East School of
    Engineering (EEBE), Universitat
    Politècnica de Catalunya (UPC),
Barcelona 08019, Spain

Niroj Maharjan
Advanced Remanufacturing and
    Technology Centre, Agency for
    Science Technology and
    Research,
Singapore

Harlal Singh Mali
Malaviya National Institute of
    Technology,
Jaipur, India

S. Martinez
CFAA – University of the Basque
    Country,
Zamudio, Spain

Anoj Meena
Malaviya National Institute of
    Technology, Jaipur, India

Abeer Mithal
Nanyang Technological
    University,
Singapore
and
Advanced Remanufacturing and
    Technology Centre, Agency for
    Science Technology and
    Research
Singapore

B.N. Mordyuk
G.V. Kurdyumov Institute for
  Metal Physics of the NAS of
  Ukraine,
Kyiv, Ukraine

O.O. Pedash
MOTOR SICH JSC,
Zaporizhzhia, Ukraine

Ana Pilar Valerga Puerta
Universidad de Cádiz, Puerto Real,
Cádiz, Spain

Binnur Sagbas
Yildiz Technical University,
Istanbul, Turkey

Pankaj Kumar Singh
IIT (BHU),
Varanasi, U.P, India

Jose Antonio Travieso-Rodríguez
Department of Material Science and
  Metallurgical Engineering,
  Barcelona East School of
  Engineering (EEBE), Universitat
  Politècnica de Catalunya (UPC),
Barcelona 08019, Spain

Arvind Ganesh a/l Vasuthaven
USM Engineering Campus
Nibong Tebal, Penang, Malaysia

Govind Kumar Verma
IIT (BHU),
Varanasi, U.P. India

Zarirah Karrim Wani
USM Engineering Campus,
Nibong Tebal, Penang, Malaysia

Chapter 1

Additive manufacturing and post-processing

An introduction

Faiz Iqbal
University of Lincoln, Lincoln, United Kingdom

Zafar Alam
Indian Institute of Technology (Indian School of Mines), Dhanbad, India

Dilshad Ahmad Khan
National Institute of Technology, Hamirpur, India

CONTENTS

1.1 INTRODUCTION

Additive Manufacturing (AM) technology has also been referred to as Rapid Prototyping (RP) or 3D printing. The first emergence of AM was in late 1980s. It gathered strong interest in the manufacturing industry and has been developing swiftly ever since. Its initial applications were to facilitate design and validation of engineering ideas and concepts. Over time, it has surmounted many challenges, from costs to tooling, and the amenities required for such a concept.

The many benefits of AM include:

- Exploration and validation of ideas.
- Refining the above.
- Platform for continuous improvements.
- Effective communication between designers and validators at idea conceiving stage.
- Optimum material consumption.
- Optimal use of resources.

For many years now, AM has been vital segment of many industrial sectors and is a key role player in Industry 4.0. It is important to note that the widespread use of RP has also sparked the concurrent creation of construction materials and software tools with application-specific functionality.

1.1.1 Working principle

AM involves the creation of a product using a platform without fixtures and predefined slices or layers. The items created by AM technologies follow an "Additive" principle, which is fundamentally different from conventional/subtractive manufacturing methods where the process begins with a block of material and undesired material is gradually eliminated until the desired part is formed.

As a result, the process is referred to as "Additive Manufacturing" and the raw feedstock material has been carefully put in a predetermined configuration. Even while the fundamentals of additive manufacturing are not new, the mechanism by which the material is treated utilizing a digitally defined blueprint itself is. Physical components are created using additive manufacturing techniques from virtual computer-aided models. Additionally, AM begins with nothing, and creates a part layer by layer, depositing each new layer on top of the one before it, until the part is finished. The layer thickness ranges between a few μm up to around 0.25 mm [1]. The uniqueness in the working principle of AM and its manufacturing ability provides various merits and demerits in comparison to the traditional manufacturing processes [2].

In addition to these features, AM enables a computer-aided design (CAD) to be directly transformed to a product by eliminating some of the necessities of subtractive manufacturing processes, such as fixtures and cutting tools. Most importantly, it also enables the production of difficult to manufacture (by traditional machining processes) products and fabrication use of hard-to-machine metals and alloys. In a nutshell, AM helps environmental friendliness, flexibility, and lean production. A need exists for support material that makes it easier to put layers for the overhanging product sections in some AM techniques. These support structures are also laid down layer by layer using the AM techniques and they are usually a different material.

1.1.2 Procedure of AM

The process of additive manufacturing involves several steps where digital data is converted from virtual prototype to solid physical item of use. To achieve the desired quality features of the finished product, however, each stage is crucial. The methodical techniques used by AM technologies, as schematically put in Figure 1.1, are explained below:

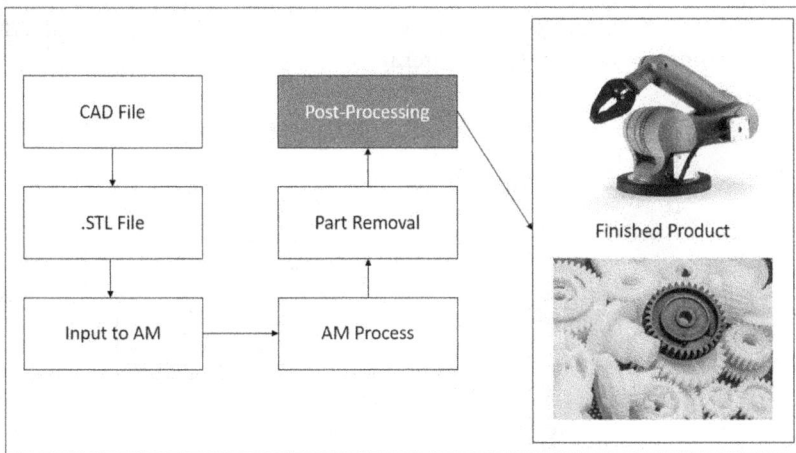

Figure 1.1 Procedural steps of AM.

- **Step 1**: The process begins with an outline of product to be manufactured using AM, typically this outline is a CAD model.
- **Step 2**: A ready-to-use CAD model is converted into the .stl format (preferable by systems controlling AM) which can be obtained directly from the CAD modelling software packages. Solid CAD models are transformed into slices of regular shapes that are connected all over the model. Since stereolithography-based AM systems began using this unique format in the late 1980s, the .stl file is also referred to as stereolithography file format.
- **Step 3**: The .stl file thus obtained from CAD software are transferred to the system controlling AM which further processes it.
- **Step 4**: In the further processing, the tool path is developed. A variety of parameters are controllable by the system as the requirements of the final product which includes the critical parameters viz. layer thickness, speed, and in-fill density. The activity of utmost importance in AM is slicing of the .stl model received from CAD software as per the required thickness of the layer. For instance, a 10 mm thick product sliced by selecting a layer thickness of 0.25 mm will result in 40 slices by the system controlling AM. Figure 1.2 schematically represents slicing phenomena in AM.
- **Step 5**: On the completion of the product, it is detached from the build base and unwanted stuck material is manually removed.
- **Step 6**: The physical model thus additively manufactured is then postprocessed making it suitable as per the end-use application. Many post-processing activities are used, few of which are pressurized air cleaning, tumbling wash, sanding, polishing, and painting.

The final step (6) is a key consideration of this book!

Apart from the above stepwise procedure which mostly apply to all AM systems, some application-specific iterations may also be there.

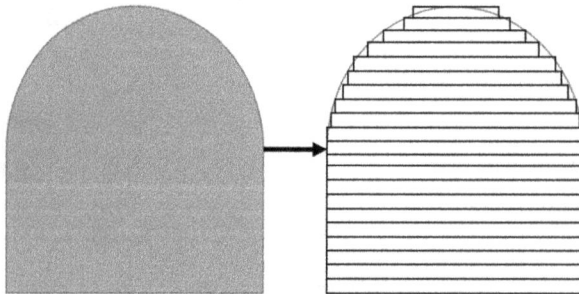

Figure 1.2 CAD model slices.

1.2 AM CATALOGUING AND FEED MATERIALS

The AM technology cataloguing/classification has drastically changed since late 1980s owing to the massive transformation in the industrial setup over the last four decades. AM techniques have been classified based on physical state of the feed materials [8] and other useful ways have been used for such a classification. A whole family tree was given for similar classification [3].

Feedstock materials are based mainly on polymers, ceramics, and/or metals. However, some specialized schemes are there for fabrication of glass, wood, sand, fibres, and other bio-based materials typical for research applications. It can be seen in Figure 1.3 that the most technology numbers in AM have polymer-based feed materials. This is because of they are cost efficient and easy to process. From these, fused deposition modelling (FDM) has primarily been adopted for general prototyping.

The techniques based on metals are there for practical engineering applications such as in aerospace and automotive sectors. Some current applications in the metal AM category are medical implants and tools. The least practiced engineering applications from AM lie in ceramic-based techniques; however, they do have merit and have been considered for development of metal casting moulds and application-specific scaffoldings.

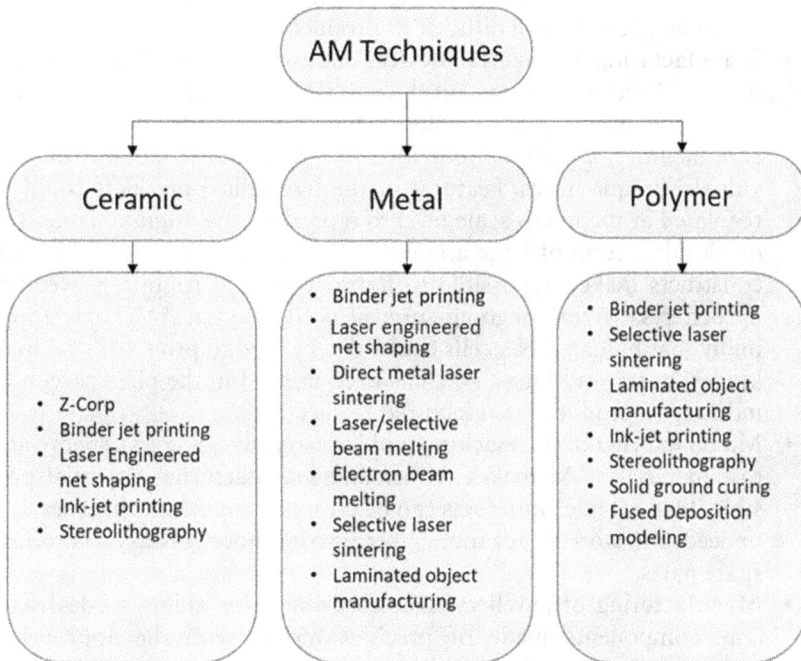

Figure 1.3 Classification of AM techniques.

1.3 APPLICATIONS

The AM technology applications have been increasing rapidly day by day, especially the industrial applications while merging into the vast and cross-disciplinary domains of science and engineering. In today's world, a broad range of AM technologies exist, and the development of these in some protuberant industrial sectors are due to reasons described in [4]:

- **Manufacturing in the automotive sector:** By reducing time and development costs, reverse engineering support from AM enables the industry to produce new products. Automobile manufacturers are embracing additive manufacturing to broaden their selection of parts, including engines and car bodywork. It is increasingly used to create small quantities of structural and functional components from lightweight metals and carbon fibres.
- **Manufacturing in the aerospace sector:** By removing assembly features and enabling the creation of internal functional channels and cooling manifolds, for example, additive manufacturing (AM) technologies produce highly complex and high-performance aerospace goods. Additionally, with additive manufacturing (AM), it is possible to create parts made of complicated geometries in sophisticated materials (viz. titanium, high-temp. ceramics, and nickel). Manufacturing of goods with these materials without AM has proved to be very time-consuming, costly, and difficult to produce.
- **Manufacturing in medical devices industries:** Using CT/MRI scan data, AM technology can produce reliable medical goods with ease. AM is a viable strategy for offering high-quality, cost-effective healthcare facilities, as well as individualized care that is tailored to individuals' unique needs. Features of the biomedical products could be regulated at the micro-scale level to reproduce the original tissue. The quick fabrication of large amounts of samples using tailored medical constructs makes it possible to improve clinical routine procedures by better satisfying ongoing surgical needs. Recent AM applications imply that human DNA cells could also be used to print artificial liver, kidney, heart, and lungs. AM has been utilized in the pharmaceutical industry to produce manufactured medications.
- **Manufacturing in the machine tool industry:** By using less energy and raw materials, AM makes on-demand manufacturing potential possible. True 3D microproducts can be created using AM techniques. It is unneeded to store legacy tooling because it is not necessary to produce spare parts.
- **Manufacturing of jewellery:** Due to its superior ability to deal with core components made of precious metals with the appropriate strength as compared to welding, additive manufacturing (AM) techniques can be used to produce a variety of jeweller pieces using CAD

software. Additionally, it is now possible to create designed jeweller architectures that would otherwise be impossible for even an experienced jeweller to create.

1.4 NEED FOR POST-PROCESSING OF ADDITIVELY MANUFACTURED PARTS

1.4.1 Fused deposition modelling

Fused Deposition Modelling (FDM) is the most popular and flourishing form of Additive Manufacturing owing to its simplicity and flexibility in manufacturing as compared to other AM techniques. It is a material extrusion method of additive manufacturing where printed materials (generally polymers) are fed into and extruded through a heated nozzle, in the form of thin filaments, onto a build platform. This nozzle moves in an X–Y direction while heating the material above its recrystallization temperature, thus joining the individual filaments together. After a layer is fully printed, the build platform moves down by one-layer thickness upon which the next layer is printed, thus, creating 3D objects layer by layer. Critical process parameters include nozzle diameter, extrusion temperature, layer thickness, and build orientation. However, due to limitations in material property and machine accuracy, FDM printed parts suffer certain mechanical disadvantages and shortcomings like poor dimensional accuracy, part strength, porosity, and surface finish.

1.4.1.1 Defects occurring in FDM parts

1. Porosity and density defects: Uncontrolled changes in the temperature and humidity during the fabrication of parts using FDM result in changes in density and porosity of the part build. Due to this, parts become porous and considerably less dense, which poses a problem regarding these parts' strength and functionality.
2. Defects due to temperature inconsistency: A temperature gradient between the bed and the first layer extruded, resulting in weaker adhesion between the bed and this layer. Hence, warpage occurs. Warpage can be eliminated by preheating the bed to a suitable temperature and implementing post-processing techniques. Similarly, inconsistency in temperature occurs between two subsequent layers leading to weak adhesion between these layers. Weak adhesion between layers results in the debonding and delamination of layers.
3. Staircase defect: The staircase effect occurs due to the discretized nature of the layered structures. When there is a substantial difference in horizontal slopes of the layers printed, there happens to be a certain

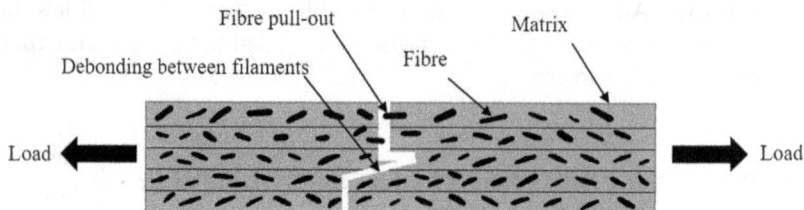

Figure 1.4 Defects occurring in FDM build composites.

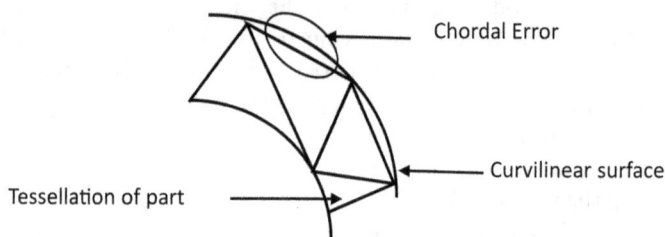

Figure 1.5 Chordal error.

approximated deviation from the actual model to the printed. These results in ladder-like defects, therefore, termed staircase defects.

4. Defects occurring in composites printed using FDM: Due to improper feeding of composite materials and the variation in parameters during manufacturing, a local alignment deviation occurs in placing fibres of composites into its matrix. This localized misalignment of fibres is termed a waviness flaw, which significantly affects the mechanical properties of the composite materials [5]. Also, temperature inconsistency between two filaments results in weaker adhesion; hence, debonding and fibre pull-out occur in FDM-formed composites, as shown in Figure 1.4.

5. Chordal error: Chordal is defined as the difference between the actual tessellation path and the curvilinear path desired, as shown in Figure 1.5.

1.4.2 Powder bed fusion

Powder Bed Fusion (PBF) consists of spreading a thin layer of metallic or polymeric powder particles onto a flatbed, which is flattened using a roller (heated or otherwise). Heat is applied to these powders in the form of a laser or electron beam in a specified scanning area (layer cross-section), which causes partial or complete melting of these powders, and they fuse to form a single layer of the part. Another powder layer is then spread on top of this fused layer, and the process is repeated. The heat source generates enough energy to penetrate further down the top layer to inhibit the fusion of inter-mediate layers, thus forming the entire part layer by layer. The extra powder material is easily removed during post-processing.

Depending on the type of energy source and fusion mechanism, PBF is sub-divided into various types: Direct metal laser sintering (DMLS), Electron beam melting (EBM), and Selective heat sintering (SHS), Selective laser melt-ing (SLM) and Selective laser sintering (SLS).

1.4.2.1 Defects occurring in PBF parts

The probable defects occurring in PBF are discussed below:

1. Process-induced defects: Defects such as material expansion due to thermal stresses during heating, formation of the unstable melt pool, vaporization of material at confined zones and electrostatic repulsion.
2. Power handling defects: This generally happens due to feeding the improper ratio of reusable sintered powder with fresh powder, wrap-ping powdered layer, and forming voids.
3. Distortions, burns, and curling: The main problem for SLM parts is the distortions, burns, and curling due to high-temperature gradients developed due to the high energy density and improper scanning speed of lasers.
4. Extended grain size: There exists a thermal gradient in the building directions in SLM. Owing to this, unidirectional solidification of fused material takes place, resulting in extended grain size and columnar structure [6].
5. Other defects: Several other defects are also probable in PBF due to variations in scanning speeds, thermal gradients, different cooling rates, amongst others. These defects include balling melts, bumps and marks appearing on surfaces, and staircase defects.

1.4.3 Laminated object manufacturing

Laminated object manufacturing (LOM) is an Additive Manufacturing pro-cess that utilizes adhesive bonding, chemicals, or heating to join thin sheets of paper, plastics or metal laminates layer by layer and then cut into the desired shape (the reverse, cutting then joining, is also possible). LOM was initially developed to create paper-based rapid prototypes, using a standard CNC cutter to cut the cross-sectional profile of each layer and stick them together using adhesives. LOM is often considered a hybrid manufactur-ing process as both additive, and subtractive techniques are used. Critical process parameters include cutting tool type and cutting time and roller temperature, speed and indentation.

LOM offers advantages like quick and inexpensive RP, the flexibility of creating solid and hollow structures and ease of post-processing by basic machining. However, there are also disadvantages like difficulty in producing complex internal structures owing to the need for material removal, low dimensional accuracy, and poor surface finish. Another problem with paper-based LOM parts is the low moisture resistance.

1.4.3.1 Defects occurring in LOM manufactured parts

LOM manufactured parts are subjected to de-cubing. Each component layer manufactured using LOM is visible to the naked eye. This layered structure exists distinctly and has weak interlaminar bonding between them. The occurrence of large voids and gaps is expected due to the formation of an air bubble within the layer. Hence, post-processing must bind these layers together and densify the structure. This densification is usually done by compressing the manufactured part between heated flat plates and then compacting it. The excess materials for de-cubing are then removed using a CO_2 laser to give proper shape to the material.

1.4.4 Wire arc additive manufacturing

Wire Arc Additive Manufacturing (WAAM) is an AM process most commonly employed for the manufacturing or repair of metal parts. It belongs under the Direct Energy Deposition (DED) family of AM processes. WAAM was developed as a 3D printing process from a fundamental welding process; Gas Metal Arc Welding (GMAW). GAMW is one of the most popular welding techniques used in the industry, and it uses consumable metal electrodes (the same as base material and shielded from oxidation by using an inert gas, e.g. He or Ag) to create an electric arc with the workpiece; this extreme heat causes the electrode to melt and deposit onto the work surface. WAAM is an integration of GMAW with CNC-based scanning and feed rate control to optimize the deposition rate and layer thickness. Critical process parameters include wire diameter, wire feed rate, current and voltage of electric arc.

WAAM has various advantages that outperform conventional manufacturing processes and other DED techniques like the ability to manufacture large parts, low setup and energy costs, wide availability, and improvement in material properties. The main limitations of WAAM are the residual stresses in printed parts and bad surface finish which requires extensive post-processing.

1.4.4.1 Need for post-processing of WAAM manufactured parts

WAAM is inclined to certain imperfections which should be addressed to get a completed part with wanted material qualities. The most common defects that occur in WAAM manufactured parts are the presence of residual stress and degraded roughness [7]. These residual stresses present in the component are reduced using stress relief treatments such as preheating. This reduces the risk of pre-mature crack generation and consecutively improves the component's life span – the second most crucial step in improving the component's surface roughness. Since the component manufactured are developed layer by layer, these layers are visible on the surface. Hence, post-processing is done using either milling or grinding operation to improve the component's tensile behaviour and make it resistant to corrosion.

1.5 POST-PROCESSING TECHNIQUES

In spite of the many advantages of additive manufacturing (AM), this process exhibits some drawbacks. One of the most important drawbacks of AM technology is to produce parts with inferior surface bulk properties. The parts produced by different AM technologies result in different values of surface roughness. As a result, AM technology alone cannot produce parts that simultaneously fulfil the demands for mechanical characteristics and surface roughness. The parts produced by the AM technologies cannot be used as a final product due to higher surface roughness, low dimensional accuracy, attached support structures, and staircase effects [8, 9]. To improve the surface and bulk properties, the post-processing of additively manufactured parts is essential to make the parts ready to use. A single post-processing technique is not enough to improve the quality of all types of parts and is produced by all types of AM techniques, so different types of post-processing techniques or the combination of more than one technique are used to achieve the desired level of quality of the parts. On the basis of the types of processes used, post-processing techniques can be categorized as follows:

1. Mechanical post-processing techniques.
2. Chemical and electrochemical post-processing techniques.
3. Thermal post-processing techniques.
4. Hybrid post-processing techniques.
5. Post-processing techniques for properties enhancement.

1.5.1 Mechanical post-processing techniques

For post-processing of AM components, both conventional and non-conventional mechanical techniques have been used due to their fast and mature nature, and good accessibility. In the conventional category of mechanical post-processing, the researchers have employed CNC milling [10], ultra-precision machining (UPM) [11, 12], vibration-assisted machining [13], grinding etc. to finish the AM components. In the advanced category of mechanical post-processing, the researchers have employed Magnetic Abrasive Finishing (MAF) [14], micro blasting with high air pressure [15], abrasive flow machining [16, 17], ultrasonic assisted abrasive polishing [18], the hybrid process of grinding and MAF [19] to remove loose and un-melted particles, balling effects and to finish the outer appearance of the component.

1.5.2 Chemical and electrochemical post-processing techniques

The chemical post-processing techniques are featured as affordable and simple to use [20]. Biomedical industry parts are mostly surface cleaned and polished using chemical post-processing techniques. Since chemical solutions can dissolve the plastic parts more, chemical post-processing techniques are typically employed on ABS (acrylonitrile butadiene styrene) parts generated with fused deposition modelling (FDM). Since in chemical post-processing, tools are not in touch with the surface of the part, it can improve dimensional and geometric stability. However, it has been noted that various chemical post-processing techniques decrease the part's tensile strength and cause deterioration in the final geometry of the part [21]. The following techniques of chemical and electrochemical are being used for the post-processing of AM components:

1.5.2.1 Chemical polishing

In chemical polishing, the first stage involves surface preparation methods, employed to clean the surface of the AM part. The part is cleaned with soapy water, isopropyl alcohol, ethanol, or distilled water etc. to remove dust, oil, and loose powder particles that may hinder the chemical process. In the second stage the part is immersed in the bath of chemical solution at a particular temperature for a specific time. The temperature of the bath and the time of immersion of part are decided by the amount of chemical present in the bath. For glossy, smooth, and oxide-free surface, the process parameters should be controlled very precisely. In the final step, the chemical residue is removed from the finished AM part and the part is further dried in the air.

1.5.2.2 Electrochemical polishing

Electrochemical polishing is a dissolution process that is carried out in the electrolytic cell. An electrochemical reaction takes place between an electrode (cathode) and the workpiece produced by AM (anode). The potential difference between cathode and anode is kept as 2 to 20 V DC [22]. The material is removed in the form of micron-sized particles from the AM part. In this process also there is no physical contact between the tool and workpiece which makes this process suitable for processing of complex, fragile, and hardened materials that cannot be processed by mechanical methods of post-processing for AM. The electrolyte and the electrode (cathode) are decided based on the material of AM part.

1.5.2.3 Electroplating

The process of electroplating involves using electrochemical processes to form a metallic layer on the surface of a metallic or non-metallic material. In our daily lives, we use a variety of items, many of which have electrolytic processes applied to their surface. The electroplating gives off some of the desired surface and mechanical qualities that other post-processing techniques are unable to produce. The electroplating increases the parts' Young's modulus, enhancing their durability, reducing corrosion, and allowing for the usage of the parts for aesthetic purposes [23–25]. As a post-processing technique, this technology has also been widely utilized in additive manufacturing. To improve the strength and conductivity of parts, it is typically applied to polymeric material components [26]. Gold, copper, silver, chrome, nickel, zinc, brass, and other metals are frequently utilized as coating materials in this technique.

1.5.2.4 Chemical etching

In the medical field, additive manufacturing techniques have several advantages over conventional ones for creating intricate porous structures that are customized for the patient. The patient's health could be at danger if particles that adhere to the surface of AM-produced parts cause the surface's quality to decline. Chemical etching is the post-processing method to remove these particles loosely held with the AM product and hence to impart better surface quality. In this process, the chemical reaction takes place between the chemical solution and the AM component to be processed for bringing the surface finish to an acceptable level. In chemical etching following chemicals are generally used: diluted and concentrated aqueous hydrochloric acid (HCl), nitric acid (HNO_3), sulfuric acid (H_2SO_4), and hydrofluoric acid (HF), concentrated aqueous solutions such as oxalic acid ((COOH) 2) and aqua regia (HCl + HNO3) chemical solutions [27].

1.5.2.5 Vapour smoothing

In the vapour smoothing process, the hot chemical vapour is utilized to smoothen the surfaces of the components. Some thermoplastics which are generally used in additive manufacturing such as acrylonitrile butadiene styrene (ABS), poly lactic acid (PLA) are polished by acetone, dichloroethylene, tetra hydro fluoride, ethyl acetate etc [28]. The process parameters in the vapor smoothing process are concentration of solution, the interaction of concentration of solution and temperature, workpiece time of exposure to chemical vapor and initial roughness of the workpiece. Due to the absorption of chemical vapor in the workpiece, the weight of the workpiece after the polishing increases.

1.5.2.6 Acetone dipping

Acetone is a quick and affordable procedure that can greatly enhance the surface quality of ABS products fabricated by FDM process. The surface quality of the parts is improved by immersing parts for a predetermined amount of time in a bath of acetone solution (90% dimethyl ketone + 10% water). It has been reported that immersion of ABS part in acetone can reduce the surface roughness by 90% [29]. The application period and acetone content of the solution are crucial factors in this post-processing technique.

1.5.3 Thermal post-processing techniques

Thermal post-processing can dramatically minimize residual stresses, cracking, and microstructure heterogeneity in the AM parts. Thermal post-processing techniques include solution heat treatment (SHT), hot isostatic pressing (HIP), and T6 heat treatment (T6 HT) for AlSi10Mg which can significantly improve the quality of the part fabricated. In HIP the manufactured component is pressed evenly from all directions at a high temperature. This results in greater density, better uniformity, and outstanding performance. This thermomechanical technique is utilized more often in for the post-processing of AM components. In this technique, the temperature reaches as high as 1000–2000°C. working pressure can approach as high as 200 MPa through an inert gas filled in a container. The HIP can treat or eliminate the pores and intrinsic flaws in parts made by the powder bed fusion (PBF) method [30, 31]. Other thermal post-processing techniques are laser-based techniques such as laser shock peening, laser polishing etc.

1.5.4 Hybrid post-processing techniques

In hybridization, two separate tools, methods, or energy systems are combined in a single system. In most of the studies on AM, the CNC milling and AM have been combined to get the benefit of both processes. In this hybrid process, the requirement of material deposition and removal are taken care of in a single station and called hybrid rapid prototyping (RP) or ECLIPSE-RP [32]. Based on the secondary processes used with AM, the hybrid AM processes can be categorised as:

1. **HAM by machining:** In this process, machining techniques like milling or turning are coupled with AM process to bring the desired product. Near-net-shape items are produced using AM, whereas machining at successive layer intervals improves surface finish and geometrical precision. Some of the additive technologies that have been employed in combination with machining include sheet lamination, plasma deposition, laser deposition, laser cladding, and laser melting.

2. **HAM by rolling and burnishing:** The two significant issues that were present during AM are resolved by rolling HAM (RHAM). First, rolling can be used to address dimensional irregularities caused by overlapping grains or layers in metal additive manufacturing. Second, residual stress created during AM causes the finished object to deform or stretch. RHAM achieves geometric accuracy and stress reduction without causing any material loss. The process of burnishing involves compacting the metal's surface by rubbing it with a small, hard tool. It is a method of surface treatment comparable to rolling that reduces residual stress and increases surface roughness. Additionally, it aids in enhancing hardness and microstructure.

3. **Friction stir AM:** Friction stir AM is a sequential HAM technique. Friction stir welding (FSW) is used to permanently join two surfaces. A non-consumable spinning tool made of refractory material is introduced into the workpiece. As a result of the pressing of the tool, heat is produced and a large quantity of permanent deformation occurs. In most cases, this causes recrystallization and substantial grain refinement in metals. These metallurgical developments have accelerated fatigue and mechanical performance. Polymer and composite materials may also be processed in a similar manner.

4. **HAM by remelting:** This procedure involves employing an energy source to remelt previously fused material. A laser or an electron beam can serve as the source. No material is taken out; rather, it is changed. Remelting raises previously set material to the melting point to better bond layers together. During 3D printing, pores are created, and the melted material fills them to produce components with a density of over 99%. After a layer or a few of layers, remelting may be performed. A specific material's remelted region experiences changes in its mechanical, physical, and chemical properties. These adjustments depend on a number of factors, including scanning pattern, power, and speed of the laser.

5. **HAM by Ablation or Erosion:** In this procedure, the top layer of accumulated material is removed or ablated using a laser or an electron beam. An integrated non-contact method called ablation between construction layers improves the product's performance and quality. By removing material, this subtractive technique, which can be compared to HAM by machining, can produce smooth and precise surfaces.

6. **HAM by Laser-assisted Plasma deposition:** The laser simultaneously applies an additional energy source to the same spot as the plasma deposits the deposited material, accelerating the building process and improving the printing quality. The constructing process is not exactly aided by the LAPD laser in the same way that laser-assisted turning and cutting are. The laser immediately aids the cutting process in laser-assisted turning by extending tool life or reducing cutting pressure through work softening.

7. **HAM by peening:** Laser shock peening, ultrasonic impact treatment, and shot peening are used to harden 3D printed components such as turbine blades. In laser shock peening, shock waves from intensifying plasma are applied to a workpiece to permanently deform it. Plasma is produced when a workpiece strikes a pulsed laser with a nanosecond pulse duration. Laser shot peening is a sequential process, just like other HAM operations. An electromechanical transducer is used in ultrasonic peening to deploy ultrasonic energy to a workpiece in a multiple or layer-by-layer fashion. An ultrasonic impact treatment, which is a surface treatment method, can improve microstructural grain, lower tension, and reduce residual stress in compression. Ultrasonic peening can enhance the fatigue, corrosion, and tribological performance of AM components. Shot peening is another process that is utilized for surface treatment. The mechanical characteristics of a near-surface layer are improved by directing a stochastic stream of beads at very high speeds under controlled coverage conditions. As the bead collides with the surface, plastic deformation results. Due to which residual compression and strain hardening take place. Glass, metal, or ceramic can all be used to make beads.

8. **Pulsed laser deposition:** Pulsed laser technology is being used more frequently in additive manufacturing. In order to print thin layers of components onto a substrate, these powerful lasers are utilized. A pulsed laser has impinged on the powder throughout this operation. Due to the fast heating, this makes the powder vaporize. Plasma plumes are produced as a result of vaporization. During printing, this plume creates a shock wave that permanently deforms the surface. Both pulsed laser deposition (PLD) and laser shot peening operate via essentially the same mechanism. The sole distinction is that PLD is paired with both peening and printing simultaneously in a single laser source. PLD facilitates desirable compressive residual stresses.

1.5.5 Post-processing techniques for properties enhancement

Internal flaws including balling, porosity, fractures, powder aggregation, and thermal stress would manifest themselves between various printing layers during the manufacturing operations in AM. The interior microstructure and mechanics of the finished products are significantly impacted by these flaws. To improve the quality of the product for its intended functionality and durability, it is necessary to eliminate these flaws, either at the time of manufacturing or at the end of manufacturing. The various post-processing techniques, which were covered earlier in their respective sections, are used to improve bulk properties, such as increasing component density, reducing porosity, increasing hardness, reducing thermal fractures, releasing residual stresses, etc.

REFERENCES

[1] Gibson, I., Rosen, D.W., & Stucker, B. (2014). *Additive Manufacturing Technologies*. New York: Springer.

[2] Huang, S.H., Liu, P., Mokasdar, A., & Hou, L. (2013). Additive manufacturing and its societal impact: A literature review. *The International Journal of Advanced Manufacturing Technology*, 67(5–8), 1191–1203.

[3] Koukka, H. The RP family tree, Helsinki University of Technology, Lahti Centre. URL: http://shatura.laser.ru/rapid/rptree/rptree.html

[4] Campbell, R.I., De Beer, D.J., & Pei, E. (2011). Additive manufacturing in South Africa: Building on the foundations. *Rapid Prototyping Journal*, 8(17), 156–162.

[5] Senthil, K., Arockiarajan, A., Palaninathan, R., Santhosh, B., & Usha, K.M. (2013). Defects in composite structures: Its effects and prediction methods – A comprehensive review. *Composite Structures*, 106, 139–149.

[6] Vrancken, B., Thijs, L., Kruth, J.P., & Humbeeck, J.V. (2012). Heat treatment of Ti6Al4V produced by selective laser melting: Microstructure and mechanical properties. *Journal of Alloys and Compounds*, 541, 177–185.

[7] Kawalkar, R., Kumar Dubey, H., & Lokhande, S.P. (2021). Wire arc additive manufacturing: A brief review on advancements in addressing industrial challenges incurred with processing metallic alloys. *Materials Today: Proceedings*, 50, 1971–1978.

[8] Tan, H., Fang, Y., Zhong, C., Yuan, Z., Fan, W., Li, Z., Chen, J., & Lin, X. (2020). Investigation of heating behavior of laser beam on powder stream in directed energy deposition. *Surface and Coatings Technology*, 397, 126061.

[9] Pyka, G., Kerckhofs, G., Papantoniou, I., Speirs, M., Schrooten, J., & Wevers, M. (2013). Surface roughness and morphology customization of additive manufactured open porous Ti6Al4V structures. *Materials*, 6(10), 4737–4757.

[10] Bai, Y., Chaudhari, A., & Wang, H. (2020). Investigation on the microstructure and machinability of ASTM A131 steel manufactured by directed energy deposition. *Journal of Materials Processing Technology*, 276, 116410.

[11] Ni, C., Zhu, L., Zheng, Z., Zhang, J., Yang, Y., Yang, J., Bai, Y., Weng, C., Lu, W.F., & Wang, H. (2020). Effect of material anisotropy on ultra-precision machining of Ti-6Al-4V alloy fabricated by selective laser melting. *Journal of Alloys and Compounds*, 848, 156457.

[12] Ni, C., Zhu, L., Zheng, Z., Zhang, J., Yang, Y., Hong, R., Bai, Y., Lu, W.F., & Wang, H. (2020). Effects of machining surface and laser beam scanning strategy on machinability of selective laser melted Ti6Al4V alloy in milling. *Materials & Design*, 194, 108880.

[13] Bai, Y., Shi, Z., Lee, Y. J., & Wang, H. (2020). Optical surface generation on additively manufactured AlSiMg0. 75 alloys with ultrasonic vibration-assisted machining. *Journal of Materials Processing Technology*, 280, 116597.

[14] Zhang, J., Chaudhari, A., & Wang, H. (2019). Surface quality and material removal in magnetic abrasive finishing of selective laser melted 316L stainless steel. *Journal of manufacturing processes*, 45, 710–719.

[15] Zhang, J., & Wang, H. (2019). Micro-blasting of 316L tubular lattice manufactured by laser powder bed fusion. In *Proceedings of the 19th International Conference of the European Society for Precision Engineering and Nanotechnology EUSPEN*.

[16] Guo, J., Song, C., Fu, Y., Au, K. H., Kum, C. W., Goh, M. H., Ren, T., Huang, R., & Sun, C. N. (2020). Internal surface quality enhancement of selective laser melted inconel 718 by abrasive flow machining. *Journal of Manufacturing Science and Engineering*, 142(10), 101003.

[17] Han, S., Salvatore, F., Rech, J., Bajolet, J., & Courbon, J. (2020). Effect of abrasive flow machining (AFM) finish of selective laser melting (SLM) internal channels on fatigue performance. *Journal of Manufacturing Processes*, 59, 248–257.

[18] Wang, J., Zhu, J., & Liew, P. J. (2019). Material removal in ultrasonic abrasive polishing of additive manufactured components. *Applied Sciences*, 9(24), 5359.

[19] Teng, X., Zhang, G., Zhao, Y., Cui, Y., Li, L., & Jiang, L. (2019). Study on magnetic abrasive finishing of AlSi10Mg alloy prepared by selective laser melting. *The International Journal of Advanced Manufacturing Technology*, 105(5), 2513–2521.

[20] Galantucci, L. M., Lavecchia, F., & Percoco, G. (2009). Experimental study aiming to enhance the surface finish of fused deposition modeled parts. *CIRP Annals*, 58(1), 189–192.

[21] Havenga, S., De Beer, D. J., Van Tonder, P. J. M., & Campbell, R. I. (2015, November). Effectiveness of acetone postproduction finishing on entry level FDM printed ABS artefacts. In *RAPDASA Conference Proceedings 2015* (pp. 4–6).

[22] Ramasawmy, H., & Blunt, L. (2007). Investigation of the effect of electrochemical polishing on EDM surfaces. *The International Journal of Advanced Manufacturing Technology*, 31(11), 1135–1147.

[23] Kannan, S., & Senthilkumaran, D. (2014). Investigating the influence of electroplating layer thickness on the tensile strength for fused deposition processed ABS thermoplastics. *International Journal of Engineering and Technology*, 6(2), 1047–1052.

[24] Saleh, N., Hopkinson, N., Hague, R. J., & Wise, S. (2004). Effects of electroplating on the mechanical properties of stereolithography and laser sintered parts. *Rapid Prototyping Journal*, 10(5):305–315.

[25] Kumar, T. N., Kulkarni, M., Ravuri, M., Elangovan, K., & Kannan, S. (2015). Effects of electroplating on the mechanical properties of FDM-PLA parts. *i-Manager's Journal on Future Engineering and Technology*, 10(3), 29.

[26] Nikzad, M., Masood, S. H., & Sbarski, I. (2011). Thermo-mechanical properties of a highly filled polymeric composites for fused deposition modeling. *Materials & Design*, 32(6), 3448–3456.

[27] Pyka, G., Burakowski, A., Kerckhofs, G., Moesen, M., Van Bael, S., Schrooten, J., & Wevers, M. (2012). Surface modification of Ti6Al4V open porous structures produced by additive manufacturing. *Advanced Engineering Materials*, 14(6), 363–370.

[28] Rajan, A. J., Sugavaneswaran, M., Prashanthi, B., Deshmukh, S., & Jose, S. (2020). Influence of vapour smoothing process parameters on fused deposition modelling parts surface roughness at different build orientation. *Materials Today: Proceedings*, 22, 2772–2778.

[29] Colpani, A., Fiorentino, A., & Ceretti, E. (2019). Characterization of chemical surface finishing with cold acetone vapours on ABS parts fabricated by FDM. *Production Engineering*, 13(3), 437–447.

[30] Tillmann, W., Schaak, C., Nellesen, J., Schaper, M., Aydinöz, M. U., & Hoyer, K. P. (2017). Hot isostatic pressing of IN718 components manufactured by selective laser melting. *Additive Manufacturing*, 13, 93–102.

[31] Rosenthal, I., Tiferet, E., Ganor, M., & Stern, A. (2015). Post-processing of AM-SLM AlSi10Mg specimens: Mechanical properties and fracture behaviour. Annals of "Dunarea de Jos" University of Galati. *Fascicle XII, Welding Equipment and Technology*, 26, 33–38.

[32] Hur, J., Lee, K., & Kim, J. (2002). Hybrid rapid prototyping system using machining and deposition. *Computer-Aided Design*, 34(10), 741–754.

Mechanical post-processing techniques for metal additive manufacturing

Beyza Gavcar and Binnur Sagbas
Yildiz Technical University, Istanbul, Turkey

CONTENTS

2.1 INTRODUCTION

Parts with complex structures and new functional characteristics are increasingly manufactured from 3D CAD model data by different sectors utilizing metal additive manufacturing (AM) processes due to the challenge of manufacturing by utilizing subtractive manufacturing (SM) methods, like CNC machining, casting, cutting, forming, grinding, and moulding. The metal powders are employed as materials' sequent layers till a final part is obtained. AM can provide material savings by reducing material loss and design flexibility in comparison with SM. The mechanical performance, material behaviour, porosity formation, and surface quality are affected by the small process variations (method type, process parameters) and size of the metal powder in metal AM, especially in aerospace and medical applications. But, additively manufactured products' surface quality is low. Products obtained through a powder bed-based process have rough surfaces that is resulting from powder particles adhering to the contour of the molten surface during the manufacturing process. Thus, non-reproducible and

DOI: 10.1201/9781003288619-2

inhomogeneous morphologies are obtained. Due to the roughness, crack initiation and extra notch effects can occur, and fatigue strength can decrease. As a result, the surface roughness should be reduced through an appropriate surface finishing method. Thus, post-processing is necessary to increase surface quality [1–10].

Different mechanical surface treatments have been applied such as abrasive flow machining (AFM), vibratory finishing (VF), shot peening (SP), and micromachining for realizing the better performance of the additively manufactured components. Abrasive flow machining (AFM) method is usually employed for finishing the complex parts that can be challenging for finishing by the conventional surface treatments, for example, grinding, honing, and lapping. Another technique is vibratory finishing that is an industrial process for cleaning, deburring, and providing isotropic finishes on metal components with little manual labour intervention. Since a tool with fixed geometry is not needed, vibratory finishing is widely applied in the finishing of complex freeform parts like orthopaedic implants, propellors, and turbine blinks. Ultrasonic abrasion finishing process is limited to flat surfaces. Surface roughness can be decreased by hot cutter machining (HCM) process. Magnetic field-assisted finishing (MAF) permits the treatment of not only complex external structures but also intrinsic properties. In addition, MAF is suitable for reducing the surface roughness of AM parts. This process can finish various materials, including ceramics, steels, and titanium alloys. In this chapter, various mechanical post-processes used to improve the surface quality of additively manufactured parts are reviewed.

2.2 MECHANICAL TECHNIQUES FOR POST-PROCESSING OF METAL AM PARTS

In despite of various advantages of AM processes, these techniques have some disadvantages on parts including anisotropy, cracks, distortion, mechanical properties at undesired level, poor surface quality, porosity, and residual stress. There are various defects in metal AM parts, which do not generally exist in conventionally manufactured metal parts, and are unsuitable for most applications. There is a requirement to apply post-processes to additively manufactured metal parts. Mechanical techniques are a subcategory of post-processes and defined as the processes in that the surface plastic deformation or mechanical properties and quality improvement are carried out through abrasives, shots, or tools. These finishing processes improve the surface characteristics, quality and mechanical properties of additively manufactured components. In this chapter, abrasive flow machining, vibratory finishing, shot peening, blasting, ultrasonic abrasion finishing, hot cutter machining, micromachining, and magnetic field-assisted finishing methods are clarified.

2.2.1 Abrasive flow machining

AFM is a flexible, modern, and non-conventional post-processing method that can be applied to the surfaces of complex structures. The advantages of AFM technique are high efficiency of polishing, low cost of manufacturing, operation suitability, and processing of complex shapes. AFM process provides machine unreachable fields in parts, which are very tough for finishing with traditional methods [11]. It can be employed to deburr, radius, polish, surface refinement and remove the complex shapes, inaccessible areas, and recasting layers for hard parts [12]. The surface smoothness can be increased up to 10 times by AFM [13]. AFM finishes various materials covering ferrous, non-ferrous, hard, soft, metallic, and non-metallic materials. In this process, fluids having different viscosities depending upon the components' sizes are employed for carrying abrasive particles. For finishing small parts and forming radius edges, fluids of low viscosity are used; otherwise, viscoelastic polymer fluids and high viscosity medium are required to polish stabile. In Figure 2.1, AFM process's principle consists of squeezing out an AFM media back and forth along the workpiece passageway, which includes different abrasives, polymers, processing oil, and additives, through a channel is shown [14]. Polymer-based abrasive media flows by way of the channel in the workpiece as well as the surface of the channel is scratched with abrasives by the pressure during the process. Different components like an implant, a gear, and a turbine blade can exist in the fixture of the workpiece and be processed by AFM medium [15, 16]. The process occurs in the forms of surface cutting, deburring as well as polishing because of the interaction of AFM medium (abrasive suspension) and the part's surface [17]. Abrasive medium has an important effect in AFM as it enables precise polishing of the chosen surfaces through the flow passage and can be distorted to any shape by applying high extrusion pressure. Alumina (Al_2O_3), diamond, silicon carbide (SiC), cubic boron nitride (CBN), and boron carbide are widely utilized as abrasives [18, 19]. To bind abrasives, media is preheated since the polymer's molecular chains require to be extended. This heating step also allows the elimination of air bubbles in suspension. As a result of the interaction between the surface and media, the surface finish is improved [20–22]. Due to the geometric challenges, it is more difficult that polish additively manufactured metal parts with conventional processes such as belt grinding, manual polishing, and sandblasting [23–25]. In AFM's polishing process, abrasive medium performs like the continuously deforming cutting tool including abrasives, polymer, et al. Micro-chips, so material removals occur by mechanical effect because of free rubbing, micro-cutting, and micro-ploughing [26, 27]. In brittle materials, micro-cutting performs, but in ductile materials, micro-ploughing and micro-cutting take place [28].

AFM process is affected by different parameters that are the process cycle number, piston speed and pressure, media flow rate, preparation method, temperature, and composition, which includes the abrasives' size, kind and

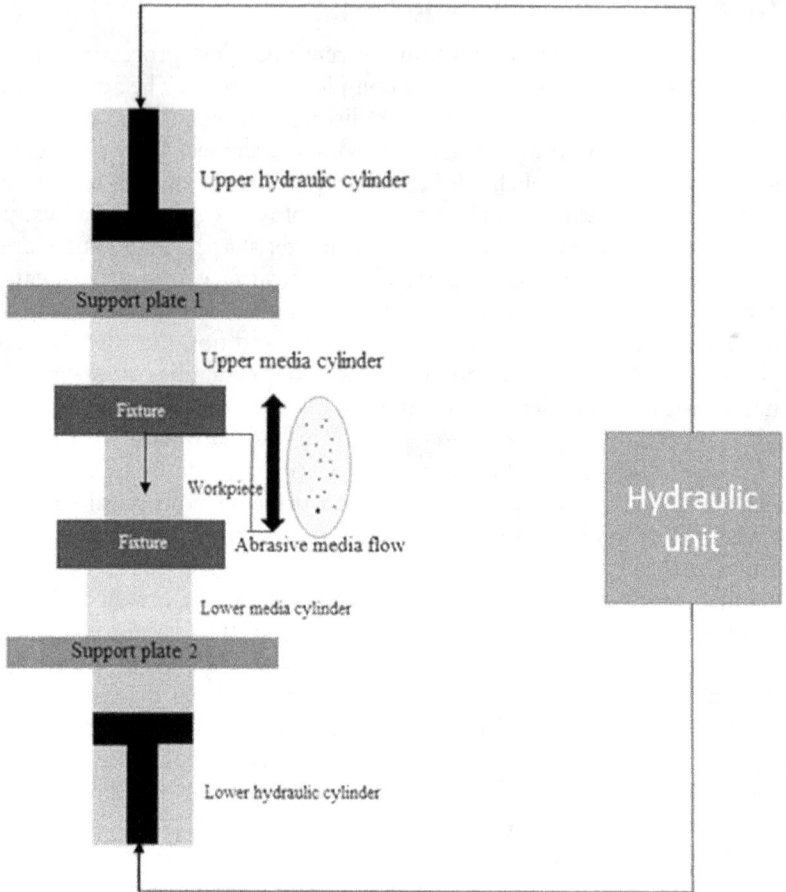

Figure 2.1 Schematic diagram of AFM process. (Reproduced with reference [29] which is Under Creative Commons CC BY 4.0 licence.)

amount, and carrier's texture. The temperature should have a constant value in the range of 25° C and 40° C. The viscosity depends on the temperature.

The staircase effect is intrinsic with additively manufactured parts especially obtained by Stereolithography (SLA), and Fused Deposition Modelling (FDM) creates a more uneven surface due to varying layer thickness [30]. AFM allows improvement of the surface quality and elimination of this effect from additively manufactured parts [31]. Ball and powder adhesion are some shortcomings that exist in Selective Laser Melting (SLM) built parts and can also be eliminated by applying AFM finishing [32, 33]. Tensile residual stresses are found in as-built parts and are also relieved by AFM by producing compressive residual stresses. Han et al. presented a new approach to determine SLM-manufactured maraging steel 300 channel's surface integrity by AFM post-treatment as well as this integrity's impact on the fatigue strength achieved by a rotating–bending fatigue test. Because of the result of abrasive particles on the roughness peaks, the roughness was reduced by AFM [34]. Uhlmann et al. used AFM process on SLM-built parts with very rough surfaces. Computational Fluid Dynamics (CFD) simulations are performed for the turbine blade's fixture and smooth reference parts to examine material removal mechanisms [35]. Peng et al. studied the effectiveness of AFM process for AM parts' finish and surface integrity. Surface quality is improved after AFM process by reducing roughness from the initial Sa 13~14 μm to the final 1.8 μm and removing surface defects because of the inherent effect of balling, and adhesion of powders. It was verified that AFM provides an enhancement in surface integrity and efficient finishing of AM components with inner surfaces of high complexity [36]. Guo et al. examined the effect and feasibility of AFM on SLM Inconel 718 parts' internal surface quality increase. They obtained good final surface roughness at the ambient factors of low temperature and extrusion pressure, large abrasive particle size, and high medium viscosity [37]. Hashmi et al. employed AFM on FDM manufactured hollow truncated cone surface to examine experimentally the internal finishing. They found that finishing time and medium viscosity are the most effective process parameters. The maximum improvement was achieved at the process parameters of AFM of medium viscosity with 2.10 Pa-sec, time of 30 minutes, and part's layer thickness with 0.2 mm. The surface roughness was decreased from 20.93 mm to 1.20 mm by an increase of 94.26% [38].

2.2.2 Vibratory finishing

VF is a mass-finishing method utilized to burnish, clean, deburr, harden and polish the parts' surfaces as well as achieve isotropic finishes on metal parts with low intervention [39, 40]. As VF does not need a tool with fixed geometry, it is widely used to finish complex parts like turbine blisks, propellers, and implants [41]. VF machine is presented in Figure 2.2.

This process occurs with a vibrating granular medium's movement. The medium includes cutting abrasives like blunt grains and pyramids with sharp edges as well as hardens and burnishes the surfaces (Figure 2.3). The granular medium becomes fluid with the vibration, and workpieces are dragged around the container with the bulk flow. In the abrasive medium's vibrating bowl, workpieces are eroded by blending.

The most significant VF parameters affecting the surface quality are machine design and dimension, medium amount, composition, form, type and velocity, the volume ratio between component and medium, abrasive amount, composition and type, process time, and water volume [42]. Surface quality describes the surface smoothness that includes defect, lay, roughness, and waviness.

Figure 2.2 VF machine.

Figure 2.3 Different abrasive mediums of vibratory polishing (Left – polymer; right ceramic).

VF is more suitable for small workpieces with a high amount due to drag and centrifugal finishing requiring a workpiece fixture [43]. It has a high process control and material removal rate. This process is the commercially existing technique applied relatively easily. Cemented carbide products like cutting tools and wear parts are usually in high amounts and need a high surface finishing. Thus, VF is a suitable option for the secondary treatment of these parts. Its applicability, low cost, and simpleness make it an appropriate treatment for additively manufactured components. Vibratory polished additively manufactured AlSi10Mg part image is illustrated in Figure 2.4.

Seltzman et al. applied VF to SLM-built Glen Research Copper (GRCop-84) to eliminate adhered powder from this material and improve the surface quality [44, 45]. VF resulted in the roughness of Ra = 0.45 µm and provided a smooth surface by removing peaks. In another study, Bernhardt et al. used VF to laser powder bed fusion (LPBF) manufactured Ti-6Al-4V ELI parts to

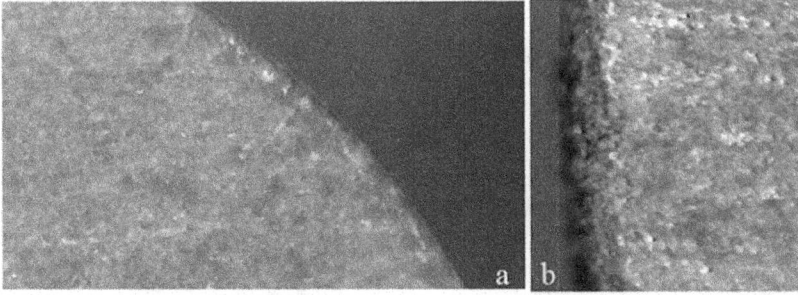

Figure 2.4 Microscopic image of vibratory polished (a) and as machined (b) AlSi10Mg AM part surfaces.

determine surface roughness. VF has a relatively long processing time. The obtainable surface roughness for VF strongly depends upon the initial surface roughness [46]. Uhlmann et al. compared the fatigue life of SLM-printed stainless steel 316L samples applied to VF and not VF to present this post-process's effectiveness. VF applied specimens presented a large scatter in the fatigue life and relatively smooth surface [47]. Kaynak et al. utilized VF and drag finishing (DF) to SLM-built 316L stainless steel specimens to improve their surface. They reported the effectiveness of DF processes on decreasing surface roughness, providing a more consistent and smooth surface finish in comparison with VF [48]. Ginestra et al. used VF to SLM-produced Ti6Al4V implants to change the surface topography and morphology. VF caused hydrophilic surfaces with low contact angle (CA) [49]. Chastand et al. applied VF to SLM-printed Ti6Al4V specimens to polish, so it was desired that a surface roughness of Ra ≈ 0.2 μm [6].

2.2.3 Shot peening

Shot peening (SP) is flexible and effective post-processing and cold working process used to enhance the fatigue strength of the parts and depends on the bombardment of a metal surface with a random flow and a large number of hard small shots that are accelerated by compressed air to achieve specific velocity (generally 50–100 m/s) [50]. These shots typically have spherical forms and dimensions in the range of a few tenths of a millimetre. Ceramic, glass, and steel are often used as materials for shots. Bombardment of shots onto the surface is repeated until the whole surface is treated uniformly and leads to severe plastic deformation bringing about the surface morphology's improvement. Each shot behaves like a small peening hammer inducing the surface grains' recrystallization and hardening of the surface [51]. This process gives rise to amorphization of the surface that prevents crack propagation through preferred crystallographic directions and placement of the peening media transforming subsurface inclusions. It decreases surface defects responsible for crack initiation [52].

During this treatment process, balls bombard the target part, and the surface layer is thus exposed to cold working that causes compressive stresses with a thickness of a few tenths of a millimetre at the surface depending on the application. Compressive stresses occur because of plastic and elastic strain's inhomogeneous distribution stemming from contact pressure between the shot and target as well as lateral plastic flow [53–55]. These stresses enhance the fatigue life of metal parts [56, 57]. Also, there have been many cases where parts' service life is increased by applying SP. This surface treatment is beneficial for improving the fatigue strength when there is a stress raiser like artificial defects, turbine blades' fir tree root, the root of gear tooth, and steep stress gradients resulting from contacts of fretting fatigue [58–62]. SP has various applications in the engineering sector like gear manufacturing and turbomachinery. This process can provide improvements in corrosion resistance and microhardness, grain refinement as well as cause residual stress [63–68]. The main restriction for SP is the sightline need that restrains the part's design freedom. Hence, a suitable surface treatment is essential for the finishing of the component's critical areas accurately [69].

There are various factors to control SP including air amount and pressure, Almen intensity, peening angle, speed, and time, shot material and size, and surface coverage that are significant in obtaining effective surface properties [70–82]. To obtain the peening effect accurately and prevent the shot deformation, shot material should be harder than the surface. Also, the peening angle is preferred as 90° [83]. For the additively manufactured components, the surface quality is improved by reducing the surface roughness during SP.

Salvati et al. studied the effect of SP on IN718 Ni-base superalloy parts obtained by Laser Metal Deposition (LMD). SP treatment produced additional tensile residual stress in the bulk material, thus failure initiation can occur far from the free surface [84]. Hadidi et al. investigated the effect of SP on the low-velocity impact features of P430 acrylonitrile butadiene styrene (ABS) samples built by FFF. This study demonstrated that additively manufactured polymer specimens' mechanical behaviour is highly affected by the frequency of layer peening in hybrid-AM. Better impact result was obtained at peening with less frequency [85]. Uzan et al. studied the effect of SP on SLM-built AlSi10Mg samples' fatigue resistance. SP applied surfaces' optimal electrochemical and mechanical polishing removed 25–30 µm from the surface-induced to important increases in fatigue resistance, particularly for a fatigue regime with high cycle. The fatigue crack initiation's depth for shot-peened parts was deeper than that for AM parts before SP. The fracture area of SLM-built AlSi10Mg parts before and after SP had a ductile fracture with relatively deep dimples [78]. Khajehmirza et al. developed a detailed numerical model for estimating LPBF manufactured AlSi10Mg parts' surface morphology after SP. SP applied parts' areal roughness parameters obtained from the provided numerical model exhibited good accordance with those of the experiments. The application of SP lead to an overall

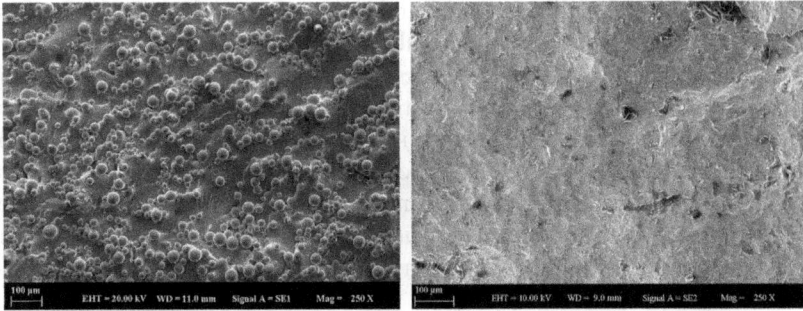

Figure 2.5 **LPBF manufactured metal surfaces without any post-processes and after shot peening.**

decrease of 40–50% in these parameters for LPBF manufactured parts [86]. An example of metal surfaces is shown in Figure 2.5. AlMangour et al. presented an approach resulting in grain refinement by SP process for Direct Metal Laser Sintering (DMLS) 17–4 stainless steel specimens to enhance their mechanical and physical characteristics. DMLS was successfully applied to manufacture these parts. SP improves the compressive yield strength, hardness, roughness, as well as wear resistance of the samples. These improvements can depend on the strong fine surface layer (i.e. grain refinement) and work-hardening arising from severe plastic deformation (SPD) [87]. Salvati et al. developed a method to investigate LMD produced nickel superalloy compressor blade's residual strain. This study concluded that the residual elastic strain simulation resulting from SP can be carried out for any 2D model [88]. Slawik et al. presented a microstructural analysis of SLM-printed Ti6Al4V treated by both SP and laser peening (laser shock peening, LSP). LSP and SP deeply affect the macro stress state of Ti6Al4V, whereas LSP exhibits much deeper stresses (up to 2.3 mm) in comparison with SP (~443 μm) which can depend on the deformation properties of the shock wave. The micro and macro stress depth were consistent. Measurements for surface roughness presented just a little roughness change of the surface treated by LSP in comparison with the reference, whereas the roughness was increased by SP strongly. The highest improvement in fatigue properties was obtained in the LSP-treated parts [89].

2.2.4 Blasting

Blasting is a mechanical treatment utilized for additively manufactured parts and has various subclasses including grit, sand (abrasive or dry abrasive), and shot blasting. In this process, there is no material removal, thus it can be described as a non-subtractive approach. This post surface treatment can provide a uniform surface roughness and gradually decrease the roughness in function of time treatment [45, 90, 91].

Blasting distance, media, pressure and time are parameters in this process [92]. The rough blasting media results in a high roughness surface with sharp edges [3]. The surface roughness and erosion rate as well as plastic deformation increase with the blasting pressure increase [93]. Another parameter affecting the surface finish is the blasting time, which is the duration for the exposure of part surface to the abrasives. The surface roughness also increases with the blasting time increase. The tribological behaviour of parts is improved, the friction coefficient of them decreases by the increase of this parameter [94]. Sandblasting machine and application are presented in Figure 2.6.

Abrasives, nut shells, and sand like corundum, corundum sand, glass beads, SiC, ceramic beads are employed as the blasting media that is thrusted by pressurized fluid or air. In Figure 2.7, SEM images of glass bead and Al_2O_3 are illustrated.

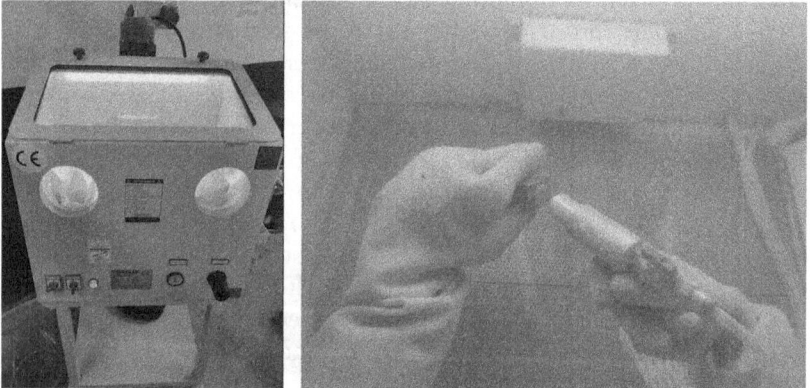

Figure 2.6 Sandblasting machine (left), an application of sandblasting (right).

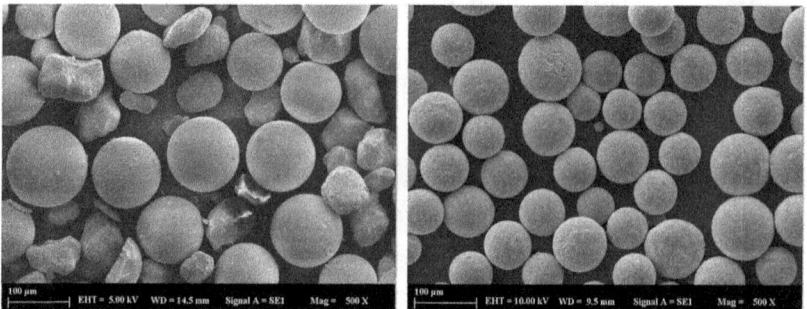

Figure 2.7 SEM images of sandblasting abrasives. Glass bead (left) and Al_2O_3 (right).

Sandblasting can enhance the surface roughness and fatigue characteristics of parts that effectively eliminates attached powders and causes the formation of a nanocrystalline structure. In powder-based AM processes, sandblasting can eliminate the unmelted adhered powder by repetitively blasting the manufactured parts with high-speed sand particles. Also, a compressive residual stress layer occurs in the subsurface region. This layer can efficiently prevent the initiation and growth of cracks [95]. Sandblasting is also used to enhance the metal surfaces' adhesive properties [96].

Abrasive blasting is used to the outer surface of additively manufactured parts for removal of the loose powder from this area [97]. This method is utilized extensively in industry to clean surfaces, engrave, and deburr [98]. Abrasive blasting can efficiently decrease the roughness by 50–70% with a lowest Ra of less than 1 μm. In spite of the restriction of method repeatability, this treatment is widely applied for finishing the microproducts, which are manufactured in different sectors like jewellery and dentistry, due to its advantages including method cost, simplicity, versatility, and cycle time [90].

Longhitano et al. presented a comparison of various surface treatments (electropolishing, chemical etching, blasting) by applying DMLS manufactured Ti6Al4V samples. As-printed parts` surface roughness was compared with that of after these modification methods and a combination of them. The combination of chemical etching and blasting caused the lowest surface roughness. Blasting treatment was applied to obtain a homogeneous surface quality, while chemical etching provided surface cleanness. These methods gradually decreased the surface roughness [99]. Bouzakis et al. investigated the effect of glass-blasting on SLM-built parts` surface roughness. After the blasting process, the surface roughness was decreased from Ra = 11.8 μm to Ra = 6.0 μm. The smoothing effect was achieved by the mechanisms of micro-forming arising during the process [100]. Seo et al. used blasting to decrease the surface roughness of SLM-printed parts. The surface roughness was obtained by about 3 μm by this effective process. The blasting method was used to decrease coarseness, provide a more uniform and blunter surface [101]. Park et al. presented a study of the degradation of CP-Ti powder by the pre-heating and blasting during Electron Beam Melting (EBM). During the blasting process, light elements' (e.g. N and O) quantity was increased by ~15% in the recycled powder. This study presented the refining effect of EBM by improving Ti parts' purity [102].

2.2.5 Ultrasonic abrasion finishing

Ultrasonic abrasion finishing is an advanced polishing process. This surface treatment method finishes part surfaces efficiently in comparison with other conventional processes. Material removal occurs from the part surface by low amplitude, high frequency tool vibrations against the surface of material in the case of the existence of fine abrasives. This method can be used to process both metals and polymers. This method can achieve a surface roughness decrease of 129 μm/min providing removal between 3 and 4 μm in 2 seconds. A final roughness between 0.3 and 0.8 μm can be obtained in finished products [103]. The requirement of the precise treatment for the waste and abrasive materials is the major disadvantage of this method [104]. Ultrasonic abrasion finishing is limited with smooth surfaces [105]. It has three different methods (oblique, longitudinal, and latitudinal) depending on the vibration direction as shown in Figure 2.8.

This process can be applied to additively manufactured parts [107]. EBM, SLA and Selective Laser Sintering (SLS) parts' surface finishing can be carried out with this technique. External and internal sections can be processed [108].

Spencer studied this finishing method and concluded that this technique provides an allowable surface roughness for SLA products. An improvement in surface finish of 66% has been accomplished with little deterioration to the part form at a very short processing time of 2 seconds [103].

2.2.6 Hot cutter machining

HCM is a conventional material removal technique and decreases surface roughness. In this technique, the heated tool moves along the specimen and provides a smoother surface. The main disadvantages of HCM are that complex additively parts' internal areas are unreachable and only flat surfaces are processed.

In this post-processing method, the cutting direction, speed, and rake angle of the cutter are important parameters [109]. The blade-like cutter cannot be applied on freeform surfaces having complex or convex structures [110].

HCM is considered a beneficial process to improve the surface finish of FDM components [111]. Pandey et al. applied HCM to increase the surface quality of FDM specimen up to Ra = 0.5 μm [112]. Pandey et al. applied HCM for machining ABS specimen's build edges. It was aimed to increase the freeform axisymmetric FDM specimen's surface quality. For this, a

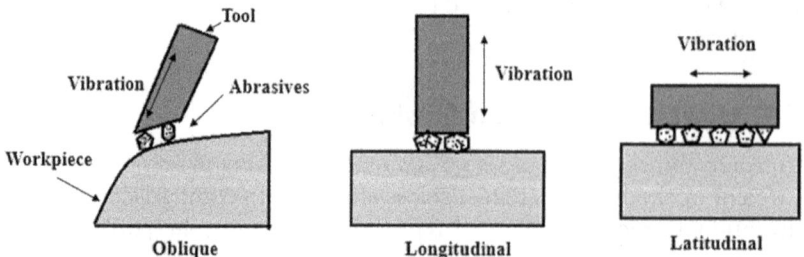

Figure 2.8 Ultrasonic abrasion finishing methods. (Reproduced with the reference of [106], which is Under Creative Commons CC BY 2.0 licence.)

virtual hybrid FDM system to specify the variation of the surface roughness before and after HCM was developed. The surface quality can be improved by this system [113].

2.2.7 Micromachining

As a result of advances in precision manufacturing, micromachining's applications are increasing in the manufacturing of small components. Because of the outstanding properties such as low set-up cost, method flexibility, usage easiness, and various part materials, micromachining is the one of promising processes for the mass production of the structures having intricate 3D contours for example medical products, micro-dies, moulds, and sensors [113]. Micromachining is also applied to additively manufactured metal parts' surfaces as a post-processing and provides good surface finish. If there is a requirement of precision and good surface quality, after AM micromachining is needed.

The micromachining process is classified as traditional, non-traditional, and hybrid processes [114]. In the traditional method, a contact between a workpiece and a tool or fine abrasive grinding wheel causes shearing, as well as material removal. Because of the interaction, tool wear and cutting force occur. These processes cover micro-grinding, micro-milling, and micro-turning [115]. The traditional mechanical micromachining possesses a higher machining efficiency; however, the smallest tool size (about 10 μm) restricts the machining resolution. The tool can wear easily, and its stiffness affects the precision [116]. Micro-milling has a high capability and accuracy on geometrically complex features.

In non-traditional micromachining, there is not a physical interaction between the workpiece and tool. These processes take place by energy transport in the forms such as arc, beam (electron, ion, or laser) energy, electrolysis, erosion energy, and vibration. Laser Beam Machining (LBM) is a non-traditional micromachining process taken advantage of for different industrial applications differing from 1D to 3D machining [117–120].

In hybrid micromachining, there is an integration of at least two traditional or non-traditional processes either in an assisted method or in a successive manner, thus multiple processes' energy sources can be applied for machining [121, 122]. The aim of utilizing hybrid processes is to improve the advantages of one process, or eliminate the disadvantages of the other, to achieve maximum benefit.

Varghese et al. examined the relation between the depth of cut, porosity, and surface integrity with cutting forces with regard to residual stress, surface finish, and microhardness when the micromachining process is applied to porous additively manufactured Ti6Al4V. As the result of porosity increase and depth of cut decrease, cutting force reduced. After the micromachining, residual stresses that are generally compressive decreased because of the deformed layer's development [123]. Lerebours et al. investigated the superfinishing of SLM-built CoCrMo parts by micromachining. This surface treatment removes a uniform material quantity and provides a smooth surface with a Ra value below 0.1 μm [124]. Ji et al. researched the machinability and microstructure of SLM-produced IN718 alloy, which is micro-milled, and compared wrought parts. Because of the micro-milling process, SLM-fabricated part has lower roughness value in comparison with the wrought alloy [125].

2.2.8 Magnetic field-assisted finishing

MAF is a non-conventional post-process that can provide required surface roughness. MAF allows both the finishing of intricate external structures and internal component properties [126, 127]. This finishing process utilizes a magnetic field to manage the finishing medium, generally including magnetic particles and non-magnetic abrasives hung in a carrier liquid [128]. Complex structures can be processed by MAF in different machining conditions by changing the magnetic tools [129]. MAF can finish various materials such as ceramics, titanium alloys, and steels. If the magnets generate a magnetic field, the brush compresses the surface of the target. Magnet rotation provides relative motion between the surface and the brush. In the polishing process, the surface roughness is changed due to the press of the abrasive by the magnetic particles and material removal from the surface by the abrasive. Ball burnishing supports the modification of the surface geometry and transmits compressive residual stress to the surface. In this process, plastic deformation and surface material flow occur by the effect of the balls. Thus, the subsurface morphology and the texture of the surface change. MAF process schematic is demonstrated in Figure 2.9.

Yamaguchi et al. studied a combination of post-processes covering polishing, sanding, and burnishing through MAF, which changes SLM-built 316L stainless steel parts' residual stress and surface roughness. The roughness Rz was reduced from over 100 μm to 0.1 μm. As a result of this study, they

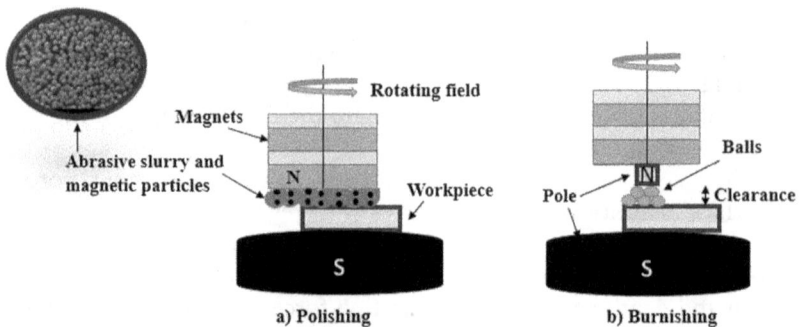

a) Polishing   b) Burnishing

Figure 2.9 MAF process schematic.

found that residual stress can be removed by adjusting the motion of the magnetic tool [130]. Wu et al. studied the result of laser scan direction with regard to SLM-manufactured 316L steel discs' surface to be treated by MAF. These parts are manufactured with different scanning approaches and orientations. The results presented that material removal rate and optimum treatment duration depend upon the orientation with regard to the scan direction [131]. Guo et al. offered an analytical and experimental study on post-processing (precision grinding with MAF) of SLS-produced Polyamide 12 (PA12) to improve surface finish and to study the connection between the parameters and surface finish. After post-processing, PA12 specimens' surface roughness Ra was decreased from over 15 µm to 0.89 µm by MAF. As a result of precision grinding, Ra = 2.85 µm was achieved. The surface treated by MAF exhibited a better tribology behaviour described by higher wear resistance and lower friction coefficient [132].

2.3 CONCLUSION

AM technology is growing rapidly in recent years. This chapter has included the mechanical post-processing techniques and their applications in AM technologies. In AM, different defect types such as cracks, residual stresses, and poor surface quality exist. To eliminate these defects and provide surface integrity, surface treatments are applied. Various post-processing methods have verified their ability to improve the surface quality and mechanical characteristics of AM parts and decrease the formation of defects. It is crucial to choose the correct combination of AM and post-processing methods.

REFERENCES

1. Sehrt, J. T. (2010). Möglichkeiten und Grenzen bei der generativen Herstellung metallischer Bauteile durch das Strahlschmelzverfahren.
2. Greitemeier, D. (2016). *Untersuchung der Einflussparameter auf die mechanischen Eigenschaften von additiv gefertigtem TiAl6V4*. Wiesbaden, Germany: Springer Fachmedien Wiesbaden.
3. Bagehorn, S., Wehr, J., & Maier, H. J. (2017). Application of mechanical surface finishing processes for roughness reduction and fatigue improvement of additively manufactured Ti-6Al-4V parts. *International Journal of Fatigue*, 102, 135–142.
4. Kasperovich, G., & Hausmann, J. (2015). Improvement of fatigue resistance and ductility of TiAl6V4 processed by selective laser melting. *Journal of Materials Processing Technology*, 220, 202–214.
5. Stoffregen, H. A., Butterweck, K., & Abele, E. (2014). Fatigue analysis in selective laser melting: review and investigation of thin-walled actuator housings. In *2014 International Solid Freeform Fabrication Symposium*. University of Texas at Austin.

6. Chastand, V., Tezenas, A., Cadoret, Y., Quaegebeur, P., Maia, W., & Charkaluk, E. (2016). Fatigue characterization of Titanium Ti-6Al-4V samples produced by Additive Manufacturing. *Procedia Structural Integrity*, 2, 3168–3176.

7. Lewandowski, J. J., & Seifi, M. (2016). Metal additive manufacturing: a review of mechanical properties. *Annual Review of Materials Research*, 46, 151–186.

8. Bagehorn, S., Mertens, T., Seack, O., & Maier, H. J. (2016, June). Reduction of the surface roughness of additively manufactured metallic parts by enhanced electrolytic smoothening. In *Rapid. Tech—International Trade Show and Conference for Additive Manufacturing* (p. 61). Munich: Carl Hanser Verlag.

9. Greitemeier, D., Dalle Donne, C., Syassen, F., Eufinger, J., & Melz, T. (2016). Effect of surface roughness on fatigue performance of additive manufactured Ti–6Al–4V. *Materials Science and Technology*, 32(7), 629–634.

10. Liu, Y. J., Li, S. J., Wang, H. L., Hou, W. T., Hao, Y. L., Yang, R., ... & Zhang, L. C. (2016). Microstructure, defects and mechanical behavior of beta-type titanium porous structures manufactured by electron beam melting and selective laser melting. *Acta Materialia*, 113, 56–67.

11. Baraiya, R., Babbar, A., Jain, V., & Gupta, D. (2020). In-situ simultaneous surface finishing using abrasive flow machining via novel fixture. *Journal of Manufacturing Processes*, 50, 266–278.

12. Wang, A. C., Cheng, K. C., Chen, K. Y., & Lin, Y. C. (2018). A study on the abrasive gels and the application of abrasive flow machining in complex-hole polishing. *Procedia Cirp*, 68, 523–528.

13. Seyedi, S. S., Shabgard, M. R., Mousavi, S. B., & Heris, S. Z. (2021). The impact of SiC, Al2O3, and B2O3 abrasive particles and temperature on wear characteristics of 18Ni (300) maraging steel in abrasive flow machining (AFM). *International Journal of Hydrogen Energy*, 46, 33991–34001.

14. Petare, A. C., & Jain, N. K. (2018). A critical review of past research and advances in abrasive flow finishing process. *The International Journal of Advanced Manufacturing Technology*, 97(1), 741–782.

15. Venkatesh, G., Sharma, A. K., & Kumar, P. (2015). On ultrasonic assisted abrasive flow finishing of bevel gears. *International Journal of Machine Tools and Manufacture*, 89, 29–38.

16. Kumar, S., Jain, V. K., & Sidpara, A. (2015). Nanofinishing of freeform surfaces (knee joint implant) by rotational-magnetorheological abrasive flow finishing (R-MRAFF) process. *Precision Engineering*, 42, 165–178.

17. Mohseni-Mofidi, S., Pastewka, L., Teschner, M., & Bierwisch, C. (2022). Magnetic-assisted soft abrasive flow machining studied with smoothed particle hydrodynamics. *Applied Mathematical Modelling*, 101, 38–54.

18. Rhoades, L. (1991). Abrasive flow machining: a case study. *Journal of Materials Processing Technology*, 28(1–2), 107–116.

19. Pusavec, F., & Kenda, J. (2014). The transition to a clean, dry, and energy efficient polishing process: an innovative upgrade of abrasive flow machining for simultaneous generation of micro-geometry and polishing in the tooling industry. *Journal of Cleaner Production*, 76, 180–189.

20. Duval-Chaneac, M. S., Han, S., Claudin, C., Salvatore, F., Bajolet, J., & Rech, J. (2018). Experimental study on finishing of internal laser melting (SLM) surface with abrasive flow machining (AFM). *Precision Engineering*, 54, 1–6.

21. Han, S., Salvatore, F., Rech, J., & Bajolet, J. (2020). Abrasive flow machining (AFM) finishing of conformal cooling channels created by selective laser melting (SLM). *Precision Engineering*, 64, 20–33.
22. Uhlmann, E., & Roßkamp, S. (2018). Surface integrity and chip formation in abrasive flow machining. *Procedia CIRP*, 71, 446–452.
23. Strano, G., Hao, L., Everson, R. M., & Evans, K. E. (2012). Surface Roughness Analysis in Selective Laser Melting. *Proceedings Innovative Developments in Virtual and Physical Prototyping*, 561–565, Boca Raton, Florida.
24. Rosa, B., Mognol, P., & Hascoët, J. Y. (2015). Laser polishing of additive laser manufacturing surfaces. *Journal of Laser Applications*, 27(S2), S29102.
25. Bhaduri, D., Penchev, P., Batal, A., Dimov, S., Soo, S. L., Sten, S., ... & Dong, H. (2017). Laser polishing of 3D printed mesoscale components. *Applied Surface Science*, 405, 29–46.
26. Jain, R. K., Jain, V. K., & Dixit, P. M. (1999). Modeling of material removal and surface roughness in abrasive flow machining process. *International Journal of Machine Tools and Manufacture*, 39(12), 1903–1923.
27. Singh, S., Shan, H. S., & Kumar, P. (2008). Experimental studies on mechanism of material removal in abrasive flow machining process. *Materials and Manufacturing Processes*, 23(7), 714–718.
28. Dixit, N., Sharma, V., & Kumar, P. (2021). Research trends in abrasive flow machining: A systematic review. *Journal of Manufacturing Processes*, 64, 1434–1461.
29. Kumar, R., Singh, S., Aggarwal, V., Singh, S., Pimenov, D. Y., Giasin, K., & Nadolny, K. (2022). Hand and Abrasive flow polished tungsten carbide die: Optimization of surface roughness, polishing time and comparative analysis in wire drawing. *Materials*, 15(4), 1287.
30. Mali, H. S., Prajwal, B., Gupta, D., & Kishan, J. (2018). Abrasive flow finishing of FDM printed parts using a sustainable media. *Rapid Prototyping Journal*, 24(3), 593–606.
31. Williams, R. E., & Melton, V. L. (1998). Abrasive flow finishing of stereolithography prototypes. *Rapid Prototyping Journal*, 4(2), 56–67.
32. Mohammadian, N., Turenne, S., & Brailovski, V. (2018). Surface finish control of additively manufactured Inconel 625 components using combined chemical-abrasive flow polishing. *Journal of Materials Processing Technology*, 252, 728–738.
33. Ferchow, J., Baumgartner, H., Klahn, C., & Meboldt, M. (2020). Model of surface roughness and material removal using abrasive flow machining of selective laser melted channels. *Rapid Prototyping Journal*, 26(7), 1165–1176.
34. Han, S., Salvatore, F., Rech, J., Bajolet, J., & Courbon, J. (2020). Effect of abrasive flow machining (AFM) finish of selective laser melting (SLM) internal channels on fatigue performance. *Journal of Manufacturing Processes*, 59, 248–257.
35. Uhlmann, E., Schmiedel, C., & Wendler, J. (2015). CFD simulation of the abrasive flow machining process. *Procedia CIRP*, 31, 209–214.
36. Peng, C., Fu, Y., Wei, H., Li, S., Wang, X., & Gao, H. (2018). Study on improvement of surface roughness and induced residual stress for additively manufactured metal parts by abrasive flow machining. *Procedia CIRP*, 71, 386–389.

37. Guo, J., Song, C., Fu, Y., Au, K. H., Kum, C. W., Goh, M. H., ... & Sun, C. N. (2020). Internal surface quality enhancement of selective laser melted Inconel 718 by abrasive flow machining. *Journal of Manufacturing Science and Engineering*, 142(10), 101003.

38. Hashmi, A. W., Mali, H. S., & Meena, A. (2022). Experimental investigation on abrasive flow Machining (AFM) of FDM printed hollow truncated cone parts. *Materials Today: Proceedings*, 56, 1369–1375.

39. da Silva Maciel, L., & Spelt, J. K. (2020). Measurements of wall-media contact forces and work in a vibratory finisher. *Powder Technology*, 360, 911–920.

40. Sood, A., & Mullany, B. (2021). Advanced Surface Analysis to Identify Media-Workpiece Contact Modes in a Vibratory Finishing Processes. *Procedia Manufacturing*, 53, 155–161.

41. Kacaras, A., Gibmeier, J., Zanger, F., & Schulze, V. (2018). Influence of rotational speed on surface states after stream finishing. *Procedia CIRP*, 71, 221–226.

42. Gillespie, L. (2006). *Mass Finishing Handbook*. South Norwalk, CT: Industrial Press.

43. Bergs, T., Müller, U., Barth, S., & Ohlert, M. (2021). Experimental analysis on vibratory finishing of cemented carbides. *Manufacturing Letters*, 28, 21–24.

44. Seltzman, A. H., & Wukitch, S. J. (2020). RF losses in selective laser melted GRCop-84 copper waveguide for an additively manufactured lower hybrid current drive launcher. *Fusion Engineering and Design*, 159, 111762.

45. Seltzman, A. H., & Wukitch, S. J. (2020). Surface roughness and finishing techniques in selective laser melted GRCop-84 copper for an additive manufactured lower hybrid current drive launcher. *Fusion Engineering and Design*, 160, 111801.

46. Bernhardt, A., Schneider, J., Schroeder, A., Papadopoulous, K., Lopez, E., Brückner, F., & Botzenhart, U. (2021). Surface conditioning of additively manufactured titanium implants and its influence on materials properties and in vitro biocompatibility. *Materials Science and Engineering: C*, 119, 111631.

47. Uhlmann, E., Fleck, C., Gerlitzky, G., & Faltin, F. (2017). Dynamical fatigue behavior of additive manufactured products for a fundamental lifecycle approach. *Procedia CIRP*, 61, 588–593.

48. Kaynak, Y., & Kitay, O. (2019). The effect of post-processing operations on surface characteristics of 316L stainless steel produced by selective laser melting. *Additive Manufacturing*, 26, 84–93.

49. Ginestra, P., Ceretti, E., Lobo, D., Lowther, M., Cruchley, S., Kuehne, S., ... & Webber, M. (2020). Post processing of 3D printed metal scaffolds: a preliminary study of antimicrobial efficiency. *Procedia Manufacturing*, 47, 1106–1112.

50. Benedetti, M., Berto, F., Marini, M., Raghavendra, S., & Fontanari, V. (2020). Incorporating residual stresses into a Strain-Energy-Density based fatigue criterion and its application to the assessment of the medium-to-very-high-cycle fatigue strength of shot-peened parts. *International Journal of Fatigue*, 139, 105728.

51. Sugavaneswaran, M., Jebaraj, A. V., Kumar, M. B., Lokesh, K., & Rajan, A. J. (2018). Enhancement of surface characteristics of direct metal laser sintered stainless steel 316L by shot peening. *Surfaces and Interfaces*, 12, 31–40.

52. Denti, L., Bassoli, E., Gatto, A., Santecchia, E., & Mengucci, P. (2019). Fatigue life and microstructure of additive manufactured Ti6Al4V after different finishing processes. *Materials Science and Engineering: A*, 755, 1–9.

53. Kobayashi, M., Matsui, T., & Murakami, Y. (1998). Mechanism of creation of compressive residual stress by shot peening. *International Journal of Fatigue*, 20(5), 351–357.
54. Kritzler, J., & Wubbenhorst, W. (2002). Inducing compressive stresses through controlled shot peening. ASM International, Member/Customer Service Center, Materials Park, OH 44073-0002, USA, 2002, 345–358.
55. Curtiss-Wright Surface Technologies. (2016). Surface technologies – shot peening applications. www.cwst.de, (Accessed 23 January 2022).
56. Lindemann, J., Buque, C., & Appel, F. (2006). Effect of shot peening on fatigue performance of a lamellar titanium aluminide alloy. *Acta Materialia*, 54(4), 1155–1164.
57. Child, D. J., West, G. D., & Thomson, R. C. (2011). Assessment of surface hardening effects from shot peening on a Ni-based alloy using electron back-scatter diffraction techniques. *Acta Materialia*, 59(12), 4825–4834.
58. Benedetti, M., Fontanari, V., Höhn, B. R., Oster, P., & Tobie, T. (2002). Influence of shot peening on bending tooth fatigue limit of case-hardened gears. *International Journal of Fatigue*, 24(11), 1127–1136.
59. Takahashi, K., Osedo, H., Suzuki, T., & Fukuda, S. (2018). Fatigue strength improvement of an aluminum alloy with a crack-like surface defect using shot peening and cavitation peening. *Engineering Fracture Mechanics*, 193, 151–161.
60. Zhang, J., Li, H., Yang, B., Wu, B., & Zhu, S. (2020). Fatigue properties and fatigue strength evaluation of railway axle steel: Effect of micro-shot peening and artificial defect. *International Journal of Fatigue*, 132, 105379.
61. James, M. N., Newby, M., Hattingh, D. G., & Steuwer, A. (2010). Shot-peening of steam turbine blades: Residual stresses and their modification by fatigue cycling. *Procedia Engineering*, 2(1), 441–451.
62. Vázquez, J., Navarro, C., & Domínguez, J. (2012). Experimental results in fretting fatigue with shot and laser peened Al 7075-T651 specimens. *International Journal of Fatigue*, 40, 143–153.
63. Bagherifard, S., Slawik, S., Fernández-Pariente, I., Pauly, C., Mücklich, F., & Guagliano, M. (2016). Nanoscale surface modification of AISI 316L stainless steel by severe shot peening. *Materials & Design*, 102, 68–77.
64. Chen, M., Jiang, C., Xu, Z., & Ji, V. (2019). Surface layer characteristics of SAF2507 duplex stainless steel treated by stress shot peening. *Applied Surface Science*, 481, 226–233.
65. Zhu, L., Guan, Y., Wang, Y., Xie, Z., Lin, J., & Zhai, J. (2017). Influence of process parameters of ultrasonic shot peening on surface roughness and hydrophilicity of pure titanium. *Surface and Coatings Technology*, 317, 38–53.
66. Pandey, V., Singh, J. K., Chattopadhyay, K., Srinivas, N. S., & Singh, V. (2019). Optimization of USSP duration for enhanced corrosion resistance of AA7075. *Ultrasonics*, 91, 180–192.
67. Altenberger, I., Scholtes, B., Martin, U., & Oettel, H. (1999). Cyclic deformation and near surface microstructures of shot peened or deep rolled austenitic stainless steel AISI 304. *Materials Science and Engineering: A*, 264(1–2), 1–16.
68. Bagherifard, S. (2019). Enhancing the structural performance of lightweight metals by shot peening. *Advanced Engineering Materials*, 21(7), 1801140.
69. Kahlin, M., Ansell, H., Basu, D., Kerwin, A., Newton, L., Smith, B., & Moverare, J. J. (2020). Improved fatigue strength of additively manufactured Ti6Al4V by surface post processing. *International Journal of Fatigue*, 134, 105497.

70. Unal, O., & Maleki, E. (2018). Shot peening optimization with complex decision-making tool: Multi criteria decision-making. *Measurement*, 125, 133–141.
71. Lin, Q., Liu, H., Zhu, C., & Parker, R. G. (2019). Investigation on the effect of shot peening coverage on the surface integrity. *Applied Surface Science*, 489, 66–72.
72. Wang, Y., Xie, H., Zhou, Z., Li, X., Wu, W., & Gong, J. (February 2020). Effect of shot peening coverage on hydrogen embrittlement of a ferrite-pearlite steel, *International Journal of Hydrogen Energy*, 45(11), 7169–7184.
73. Wagner, L. (Ed.). (2003). *Shot Peening* (Vol. 8). New Jersey, U.S.: John Wiley & Sons.
74. Hassani-Gangaraj, S. M., Cho, K. S., Voigt, H. J., Guagliano, M., & Schuh, C. A. (2015). Experimental assessment and simulation of surface nanocrystallization by severe shot peening. *Acta Materialia*, 97, 105–115.
75. Mutoh, Y., Fair, G. H., Noble, B., & Waterhouse, R. B. (1987). The effect of residual stresses induced by shot-peening on fatigue crack propagation in two high strength aluminium alloys. *Fatigue & Fracture of Engineering Materials & Structures*, 10(4), 261–272.
76. Winkler, K. J., Schurer, S., Tobie, T., & Stahl, K. (2019). Investigations on the tooth root bending strength and the fatigue fracture characteristics of case-carburized and shot-peened gears of different sizes. *Proceedings of the Institution of Mechanical Engineers, Part C: Journal of Mechanical Engineering Science*, 233(21–22), 7338–7349.
77. Benedetti, M., Torresani, E., Leoni, M., Fontanari, V., Bandini, M., Pederzolli, C., & Potrich, C. (2017). The effect of post-sintering treatments on the fatigue and biological behavior of Ti-6Al-4V ELI parts made by selective laser melting. *Journal of the Mechanical Behavior of Biomedical Materials*, 71, 295–306.
78. Uzan, N. E., Ramati, S., Shneck, R., Frage, N., & Yeheskel, O. (2018). On the effect of shot-peening on fatigue resistance of AlSi10Mg specimens fabricated by additive manufacturing using selective laser melting (AM-SLM). *Additive Manufacturing*, 21, 458–464.
79. Tan, L., Yao, C., Zhang, D., Ren, J., Zhou, Z., & Zhang, J. (2020). Evolution of surface integrity and fatigue properties after milling, polishing, and shot peening of TC17 alloy blades. *International Journal of Fatigue*, 136, 105630.
80. Vantadori, S., & Zanichelli, A. (2021). Fretting-fatigue analysis of shot-peened aluminium and titanium test specimens. *Fatigue & Fracture of Engineering Materials & Structures*, 44(2), 397–409.
81. Kirk, D. (2009). Size and variability of cast steel of shot particles. *Shot Peener*, 23(1), 24–32.
82. Wu, D., Yao, C., & Zhang, D. (2017). Surface characterization of Ti1023 alloy shot peened by cast steel and ceramic shot. *Advances in Mechanical Engineering*, 9(10), 1687814017723287.
83. Sagbas, B. (2021). Surface texture properties and tribological behavior of additive manufactured parts. *Tribology and Surface Engineering for Industrial Applications*, 3, 85.
84. Salvati, E., Lunt, A. J., Heason, C. P., Baxter, G. J., & Korsunsky, A. M. (2020). An analysis of fatigue failure mechanisms in an additively manufactured and shot peened IN 718 nickel superalloy. *Materials & Design*, 191, 108605.

85. Hadidi, H., Mailand, B., Sundermann, T., Johnson, E., Madireddy, G., Negahban, M., ... & Sealy, M. (2019). Low velocity impact of ABS after shot peening predefined layers during additive manufacturing. *Procedia Manufacturing*, 34, 594–602.
86. Khajehmirza, H., Astaraee, A. H., Monti, S., Guagliano, M., & Bagherifard, S. (2021). A hybrid framework to estimate the surface state and fatigue performance of laser powder bed fusion materials after shot peening. *Applied Surface Science*, 567, 150758.
87. AlMangour, B., & Yang, J. M. (2016). Improving the surface quality and mechanical properties by shot-peening of 17–4 stainless steel fabricated by additive manufacturing. *Materials & Design*, 110, 914–924.
88. Salvati, E., Lunt, A. J. G., Ying, S., Sui, T., Zhang, H. J., Heason, C., ... & Korsunsky, A. M. (2017). Eigenstrain reconstruction of residual strains in an additively manufactured and shot peened nickel superalloy compressor blade. *Computer Methods in Applied Mechanics and Engineering*, 320, 335–351.
89. Slawik, S., Bernarding, S., Lasagni, F., Navarro, C., Periñán, A., Boby, F., ... & Mücklich, F. (2021). Microstructural analysis of selective laser melted Ti6Al4V modified by laser peening and shot peening for enhanced fatigue characteristics. *Materials Characterization*, 173, 110935.
90. Nagarajan, B., Hu, Z., Song, X., Zhai, W., & Wei, J. (2019). Development of micro selective laser melting: The state of the art and future perspectives. *Engineering*, 5(4), 702–720.
91. Muthaiah, V. S., Indrakumar, S., Suwas, S., & Chatterjee, K. (2022). Surface engineering of additively manufactured titanium alloys for enhanced clinical performance of biomedical implants: A review of recent developments. *Bioprinting*, 25, e00180.
92. Hashmi, A. W., Mali, H. S., & Meena, A. (2021). Improving the surface characteristics of additively manufactured parts: A review. *Materials Today: Proceedings*.
93. Yetik, O., Koçoğlu, H., Avcu, Y. Y., Avcu, E., & Sınmazçelik, T. (2020). The effects of grit size and blasting pressure on the surface properties of grit blasted Ti6Al4V alloy. *Materials Today: Proceedings*, 32, 27–36.
94. Chen, L., Liu, Z., & Song, W. (2020). Process-surface morphology-tribological property relationships for H62 brass employing various manufacturing approaches. *Tribology International*, 148, 106320.
95. Yang, J., Gu, D., Lin, K., Zhang, Y., Guo, M., Yuan, L., ... & Zhang, H. (2022). Laser Additive Manufacturing of Bio-inspired Metallic Structures. *Chinese Journal of Mechanical Engineering: Additive Manufacturing Frontiers*, 1, 100013.
96. Szymczyk-Ziółkowska, P., Hoppe, V., Gąsiorek, J., Rusińska, M., Kęszycki, D., Szczepański, Ł., ... & Detyna, J. (2021). Corrosion resistance characteristics of a Ti-6Al-4V ELI alloy fabricated by electron beam melting after the applied post-process treatment methods. *Biocybernetics and Biomedical Engineering*, 41(4), 1575–1588.
97. Tyagi, P., Goulet, T., Riso, C., Stephenson, R., Chuenprateep, N., Schlitzer, J., ... & Garcia-Moreno, F. (2019). Reducing the roughness of internal surface of an additive manufacturing produced 316 steel component by chempolishing and electropolishing. *Additive Manufacturing*, 25, 32–38.

98. Hashimoto, F., Yamaguchi, H., Krajnik, P., Wegener, K., Chaudhari, R., Hoffmeister, H. W., & Kuster, F. (2016). Abrasive fine-finishing technology. *CIRP Annals*, 65(2), 597–620.

99. Longhitano, G. A., Larosa, M. A., Munhoz, A. L. J., Zavaglia, C. A. D. C., & Ierardi, M. C. F. (2015). Surface finishes for Ti-6Al-4V alloy produced by direct metal laser sintering. *Materials Research*, 18, 838–842.

100. Bouzakis, E., Arvanitidis, A., Kazelis, F., Maliaris, G., & Michailidis, N. (2020). Comparison of additively manufactured vs. conventional maraging steel in corrosion-fatigue performance after various surface treatments. *Procedia CIRP*, 87, 469–473.

101. Seo, B., Park, H. K., Kim, H. G., Kim, W. R., & Park, K. (2021). Corrosion behavior of additive manufactured CoCr parts polished with plasma electrolytic polishing. *Surface and Coatings Technology*, 406, 126640.

102. Park, H. K., Ahn, Y. K., Lee, B. S., Jung, K. H., Lee, C. W., & Kim, H. G. (2017). Refining effect of electron beam melting on additive manufacturing of pure titanium products. *Materials Letters*, 187, 98–100.

103. Spencer, J. D. (1993). Vibratory finishing of stereolithography parts. In *1993 International Solid Freeform Fabrication Symposium*.

104. Guzzo, P. L., Shinohara, A. H., & Raslan, A. A. (2004). A comparative study on ultrasonic machining of hard and brittle materials. *Journal of the Brazilian Society of Mechanical Sciences and Engineering*, 26(1), 56–61.

105. Park, S. H., Son, S. J., Lee, S. B., Yu, J. H., Ahn, S. J., & Choi, Y. S. (2021). Surface machining effect on material behavior of additive manufactured SUS 316L. *Journal of Materials Research and Technology*, 13, 38–47.

106. Zhao, J., Zhan, J., Jin, R., & Tao, M. (2000). An oblique ultrasonic polishing method by robot for free-form surfaces. *International Journal of Machine Tools and Manufacture*, 40(6), 795–808.

107. Kumbhar, N. N., & Mulay, A. V. (2018). Post processing methods used to improve surface finish of products which are manufactured by additive manufacturing technologies: a review. *Journal of The Institution of Engineers (India): Series C*, 99(4), 481–487.

108. García-Verdugo Zuil, A., & Herrero Martín, A. (2020). Additive manufacturing and radio frequency filters: A case study on 3D-printing processes, postprocessing and silver coating methods.

109. Pandey, P. M., Reddy, N. V., & Dhande, S. G. (2003). Real time adaptive slicing for fused deposition modelling. *International Journal of Machine Tools and Manufacture*, 43(1), 61–71.

110. Boschetto, A., Bottini, L., & Veniali, F. (2016). Finishing of fused deposition modeling parts by CNC machining. *Robotics and Computer-Integrated Manufacturing*, 41, 92–101.

111. Galantucci, L. M., Lavecchia, F., & Percoco, G. (2010). Quantitative analysis of a chemical treatment to reduce roughness of parts fabricated using fused deposition modeling. *CIRP Annals-Manufacturing Technology*, 59(1), 247–250.

112. Pandey, P. M., Reddy, N. V., & Dhande, S. G. (2003). Improvement of surface finish by staircase machining in fused deposition modeling. *Journal of Materials Processing Technology*, 132(1–3), 323–331.

113. Pandey, P. M., Venkata Reddy, N., & Dhande, S. G. (2006). Virtual hybrid-FDM system to enhance surface finish. *Virtual and Physical Prototyping*, 1(2), 101–116.

114. Masuzawa, T. (2000). State of the art of micromachining. *CIRP Annals*, 49(2), 473–488.
115. Saxena, K. K., Bellotti, M., Qian, J., Reynaerts, D., Lauwers, B., & Luo, X. (2018). Overview of hybrid machining processes.
116. Asad, A. B. M. A., Masaki, T., Rahman, M., Lim, H. S., & Wong, Y. S. (2007). Tool-based micro-machining. *Journal of Materials Processing Technology*, 192, 204–211.
117. Dornfeld, D., Min, S., & Takeuchi, Y. (2006). Recent advances in mechanical micromachining. *CIRP Annals*, 55(2), 745–768.
118. Pandey, A., & Singh, S. (2010). Current research trends in variants of Electrical Discharge Machining: A review. *International Journal of Engineering Science and Technology*, 2(6), 2172–2191.
119. Spieser, A., & Ivanov, A. (2013). Recent developments and research challenges in electrochemical micromachining (μECM). *The International Journal of Advanced Manufacturing Technology*, 69(1), 563–581.
120. Huang, H., Zhang, H., Zhou, L., & Zheng, H. Y. (2003). Ultrasonic vibration assisted electro-discharge machining of microholes in Nitinol. *Journal of Micromechanics and Microengineering*, 13(5), 693.
121. Gower, M. C. (1998, September). Industrial applications of pulsed laser micromachining. In *CLEO/Europe Conference on Lasers and Electro-Optics* (pp. 247–247). IEEE.
122. Rahman, M., Asad, A. B. M. A., Masaki, T., Saleh, T., Wong, Y. S., & Kumar, A. S. (2010). A multiprocess machine tool for compound micromachining. *International Journal of Machine Tools and Manufacture*, 50(4), 344–356.
123. Li, G., Chandra, S., Rashid, R. A. R., Palanisamy, S., & Ding, S. (2022). Machinability of additively manufactured titanium alloys: A comprehensive review. *Journal of Manufacturing Processes*, 75, 72–99.
124. Varghese, V., & Mujumdar, S. (2021). Micromilling-induced surface integrity of porous additive manufactured Ti6Al4V alloy. *Procedia Manufacturing*, 53, 387–394.
125. Ji, H., Gupta, M. K., Song, Q., Cai, W., Zheng, T., Zhao, Y., ... & Pimenov, D. Y. (2021). Microstructure and machinability evaluation in micro milling of selective laser melted Inconel 718 alloy. *Journal of Materials Research and Technology*, 14, 348–362.
126. Lerebours, A., Demangel, C., Dembinski, L., Bouvier, S., Rassineux, A., & Egles, C. (2020). Effect of the residual porosity of CoCrMo bearing parts produced by additive manufacturing on wear of polyethylene. *Biotribology*, 23, 100138.
127. Yamaguchi, H., Shinmura, T., & Kobayashi, A. (2001). Development of an internal magnetic abrasive finishing process for nonferromagnetic complex shaped tubes. *JSME International Journal Series C Mechanical Systems, Machine Elements and Manufacturing*, 44(1), 275–281.
128. Yamaguchi, H., & Graziano, A. A. (2014). Surface finishing of cobalt chromium alloy femoral knee components. *CIRP Annals*, 63(1), 309–312.
129. Kum, C. W., Sato, T., Guo, J., Liu, K., & Butler, D. (2018). A novel media properties-based material removal rate model for magnetic field-assisted finishing. *International Journal of Mechanical Sciences*, 141, 189–197.
130. Yamaguchi, H., Fergani, O., & Wu, P. Y. (2017). Modification using magnetic field-assisted finishing of the surface roughness and residual stress of additively manufactured components. *CIRP Annals*, 66(1), 305–308.

131. Wu, P. Y., Hirtler, M., Bambach, M., & Yamaguchi, H. (2020). Effects of build- and scan-directions on magnetic field-assisted finishing of 316L stainless steel disks produced with selective laser melting. *CIRP Journal of Manufacturing Science and Technology*, 31, 583–594.
132. Guo, J., Bai, J., Liu, K., & Wei, J. (2018). Surface quality improvement of selective laser sintered polyamide 12 by precision grinding and magnetic field-assisted finishing. *Materials & Design*, 138, 39–45.

Chapter 3

Vibratory surface finishing of additive components prepared from the laser powder bed fusion process

Hamaid Mahmood Khan and Ebubekir Koç

Fatih Sultan Mehmet Vakif University, Istanbul, Turkey

CONTENTS

3.1 INTRODUCTION

Traditionally seen as a promising technology for producing small, low-quality prototypes, 3D printing is widening its potential by becoming faster and more dependable. Instead of cutting a shape from a bigger block or casting a molten material in a mould, 3D printing entails adding items layer by layer, making the process more cost-effective and material-efficient [1]. Recent material and printing process advancements have resulted in the fast commercial expansion of 3D printer markets, making the technology more diverse and competitive than it has ever been [2]. Using the capacity to create unique designs such as lattice materials, medical implants, and conformal cooling channels, 3D printing has made a reality what was previously only a distant dream with casting or milling [3].

Powder Bed Fusion (PBF) and directed energy deposition (DED) continue to dominate the commercial side of metal-based AM processes, also extensively supporting polymers, and to a lesser extent, ceramics and composites [4]. Originally conceived as a manufacturing method to create plastic prototypes, significant modifications in AM laser induction and powder

DOI: 10.1201/9781003288619-3

deposition system led to the discovery of the dedicated metal-based selective laser melting (SLM), direct metal laser sintering (DMLS), and electron beam melting (EBM) processes [5]. A computer regulated heat source, such as a laser or electron beam, is used in the PBF system to fuse discrete powder particles dispersed in a layer in a protective environment of argon/nitrogen/vacuum to recreate a 2D contour of a 3D CAD model [4, 5].

While SLM uses a high-powered laser source to fuse powder particles in an inert environment of nitrogen or argon, EBM uses an electron beam in vacuum to build solid components. DED, on the other hand, sequentially adds feedstock material, either metal powder or wire, and energy to the build surface through a nozzle. Depending on the version, DED's energy source can be a laser, an electron beam, or an arc, which melts the powder particles in an airtight chamber prior to deposition [5, 6]. DED techniques, as compared to PBF procedures, can create big structures and repair damaged sections, albeit at a lower resolution. It is also relatively quick, resulting in efficient material usage and cost savings. However, unlike PBF processes, DEDs are not suitable to fabricate complex designs, like cooling channels and lattices. Between SLM and EBM, the SLM can handle a broad range of materials, including titanium, aluminium, steel, CoCr, Inconel, and different composites [6, 7]. In short, designers may use additive manufacturing technologies to make distinctive breakthroughs in the fundamental design of a component and build parts with sophisticated characteristics in a short period of time. PBF technologies driven by a laser source are commonly referred to as laser powder bed fusion (LPBF), and this terminology is used consistently throughout the present literature for SLM, DMLS, and selective laser sintering processes. Figure 3.1 shows the schematic model for the LPBF process.

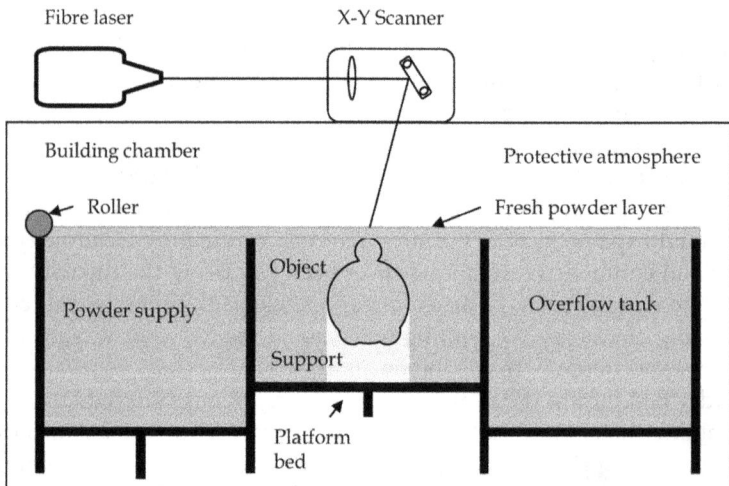

Figure 3.1 Schematic representation of a LPBF process.

Despite handling a wide range of materials and producing unique designs with properties comparable to wrought or superior to cast counterparts, metal AM has yet to make progress in overcoming challenges including poor surface roughness, staircase effect, dimensional accuracy, low fatigue strength, low wear and corrosion resistance, high structural porosity, and high residual stresses [5]. This truly confines the AM products to simple applications unless finished components are subjected to extra production steps, like thermal, chemical, or mechanical-based post-processing operations. Already, thermal processing of AM products has shown remarkable improvement in their overall mechanical and microstructural behaviours. However, heat-treatments are not fit to address challenges arising from the poor surface texture that are often linked to a low corrosion and wear performance. It is consequently essential to expose the surfaces of AM components to mechanical treatment in order to facilitate surface layer strengthening. Many research groups have worked on various post-processing solutions to alleviate these constraints [4]. Figure 3.2 compares the AM process with the existing conventional techniques in terms of product design, process, and physical, mechanical and microstructural behaviours.

There are several post-processing procedures available to inject desired mechanical and physical properties into PBF components. The final nature of the surface layer has been discovered to have a substantial impact on the mechanical performances of the treated component. Some of the most prevalent finishing methods in the market include shot peening, VSF, drag finishing, grinding, polishing, electrochemical polishing, CNC milling, abrasive flow machining, electroplating, and micro machining. VSF, a popular mass-finishing technology, is rapidly being used in a variety of commercial and industrial applications to increase mechanical strength and physical

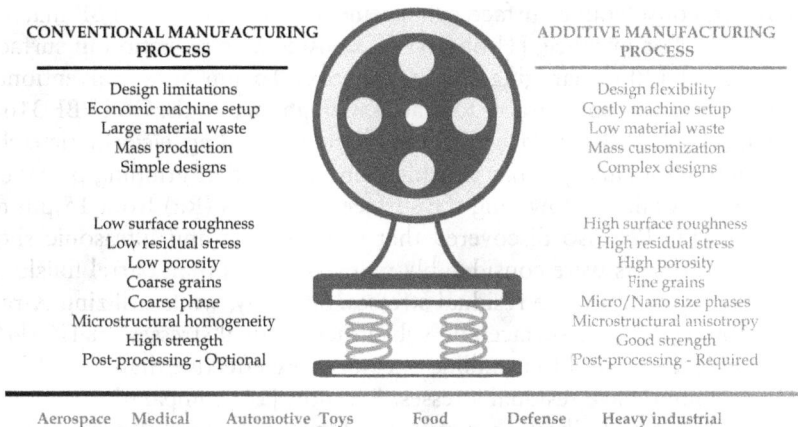

CONVENTIONAL MANUFACTURING PROCESS	ADDITIVE MANUFACTURING PROCESS
Design limitations	Design flexibility
Economic machine setup	Costly machine setup
Large material waste	Low material waste
Mass production	Mass customization
Simple designs	Complex designs
Low surface roughness	High surface roughness
Low residual stress	High residual stress
Low porosity	High porosity
Coarse grains	Fine grains
Coarse phase	Micro/Nano size phases
Microstructural homogeneity	Microstructural anisotropy
High strength	Good strength
Post-processing - Optional	Post-processing - Required

Aerospace   Medical   Automotive Toys   Food   Defense   Heavy industrial

Figure 3.2 Comparing additively manufactured components with the traditional ones.

properties of both commercial and PBF components. VSF consists of a large vibrating container containing multiple workpieces, one or more types of abrasive and additive media, and lubricating compound depending on the application to change the surface layer for enhanced mechanical strength and aesthetic goals. The VSF process exposes the workpiece surface to the moving abrasive media so that surface grinding, plastic deformation, and material erosion can all occur at the same time to flatten the workpiece surface and strengthen the surface layer with in-depth compressive residual stresses for improved fatigue performance.

This chapter aims to discuss the overall influence of the VSF technology on the mechanical and electrochemical performance of PBF components. Scholarly literature on VSF treatment of PBF parts are scarce, so this chapter has extensively studied the impact of VSF and other mass finishing methods on traditional components to comprehend the VSF mechanism, such as the relative motion of the media particle with respect to the workpiece surface, other media abrasives, and container walls. The influence of vibrating frequency, media rolling speed, eccentric weights, workpiece fixture, and other factors are also considered to evaluate the physical, mechanical, and electrochemical performances of VSF treated components. This chapter will cover both experimental and numerical analyses of VSF components in order to offer a comprehensive picture.

3.2 EXISTING MECHANICAL POST-PROCESSING METHODS AND THEIR LIMITATIONS

Traditional procedures such as glass bead blasting, conventional milling, electrical discharge machining (EDM), and surface grinding are available to provide desired surface finish to AM components. Several researchers have discovered considerable surface enhancement in a variety of LPBF materials [1, 8, 9]. For instance, [1] observed a considerable reduction in surface roughness of LPBF maraging steel to approx. 0.5 μm after conventional milling, which was comparable to that of wrought materials. On LPBF 316L SS, [10] evaluated the surface finishing effects of grinding, electrolytic etching, sandblasting, and plasma polishing, and found that grinding provided the greatest results by lowering the surface roughness (Ra) from 15 μm to 0.34 μm. [11, 12] also discovered that shot peening and ultrasonic shot peening treatments were considerably more successful than barrel finishing in generating compressive residual stresses. Similarly, while utilizing X-ray diffraction to measure surface and subsurface residual stresses on Ti6Al4V blades, [13] discovered that milling was far more effective than the VSF in inducing compressive residual stresses. Recently, [14] compared the impact of shot peening and vibration surface finishing on the fatigue performance of TC17 alloy. They discovered that using VSF after shot peening improved the overall surface integrity of the TC17 alloy.

However, the bulk of these procedures were only found to be compatible with the normal planar geometry, leaving the more complicated AM profile unaffected, which is the difficulty that post-processing approaches are designed to solve. To get over these constraints, mass finishing processes are becoming more popular as a way to improve the surface roughness of geometrically complicated, high-value AM products. Although several studies have addressed the benefits of mass finishing procedures, much more research is needed to standardize the influence of mass finishing processes on AM products. [15] discovered that traditionally sintered components had greater wear resistance than unsintered samples due to their improved surface hardness. When it comes to additive samples, [16–18] conducted a series of investigations and discovered that AM samples outperformed their as-built counterparts in mechanical strength and wear resistance. [19, 20] observed improved mechanical performance in LPBF samples following the mass finishing procedure. [21] observed comparable fatigue performance in VSF treated aluminium components to shot-peened ones. Moreover, VSF samples outperformed shot-peened samples in terms of surface quality. Such observations were also confirmed by other researchers [22]. Mass finishing is suited for large and medium-sized batches due to its capacity to handle several components at once, and many of its versions do not require clamping fixtures or intricate tool path design [23].

3.3 MASS FINISHING TECHNIQUES

Mass finishing processes are widely popular in manufacturing industries for surface finishing, cleaning, deburring, descaling, edge radiusing, and stress relieving complex freeform surfaces [24]. Its basic functionality involves immersing components in a big bowl filled with an abrasive media compound. Depending on the bowl movement and media-surface interaction, mass finishing processes can be categorized into several types, such as VSF, stream finishing, centrifugal barrel finishing [25], centrifugal disc mass finishing, cascade finishing [26], and drag finishing [20]. Figure 3.3 describes the basic functionality of some of the most popular mass finishing techniques in the market. Due to their widespread appeal, ease of use, and good surface finish, mass finishing techniques are becoming more prevalent in post-processing of complex AM components. Table 3.1 provides the process detail of different types of commercially available mass finishing techniques.

The key concept of using mass finishing techniques is to eliminate the excess material from the surface to minimize the surface irregularity, like surface roughness and subsurface porosities. Out of all the mass finishing techniques, VSF is one of the most popular mechanical-based post-processing technology to modify surfaces by creating compressive residual stresses and eliminating surplus material for a smooth surface finish [29]. It consists of a large container containing media pellets, which is subjected to a

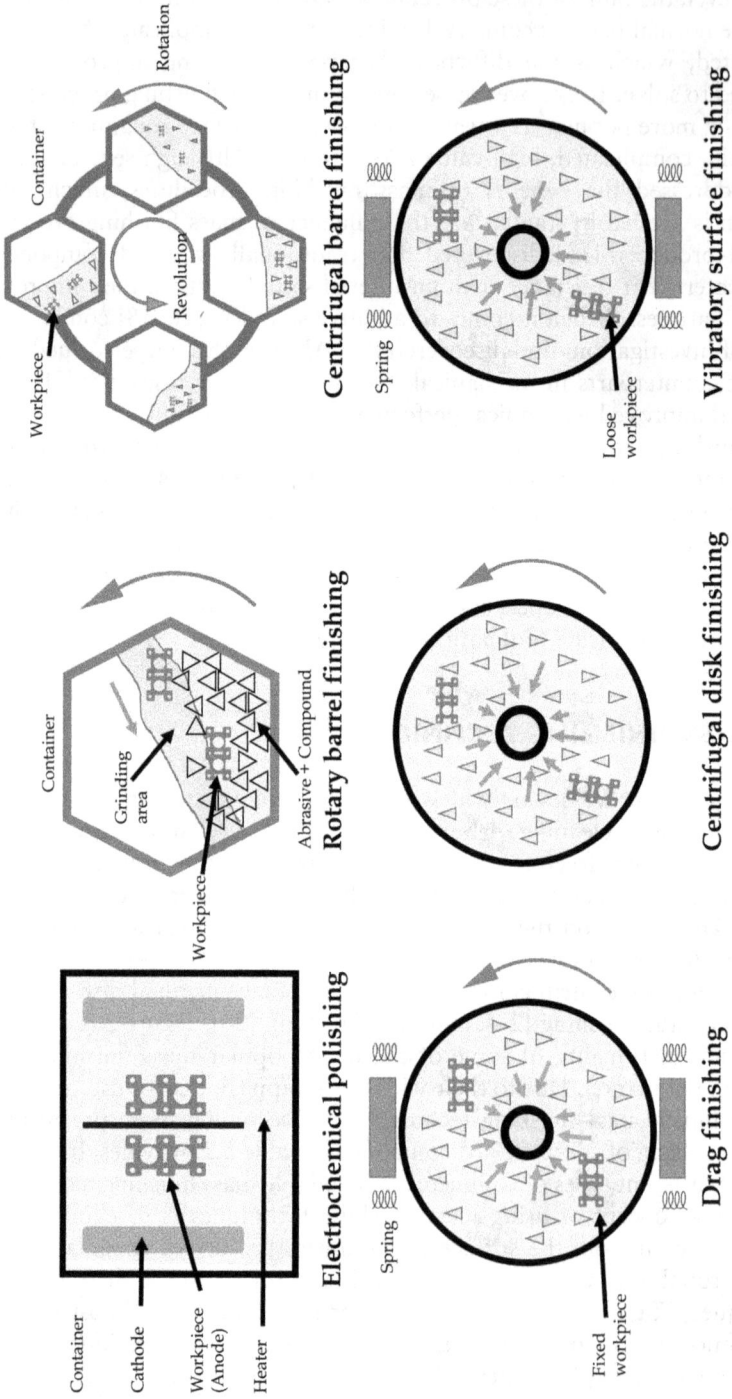

Figure 3.3 Schematic models of various mass finishing techniques.

Table 3.1 Definition and usage of a various mass finishing techniques

Mass-finishing process	Description, application, and usage
Stream finishing (SF) or Gyro finishing (GF)	A revolving workpiece is held and lowered into a rotating container containing polishing material. Short finishing times can achieve extremely low surface roughness because media rotation in SF is quicker than other mass finishing processes. SF is suitable for small batches only, with large and complex structures can be processed easily. Suitable for use in the aerospace and automotive industries [27].
Centrifugal disc finishing (CDF)	The workpiece is placed in an open drum containing polishing granule and a movable base plate to create a toroidal stream. The CDF finishing effect can produce high media-surface contact, making the process 20 times more efficient than traditional vibrators. Suitable for deburring and edge rounding of intricate workpieces.
Rotary barrel finishing (RBF)	RBF consists of a container in which a variety of workpieces interact with freely moving abrasive media. While the RBF container is positioned horizontally, it is rotated along its major axis to generate motion, allowing the abrasive material to travel at a high velocity and provide the required pressure to the workpiece. The circular motion causes the inserted workpieces to be bulk finished.
Centrifugal barrel finishing (CBF)	CBF is made up of many containers that spin and revolve around the primary axis at a rapid speed. This results in a very high centrifugal force that keeps the workpiece away from the container wall. As a result, the workpiece at the centre of the container experiences the maximum force that results in a far higher productivity for CBF than RBF.
Drag finishing (DF)	The workpiece is clamped in special holders that revolve at a high speed in a container with polishing granules. Due to the created high contact pressure between the workpiece and the media, DF can deliver high-precision results in a short period compared to VSF [20].
Electropolishing (EP)	EP removes material electrochemically. It involves workpiece (anode) immersion in an electrolyte solution to polish, passivate, and deburr surfaces. EP improves corrosion resistance and create mirror like finish.
Cascade finishing (CF)	A stream of small abrasives is supplied onto the clamped part from the outside as it oscillates in a housing. The media is pushed around the part by the housing oscillation, causing the abrasives to remove burrs with or without water. The addition of compounds or water speeds up the abrasive and cleaning action, resulting in a high rate of energy loss. Also suitable for cleaning internal channels.
Vibratory surface finishing (VSF)	VSF entails immersion of a part in a large bowl containing abrasive material. Continuous hammering of parts by a freely moving mass media, powered by bowl oscillation, results in parts' deburring, cleaning, corner radiusing, and stress relieving. Multiple materials and parts of various sizes may be handled at the same time, giving in a cost-effective and high-quality finishing approach [16, 28].

vibratory motion using a set of weights and an electric motor to intensify the media surface interaction, as shown in Figure 3.4a [30]. An aggressive movement of media pellets causes the development of normal and tangential force components on the workpiece surface. While normal impact causes compressive stresses on the surface due to the plastic deformation, transversal impact allows for a cutting action, resulting in mechanical abrasions on the component surface. Figure 3.4d shows the media behaviour on the workpiece surface. Given the dual impact of the vibrating media, components surface roughness reduces, and mechanical strength rises [19, 31, 32].

The performance of mass finishing techniques can be quantitatively assessed through parameters like material removal rate (MRR) and the final surface roughness, which are directly dependent on the selection of various parameters, as shown in Figure 3.4b. Governed by the excitation system, the net outcome of MRR and surface roughness of VSF treated components are generally found to influenced by three key variables:

  i. Media contact pressure.
  ii. Media impact and sliding velocity.
  iii. Media size and shape.

One of the thoroughly researched variables of the three is media contact pressure on a workpiece. To get a contact force data in the VSF process, force sensors have frequently been placed on a workpiece to gather the impact intensity of the abrasive media. Several studies have thoroughly examined the application of force sensors and other sensing devices on traditional structures [33]. [34] integrated the force-sensing device on the workpiece to monitor the normal acting force. Later, [35] utilized the same setup with some modifications to capture both normal and tangential force components. It was discovered that the media impact on the workpiece occurs in three distinct ways: free impacts, isolated media rolling, and neighbouring media rolling over the stationary media, Figure 3.4c. An experimental investigation by [36] shows that both the abrasive media configuration and eccentric weights are essential to define the dynamic behaviour of the VSF machine. The eccentric weights in the VSF system are installed on each end of a vertical shaft or motor. The amount of weight on the top eccentric regulates the speed of mass around the tub, while the bottom one controls the rollover rate of media mass [37]. In one relatively simplified setup, [38] tied a piezoelectric force sensor onto the workpiece, and measured the impact velocity and force signals by dropping a porcelain and steel spheres from a known height. In recent years, a research group carried out a series of experiments using a laser sensor to monitor the impact of media velocities inside the VSF tub [28, 39, 40]. By measuring the displacement signals using a laser displacement probe, they revealed that the density and elastic properties of the media and the sensor orientation can alter the velocity measurement. Lately, [41] provided a consistent and repeatable result by

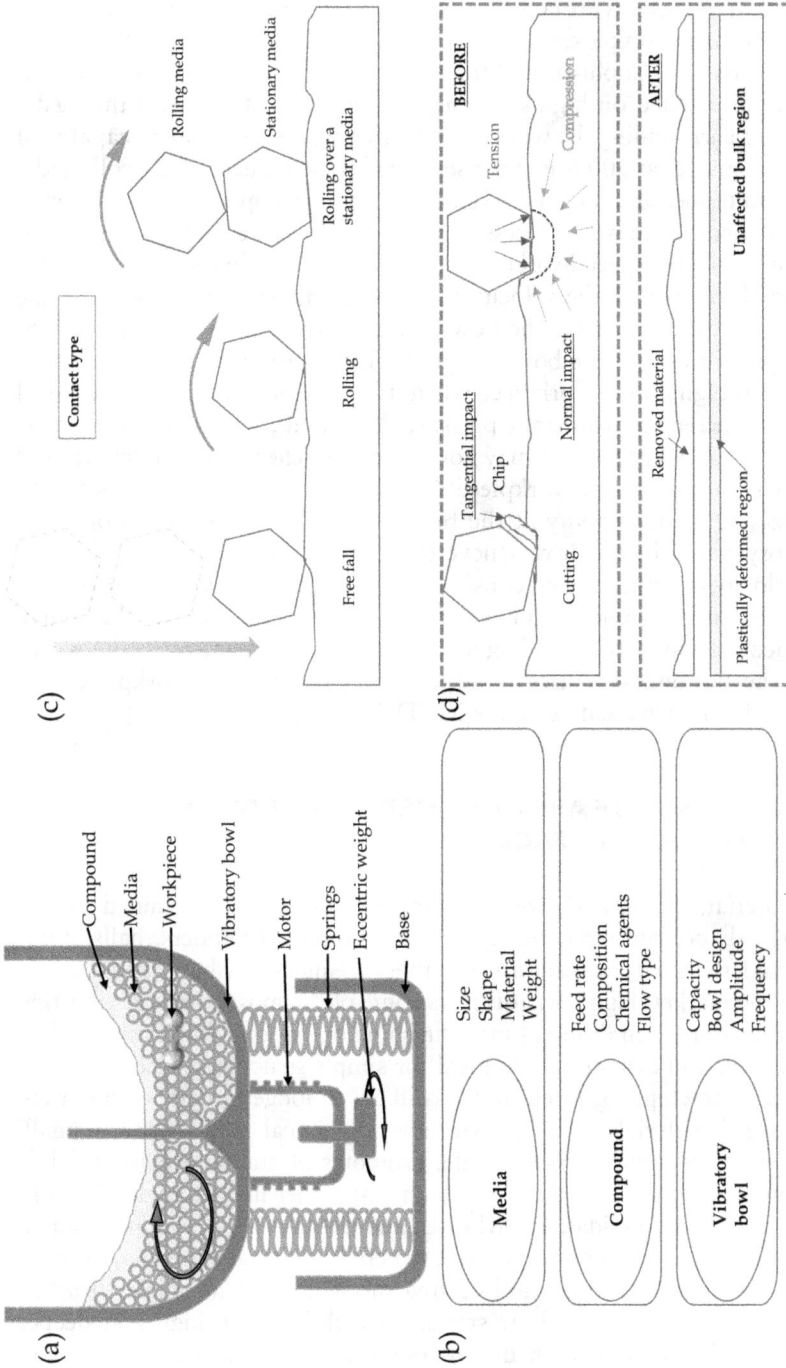

(a)

Compound
Media
Workpiece
Vibratory bowl
Motor
Springs
Eccentric weight
Base

(b)

| Media | Size
Shape
Material
Weight |

| Compound | Feed rate
Composition
Chemical agents
Flow type |

| Vibratory bowl | Capacity
Bowl design
Amplitude
Frequency |

(c)

Contact type

Rolling media

Stationary media

Rolling over a
stationary media

Free fall   Rolling

(d)

BEFORE

Tension   Compression

Tangential impact
Chip
Normal impact

Cutting

AFTER

Removed material

Unaffected bulk region

Plastically deformed region

Figure 3.4 (a) Schematic model of the VSF process; (b) various processing parameters of the VSF system; (c) different types of media-surface contacts; and (d) normal and tangential impact mechanism between media and the workpiece.

attaching a strain sensor to a workpiece holder for pressure measurement. Many of these systems are conceptually successful, but their implementation is expensive and complex.

Instead of using sophisticated force sensors or probing devices, researchers have also relied on high-speed imaging equipment to assess the media behaviour surrounding the workpiece. With a high-speed camera capable of recording images at 500 frame per second (fps), [42] successfully collected a large data to measure the media velocity for process optimization and overall assessment of the VSF process. Similarly, [43] used a 100-fps high-speed imaging through a specially built transparent window on a side of the vibratory bowl to estimate the velocity magnitude. However, in both cases, the camera was placed outside the bowl, limiting the velocity measurement to media present only at the bowl's edge. According to [44], the contact force intensity is high on the workpiece when it is positioned deeper in the barrel due to the increased hydrostatic pressure. As a result, velocity measurements of media particles at borders may not accurately reflect the local impact and slide velocities on a real workpiece. Recently, [45] used an X-ray setup to visualize the media activity at the bowl centre and to plot the workpiece trajectory inside the tub. It was, nevertheless, inadequate to accurately compute velocity magnitudes, but considerably better than previous approaches. A comprehensive wireless setup was recently developed in one of the unique approaches from [46] by adding an integrated video camera to the workpiece. The motion trajectories of the granular media and the workpiece were afterwards characterized using the MATLAB program.

3.4 INFLUENCE OF ABRASIVE MEDIA TYPE ON THE WORKPIECE SURFACE

The material, shape, and size of vibratory media are determined by the material, shape, and strength of the finishing part. Cylinders, balls, pyramids, and sharp-edged stars are the most frequent finishing media available in the market. Figure 3.5 displays some of the most common vibrating media shapes used in different mass finishing operations. Media shapes like round, oval, and cylindrical are ideal for simple structures since they have reduced wear, chipping rates, and less likely to lodge in components than sharp-edged materials [47, 48]. Moreover, spherical media, with a small/ orthogonal orientation impact angle, promotes plastic deformation, while the main mechanism becomes cutting only at higher impact angles. By contrast, triangles, pyramidal, arrowheads, and tri-star shapes are more suited for polishing tight regions of complex structures. Pyramidal media, for their high rheological behaviour and cutting mechanism at all angles, lead to much higher normal and shear stresses, which are even higher at deeper immersion depths. However, due to quicker wear, such media are more prone to chipping [49, 50]. In short, spherical media favours abrasion and micro-cutting, whereas triangular media facilitates cutting [51–53].

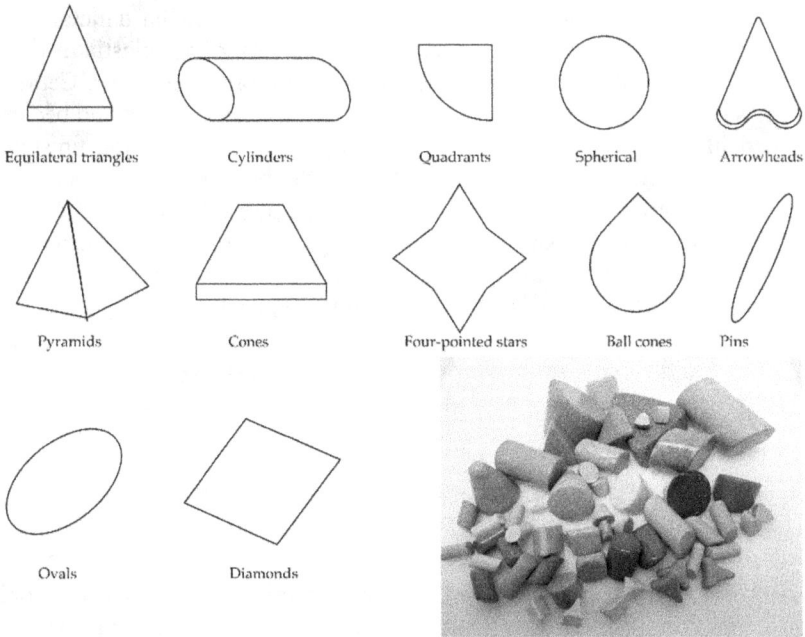

Figure 3.5 Some common types of vibrating finishing media.

In terms of size, a smaller media mass with a larger surface area provides more contact points, allowing for smoothing and fine finishing, but require longer finishing times due to the sensitive handling. Larger media, on the other hand, makes grinding more aggressive due to the high punching effect and inertia, allowing for quicker burr removal and smooth edge radiusing [25]. However, [54] noticed that the component damaging can be more severe with smaller media, when they investigated the effect of various abrasive media (size and shape) on the roughness evolution of the LPBF Ti6Al4V components. This might be attributed to the LPBF surface, which is known to have higher surface porosity and surface roughness.

[55] observed that a lack of debris filtering from the part and media particles may cause poor surface finishing. It is because the presence of debris scales has been linked to significant component distortion. According to [36], higher initial surface roughness correlates to more material removal since loose edges are weak in force and easy to remove. As a result, AM surfaces, which are higher in surface roughness can produce more debris than traditional components. More scientific investigation is thus necessary to comprehend the surface morphology of the LPBF components, so that the primary attention may be given to the subsurface porosity, high residual stresses, high surface roughness, and non-homogeneous microstructure.

For commercial systems, several media variants are available, such as ceramic, plastic, steel, and organic compounds. Ceramic and plastic media account for more than 90% of finishing media. Because of its high density, ceramic media is commonly used for grinding and polishing hard metals such as steel, stainless steel, and titanium, whereas polyester base plastic media is often used for "softer" metals such as aluminium, brass, and zinc. Ceramic media is typically strong and long-lasting; nonetheless, it can chip and become trapped in microscopic parts of metal components [47]. During finishing, both ceramic and plastic media are blended with abrasives. Silica, silicon carbide, aluminium oxide, and zirconium are examples of common abrasives. For aggressive grinding, silicon carbide and aluminium oxide are commonly employed on tougher metals. Steel media, on the other hand, is comprised of both hardened carbon and stainless steel and is typically used to deburr steel components and polish stainless steel [20]. As a result, the optimal media selection and their size optimization is important to achieve the desired result in the shortest amount of time and at the lowest possible cost.

3.5 INFLUENCE OF MEDIA ROTATION ON THE WORKPIECE SURFACE

A granular media can be fluidized by using an injected fluid flow or vibration of its container. Different factors affect the quality of fluidization in both cases such as fluid density and viscosity, media size and coefficients of restitution and friction. In the vibrational fluidization, the frequency and amplitude of vibration is effective [56]. [57] showed that higher vibration frequencies (f = 15–25 Hz) can boost impact intensity and media-surface contact density. A study from [43] demonstrated that MRR becomes aggressive at higher velocities, enabling faster finishing at higher vibration frequencies (f = 40 Hz) than traditional machines (f = 25 Hz). [58] also demonstrated that the force detected using a sensor at 75 Hz was more than that at 50 Hz, resulting in a bigger impact force and polishing effect.

[59] emphasized that the collective behaviour of rotating abrasive granules in the VSF process is comparable to that of gas molecules or atoms. As a result, the media impact on the workpiece generates radial and tangential force components, and that the impact strength varied with the rotational speed of the media particles inside the container. While [59] reported an increase in net MRR with rotating speed, [54] discovered that the surface roughness reduction was initially greater but eventually became saturated. For LPBF Ti6Al4V components exposed to the CBF process, [60] discovered that employing a faster rotating speed can generate superior surface quality. [61] also studied the effect of the rotating speed on work hardening, residual stresses, and surface topography of AISI4140 specimen using a modified Almen method. Their research demonstrated that a fast rotating speed yields deeper depths of induced compressive residual stresses and work hardening in a shorter possible time, making the process quite efficient.

It is important to highlight that the media rotation inside the container varies with its location, with media closer to the walls rotating slower than those in the middle [62]. According to recent research [41] stronger contact forces between media and the workpiece in the gyro finishing test may be produced when the workpiece is placed near the bottom of the container, as well as in the presence of a greater number of media particles. However, because of the freely floating behaviour of the workpiece inside the VSF container, its location is difficult to predict, and fluctuates throughout the operation. In cases where workpieces are secured to a fixture, such as those used in the drag finishing system, material removal rate and plastic deformation has been reported as higher due to the increased and severe media surface contact than those reported with the freely floating VSF system [16, 18]. They noticed that the surface layer became more refined with the drag finishing operation, raising the surface hardness and reducing the surface roughness significantly. Figure 3.6 describes the extent of plastic deformation by the VSF process on the LPBF prepared maraging steel surface.

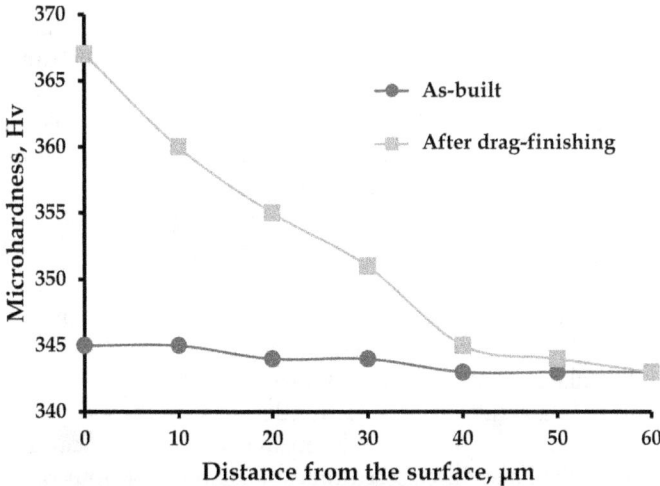

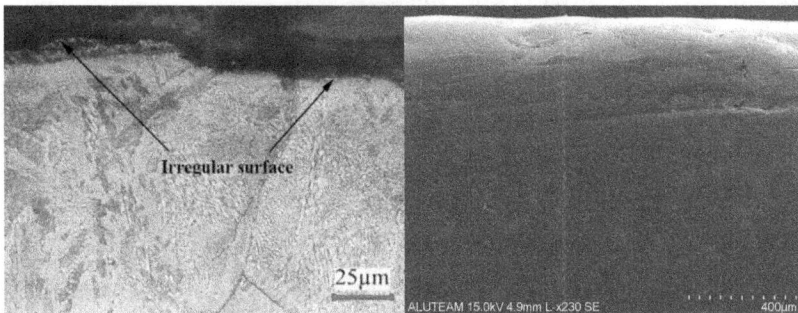

Figure 3.6 Extent of plastic deformation by the VSF process on LPBF maraging steel samples.

Commonly used fixtures are of the mechanical type where components are held on the fixture which is either rigidly bolted or freely revolving in the media; see Figure 3.3. Workpieces on a revolving fixture, such as those used in the drag finishing system, allow for double vibro polishing, in which the workpiece fixture rotate in addition to the revolving media particles in the container [63]. As a result, higher media impact intensity can be attained, and various studies have shown a considerable decrease in the surface roughness and an improvement in surface hardness of many AM metal alloys, such as LPBF 316L SS, IN718, and maraging steel alloys [16, 18, 20]. [63] were also able to reach the roughness saturation limit faster by adding the vibratory fixture to Ti6Al4V samples in a vibratory trough containing plastic media.

3.6 INFLUENCE OF THE PROCESS DURATION ON THE WORKPIECE SURFACE

The influence of the VSF process time on the workpiece surface has been numerous [19, 63, 64], and like any other parameter, it must be tuned to achieve the required surface integrity. [64] noticed significant improvements in the surface roughness of Ti6Al4V and Inconel718 geometries while working with the CBF technique for 10 hours at rotating speeds of around 100 rpm and higher. They noticed that on increasing the process time to 50 hours, no discernible effect was found. Similarly, after attaining a 70% decrease in the surface roughness of barrel finished 316L SS at 60 rpm and 40 min, [65] also did not observe any meaningful improvement in the surface roughness for longer periods. As previously stated, the media impact includes both normal and tangential force components, and when structures are exposed to rotating media, plastic deformation and material attrition of the surface begin immediately. As a result, the surface hardens due to the refined grains inside the surface layer, which lowers the media impact significantly and hence material loss. [19] also discovered that during the first hour of exposure, the hardness of the SLS PA12 increases from the as-built condition, which remained stable over the time. The material removal from the hardened surface reduces caused the MRR to be higher only in the beginning, and it then reduced continuously until a saturation limit was reached.

3.7 EFFECT OF LUBRICANTS ON THE WORKPIECE SURFACE

Lubrication has received less attention than other aspects in the VSF process. According to [33], abrasive media in the presence of a lubricant such as water imparts better surface finish to the component surface. [66] also observed that an aggressive material attrition of aluminium (Al6061T6) components occur in the presence of the chemical additive. However, water or detergent mixed abrasive exhibit less media impact and plastic deformation due to a lower dynamic coefficient of friction. As a result, the total surface roughness was found to be greatly decreased in a VSF process [34, 35]. Moreover, given the fact that high friction in dry conditions limits media flow, this allows for a low tangential impact of media particles on the surface. In contrast, lubricated compound increases the media impact in tangential motion, allowing for scratching and cutting rather than plastic deformation [67]. Furthermore, [52] demonstrates that a higher lubrication level might result in a higher degree of sliding, resulting in a smoother finish.

The leftover debris after the media impact can also pose serious issues to the final surface quality. This can be much more difficult for additive components, since evacuated debris density can be very high, potentially causing significant alteration in the media-surface interaction. Debris particles may grind polished surfaces, loosely fill fracture gaps, or partially fuse to the surface [19, 55]. Moreover, [67] noticed that the debris lodging into the workpiece surface can be severe in the dry state, but washes away in the water-sprayed condition.

3.8 NUMERICAL STUDIES

Despite the fact that the research on model training for the VSF process is sparse, it provides a comprehensive assessment of several key variables in order to understand their impact on the workpiece surface. Researchers have utilized statistical, geometrical, and numerical models to simulate the effect of VSF media, such as its speed, mass, volume, size, and shape, on the material removal rate, surface roughness reduction, and compressive stress evaluation of the workpiece surface. Their findings revealed that the nature of the media-surface interaction governs the intensity of the plastic deformation or material erosion from the workpiece surface. According to the statistical model presented by [68], the surface peening by the VSF media or the extent of plastic deformation of the workpiece surface was found to greatly influenced by the media impact frequency followed by media size and process duration. It is to note here that the surface finishing using the VSF process strongly depends on the workpiece material, its initial surface texture, and its position in the container [69]. [44] discovered that the initial surface texture is a key parameter in their mathematical model to define the extent of MRR from the workpiece surface.

Despite the greatest attempts to comprehend the VSF process using broad statistical or geometrical models employing sensors, probing devices, and high-speed cameras, a comprehensive knowledge remains elusive unless sufficient consideration is given to numerical or computational investigations. Numerical simulations are cost- and time-effective tools to simplify complex media motion surrounding the workpiece. Discrete element modelling

(DEM) can successfully model a wide range of granular flows, providing a comprehensive insight of the micro-dynamics of powder flows in various mass finishing techniques. Table 3.2 shows some of the computational based studies of different mass finishing techniques.

[70] successfully demonstrated that the DEM model is a viable alternative to the time and cost-intensive trial and error method to simulate media-surface interaction. A DEM approach to track media interactions with the workpiece within the VSF machine was utilized by [71]. Based on Hertzian contact mechanics, the DEM model was used to explore the effect of contact stiffness, friction, and damping on media motion in order to estimate normal and tangential contact forces. Similarly, [72] utilized the Hertz-Mindlin DEM model to investigate the media impact on the workpiece surface, and validated it with the experimental results. The Hertz-Mindlin contact force model is widely recognized for modelling contact between two elastic spheres or one sphere and a half space [73]. [74] used the DEM approach to simulate a material removal model based on Preston's law. Their modelling results for material removal and final surface roughness were within 12% of the experimental values on average. [75] combine high-speed imaging, particle imaging velocimetry, and computational fluid dynamics to demonstrate different media flow characteristics at the bottom than the top of the bowl.

In a micro-scale drag finishing process, [76] used a DEM approach to illustrate contact forces between the workpiece and the granular material. [49] observed that MRR in a centrifugal disc mass finishing machine is a function of density and hardness of the workpiece and the spherical granular media. In comparison to DEM simulations, a finite element (FE) continuum model of granular flows for the VSF process was found to cut computation time by around eight times [40]. [77] used Gene Expression Programming (GEP) and artificial neural network (ANN) methodologies to investigate the power consumption and MRR in the mass finishing process as a function of workpiece material, media, process time, and so on. They discovered that the media component had the greatest effect on the power consumption and MRR. [78] used the ADAMS and EDEM software to forecast MRR and final surface quality by simulating the granular media response to the vibrating flow field in a dynamic VSF model.

The fluidized granular material was modelled by [56] using DEM. When subjected to high-frequency shaking, media flow displayed waviness at their bottom walls and sharp leaps at their free surface. While media particles were found to slow down with increased vibration frequency, they were discovered to move quicker with increasing vibration amplitude due to the fluidized bed's higher kinetic energy. As a result, it was determined that increasing vibration amplitude would be more efficient than increasing vibration frequency for fluidization in processes such as mixing, segregation, and vibratory finishing. In a recent work, [79] used the VSF technique to treat the LPBF 316L SS cooling channel. They discovered that cooling efficiency improved greatly following the VSF treatment using 98.4 wt. %

Table 3.2 List of some computational studies on the VSF and other mass finishing techniques

Reference	Research model	Changing variables	Measurement variables	Effect
Hashimoto	Mathematical models – differential equations	Initial surface texture	Surface roughness and material removal	Surface roughness first reduces to a constant value "roughness limitation (R_L)", which depends on the initial surface roughness (R_i). If the difference $R_i - R_L$ is huge, MRR will be faster. At steady state, MRR is constant, which means independent of time.
[36]		Vibration frequency, workpiece mass, material, and velocity	Material removal rate	MRR is constant at a steady state. MRR depends on the bowl acceleration. However, [36] assumed that peening effect of granules was negligible on the workpiece, and that granules effect reduces over time.
[80]	DEM for 2D bed	Velocities, collision (forces) effect of media, workpiece, and container wall	Contact parameters: normal and tangential stiffness (k), normal and tangential damping (β), and sliding friction (μ)	Material removal depends on the relative motion of media and workpiece. N_V was 6–8 times T_V numerically, and normal contact force was found 10 times of tangential forces experimentally. Plastic deformation of workpiece depends on media impact.
[69]	DEM for 2D using two different media in a large container	Large setup, container depth	Surface finish	Media circulation increased with the container depth. Workpiece placed deeper in the container resulted in a better finish.

(Continued)

Table 3.2 (Continued)

Reference	Research model	Changing variables	Measurement variables	Effect
[81]	Geometry based model	Transient time	Surface roughness	MRR is proportional to surface roughness is assumed. MRR requires evaluation for different workpiece shapes, materials, initial roughness, and VSF media.
[76]	DEM and geometrical model	processing time, workpiece speed, excitation frequency, media type, and initial surface roughness	Transient MRR between R_I and R_L	For micro-cutting and predominantly abrasion mechanism: MRR depends on the relative motion between the media and workpiece. For predominantly cutting mechanism: MRR depends on the media impact. Excitation frequency given to media and workpiece speed improves surface finish for both VSF and DF processes.
[79]	DEM and VSF experimental study	Surface quality, cooling channel shapes	Surface roughness, cooling efficiency	VSF treatment improves the surface finish of the inside of cooling channel, resulting in improved cooling efficiency by reducing overall cooling time. Various cooling channel shapes and their efficiencies were modelled with the DEM method.

satellite media, 0.738 wt. % water, and 0.837 wt. % 10 μm sized SiC abrasives. Furthermore, a DEM analysis was carried out using various channel shapes to assess the surface finishing impact of the VSF process.

3.9 FUTURE POSSIBILITIES AND SUMMARY

AM is revolutionizing various sectors by enabling the manufacturing of lightweight, robust, visually appealing, and personalized components in a timely and cost-effective manner. Advances in computational and process technology/software, low-cost equipment, innovative feedstock materials, and improved automation are all expected to accelerate the AM process. However, intrinsic defects in the AM process, microstructural heterogeneity, and post-processing limits present a variety of scientific and technological issues that may prevent AM from becoming completely commercialized. As a result, despite all of the positive feedback and technological acceptance, AM structures must be upgraded to overcome process limitations in order to meet important quality criteria for real-world industrial applications.

VSF, a popular mass finishing technology, is a feasible and an efficient way to add physical and mechanical properties to AM products. VSF processes with optimal parameters, such as media type, lubricant, vibration frequency, and more, are progressively becoming a realistic alternative for a number of sectors to make ready-to-go products. Given the effective media influence on the workpiece, the resulting surface properties, such as exceptionally reduced surface roughness and higher compressive stresses, can turn any AM structure into an industrial-ready component. However, contrary to the comprehensive study on conventional components, scientific findings on additive components are limited. Given the substantial differences in the mechanical and microstructural properties of AM materials, research into the VSF method is anticipated to yield more critical insights.

Many post-processing methods can overcome AM flaws to a greater extent. However, mechanical processing of many AM materials, notably lattice materials, remains a challenge. Post-processing of lattice materials is exceptionally hard due to the micro or mesoscale specified porosities. Vibrational surface finishing can handle this issue to some extent by targeting the inner areas, which is otherwise a difficult aspect with typical machining, shot peening, or polishing methods. Although electrochemical solutions can reach inner zones, they are unfit to mechanically upgrade the lattice strut boundaries. On the contrary, lattice materials may be successfully polished using micro and mesoscale VSF media. This area is open for research to address mechanical, physical, electrochemical, and wear behaviours of VSF treated lattice materials.

Although there are many numerical studies on traditionally manufactured components, such studies on AM structures are extremely limited; in fact, no major advances have been documented in the published literature. Given the extremely high surface roughness, subsurface porosity, and extremely high residual stresses of AM structures, incorporating AM defects into geometrical and DEM models with and without lubricant can be critical to understanding the overall impact of the VSF process on AM materials.

Indeed, the VSF method is not well adapted to modifying the mechanical and physical properties of AM materials in the same way as shot peening and milling. As a result, for complex AM materials, a hybrid effect of numerous other mechanical techniques beside the VSF process could be an important study topic to enhance the mechanical properties of AM materials, like lattices. Furthermore, such combinations can also facilitate the creation of novel lattice designs. Aside from mechanical post-processing procedures, VSF may also be used in conjunction with the chemical and thermal processing techniques to provide exceptional results on a wide range of AM components.

VSF processes, as well as other variants of mass finishing techniques, have hitherto been used only with standard parameters supplied by manufacturers for broad comparison research. More insight into the selection of process parameters could be a significant step towards optimizing a wide range of AM geometries, designs, and materials. To summarize, VSF is a realistic and cost-effective option to mechanically post-process AM materials that should be investigated further to standardize the net effect of VSF on AM materials.

REFERENCES

1. Tascioglu, E., Khan, H. M., Kaynak, Y., Coşkun, M., Tarakci, G., & Koç, E. (2021). Effect of aging and finish machining on the surface integrity of selective laser melted maraging steel. *Rapid Prototyping Journal, ahead-of-p*(ahead-of-print). https://doi.org/10.1108/RPJ-11-2020-0269
2. Khan, H. M., Waqar, S., & Koç, E. (2022). Evolution of temperature and residual stress behavior in selective laser melting of 316L stainless steel across a cooling channel. *Rapid Prototyping Journal, ahead-of-p*(ahead-of-print). https://doi.org/10.1108/RPJ-09-2021-0237
3. Khan, H. M., Dirikolu, M. H., & Koç, E. (2018). Parameters optimization for horizontally built circular profiles: Numerical and experimental investigation. *Optik, 174*(August), 521–529. https://doi.org/10.1016/j.ijleo.2018.08.095
4. Khan, H. M., Karabulut, Y., Kitay, O., Kaynak, Y., & Jawahir, I. S. (2021). Influence of the post-processing operations on surface integrity of metal components produced by laser powder bed fusion additive manufacturing: A review. *Machining Science and Technology, 25*(1), 118–176. https://doi.org/10.1080/10910344.2020.1855649
5. Khan, H. M., Özer, G., Yilmaz, M. S., & Koç, E. (2022). Corrosion of additively manufactured metallic components: A review. *Arabian Journal for Science and Engineering.* https://doi.org/10.1007/s13369-021-06481-y
6. DebRoy, T., Wei, H. L., Zuback, J. S., Mukherjee, T., Elmer, J. W., Milewski, J. O., … Zhang, W. (2018). Additive manufacturing of metallic components – Process,

structure and properties. *Progress in Materials Science*, *92*, 112–224. https:// doi.org/10.1016/j.pmatsci.2017.10.001

7. Zadpoor, A. A. (2019). Mechanical performance of additively manufactured meta-biomaterials. *Acta Biomaterialia*, *85*, 41–59. https://doi.org/10.1016/j. actbio.2018.12.038

8. Khorasani, A. M., Gibson, I., Chegini, N. G., Goldberg, M., Ghasemi, A. H., & Littlefair, G. (2016). An improved static model for tool deflection in machining of Ti–6Al–4V acetabular shell produced by selective laser melting. *Measurement*, *92*, 534–544.

9. Kaynak, Y., Tascioglu, E., Poyraz, Ö., Orhangül, A., & Ören, S. (2020). *The Effect of Finish-Milling Operation on Surface Quality and Wear Resistance of Inconel 625 Produced by Selective Laser Melting Additive Manufacturing. Advanced Surface Enhancement.* Advanced Surface Enhancement. https://doi. org/10.1007/978-981-15-0054-1_27

10. Petters, R., Kühn, U., Löber, L., Flache, C., & Eckert, J. (2013). Comparison of different post processing technologies for SLM generated 316l steel parts. *Rapid Prototyping Journal*, *19*(3), 173–179. https://doi. org/10.1108/13552541311312166

11. Lesyk, D., Martinez, S., Mordyuk, B., Dzhemelinskyi, V., & Lamikiz, A. (2019). Surface finishing of complexly shaped parts fabricated by selective laser melting. In *Grabchenko's International Conference on Advanced Manufacturing Processes* (pp. 186–195). Springer.

12. Lesyk, D. A., Martinez, S., Mordyuk, B. N., Dzhemelinskyi, V. V., Lamikiz, A., & Prokopenko, G. I. (2020). Post-processing of the Inconel 718 alloy parts fabricated by selective laser melting: Effects of mechanical surface treatments on surface topography, porosity, hardness and residual stress. *Surface and Coatings Technology*, *381*, 125136.

13. Zhang, J.-Y., Yao, C.-F., Cui, M.-C., Tan, L., & Sun, Y.-Q. (2021). Three-dimensional modeling and reconstructive change of residual stress during machining process of milling, polishing, heat treatment, vibratory finishing, and shot peening of fan blade. *Advances in Manufacturing*, *9*(3), 430–445. https://doi.org/10.1007/s40436-021-00351-4

14. Shi, H., Liu, D., Pan, Y., Zhao, W., Zhang, X., Ma, A., … Wang, W. (2021). Effect of shot peening and vibration finishing on the fatigue behavior of TC17 titanium alloy at room and high temperature. *International Journal of Fatigue*, *151*, 106391. https://doi.org/10.1016/j.ijfatigue.2021.106391

15. Karthik, R., Elangovan, K., Girisha, K. G., & Shankar, S. (2022). Experimental investigation on wear behaviour of laser sintered and unsintered Inconel 718 in vibratory finishing. *Materials Today: Proceedings*. https://doi.org/10.1016/j. matpr.2022.04.201

16. Kaynak, Y., & Tascioglu, E. (2020). Post-processing effects on the surface characterisics of Inconel 718 alloy fabricated by selective laser melting additive manufacturing. *Progress in Additive Manufacturing*, *5*(2), 221–234. https:// doi.org/10.1007/s40964-019-00099-1

17. Kaynak, Y., & Kitay, O. (2018). Porosity, surface quality, microhardness and microstructure of selective laser melted 316L stainless steel resulting from finish machining. *Journal of Manufacturing and Materials Processing*, *2*(2), 36. https://doi.org/10.3390/jmmp2020036

18. Kaynak, Y., & Kitay, O. (2019). The effect of post-processing operations on surface characteristics of 316L stainless steel produced by selective laser

melting. *Additive Manufacturing*, 26(December 2018), 84–93. https://doi.org/10.1016/j.addma.2018.12.021

19. Khan, H. M., Sirin, T. B., Tarakci, G., Bulduk, M. E., Coskun, M., Koç, E., & Kaynak, Y. (2021). Improving the surface quality and mechanical properties of selective laser sintered PA2200 components by the vibratory surface finishing process. *SN Applied Sciences*, 2, 1–14.

20. Khan, H. M., Özer, G., Tarakci, G., Coşkun, M., Koç, E., & Kaynak, Y. (2021). The impact of aging and drag-finishing on the surface integrity and corrosion behavior of the selective laser melted maraging steel samples. *Materialwissenschaft und Werkstofftechnik*, 52(1), 60–73. https://doi.org/10.1002/mawe.202000139

21. Sangid, M. D., Stori, J. A., & Ferriera, P. M. (2011). Process characterization of vibrostrengthening and application to fatigue enhancement of aluminum aerospace components—part I. Experimental study of process parameters. *The International Journal of Advanced Manufacturing Technology*, 53(5), 545–560. https://doi.org/10.1007/s00170-010-2857-2

22. Canals, L., Badreddine, J., McGillivray, B., Miao, H. Y., & Levesque, M. (2019). Effect of vibratory peening on the sub-surface layer of aerospace materials Ti-6Al-4V and E-16NiCrMo13. *Journal of Materials Processing Technology*, 264, 91–106. https://doi.org/10.1016/j.jmatprotec.2018.08.023

23. Kolganova, E. N., Goncharov, V. M., & Fedorov, A. V. (2019). Investigation of deburring process at vibro-abrasive treatment of parts having small grooves and holes. *Materials Today: Proceedings*, 19, 2368–2373. https://doi.org/10.1016/j.matpr.2019.07.726

24. Li, W. H., Yang, S. Q., Li, X. H., & Li, W. D. (2014). Development status and trends of mass finishing processes. In *Key Engineering Materials* (Vol. 621, pp. 111–120). Trans Tech Publ.

25. Khorasani, M., Gibson, I., Ghasemi, A., Brandt, M., & Leary, M. (2020). On the role of wet abrasive centrifugal barrel finishing on surface enhancement and material removal rate of LPBF stainless steel 316L. *Journal of Manufacturing Processes*, 59, 523–534. https://doi.org/10.1016/j.jmapro.2020.09.058

26. Wang, X., Wang, Y., Yang, S., & Hao, Y. (2021). Analysis of velocity of granular media in cascade finishing by discrete element method with experimental validation. *Granular Matter*, 23(4), 83. https://doi.org/10.1007/s10035-021-01154-x

27. Tan, K. L., Neoh, E. T., Lifton, J. J., Cheng, W. S., & Itoh, S. (2022). Internal measurement of media sliding velocity in a stream finishing bowl. *The International Journal of Advanced Manufacturing Technology*. https://doi.org/10.1007/s00170-022-09053-y

28. Hashemnia, K., & Spelt, J. K. (2014). Particle impact velocities in a vibrationally fluidized granular flow: Measurements and discrete element predictions. *Chemical Engineering Science*, 109, 123–135. https://doi.org/10.1016/j.ces.2014.01.027

29. Ciampini, D., Papini, M., & Spelt, J. K. (2009). Modeling the development of Almen strip curvature in vibratory finishing. *Journal of Materials Processing Technology*, 209(6), 2923–2939.

30. Gillespie, L. (2006). *Mass Finishing Handbook*. Industrial Press.

31. Tong, Q., Xue, K., Wang, T., & Yao, S. (2020). Laser sintering and invalidating composite scan for improving tensile strength and accuracy of SLS parts. *Journal of Manufacturing Processes*, 56, 1–11.

32. Karthik, R., Elangovan, K., Shankar, S., & Girisha, K. G. (2022). An experimental analysis on surface roughness of the selective laser sintered and unsintered Inconel 718 components using vibratory surface finishing process. *Materials Today: Proceedings.* https://doi.org/10.1016/j.matpr.2022.04.448

33. Mediratta, R., Ahluwalia, K., & Yeo, S. H. (2016). State-of-the-art on vibratory finishing in the aviation industry: An industrial and academic perspective. *International Journal of Advanced Manufacturing Technology, 85*(1–4), 415–429. https://doi.org/10.1007/s00170-015-7942-0

34. Wang, S., Timsit, R. S., & Spelt, J. K. (2000). Experimental investigation of vibratory finishing of aluminum. *Wear, 243*(1), 147–156. https://doi.org/10.1016/S0043-1648(00)00437-3

35. Yabuki, A., Baghbanan, M. R., & Spelt, J. K. (2002). Contact forces and mechanisms in a vibratory finisher. *Wear, 252*(7–8), 635–643. https://doi.org/10.1016/S0043-1648(02)00016-9

36. Domblesky, J., Evans, R., & Cariapa, V. (2004). Material removal model for vibratory finishing. *International Journal of Production Research, 42*(5), 1029–1041.

37. Nebiolo, W. P., & Chemicals, R. E. M. (2008). An easily understood technique for measuring vibratory bowl speed and optimizing vibratory bowl processing efficiency. *Plating and Surface Finishing, 95*(4), 14.

38. Ciampini, D., Papini, M., & Spelt, J. K. (2007). Impact velocity measurement of media in a vibratory finisher. *Journal of Materials Processing Technology, 183*(2–3), 347–357.

39. Hashemnia, K., Mohajerani, A., & Spelt, J. K. (2013). Development of a laser displacement probe to measure particle impact velocities in vibrationally fluidized granular flows. *Powder Technology, 235*, 940–952. https://doi.org/10.1016/j.powtec.2012.12.001

40. Hashemnia, K., & Spelt, J. K. (2015). Finite element continuum modeling of vibrationally-fluidized granular flows. *Chemical Engineering Science, 129*, 91–105. https://doi.org/10.1016/j.ces.2015.02.025

41. Hashimoto, Y., Ito, T., Nakayama, Y., Furumoto, T., & Hosokawa, A. (2021). Fundamental investigation of gyro finishing experimental investigation of contact force between cylindrical workpiece and abrasive media under dry condition. *Precision Engineering, 67*, 123–136.

42. Fleischhauer, E., Azimi, F., Tkacik, P., Keanini, R., & Mullany, B. (2016). Application of particle image velocimetry (PIV) to vibrational finishing. *Journal of Materials Processing Technology, 229*, 322–328.

43. Pandiyan, V., Castagne, S., & Subbiah, S. (2016). High frequency and amplitude effects in vibratory media finishing. *Procedia Manufacturing, 5*, 546–557.

44. Hashimoto, F., & DeBra, D. B. (1996). Modelling and optimization of vibratory finishing process. *CIRP Annals, 45*(1), 303–306.

45. Zhang, C., Liu, W., Wang, S., Liu, Z., Morgan, M., & Liu, X. (2020). Dynamic modeling and trajectory measurement on vibratory finishing. *The International Journal of Advanced Manufacturing Technology, 106*(1), 253–263.

46. Lachenmaier, M., Brocker, R., Trauth, D., & Klocke, F. (2018). Analysis of the relative velocity and its influence on the process results in unguided vibratory finishing. *Journal of Manufacturing Science and Engineering, 140*(3), 1–9.

47. Uhlmann, E., & Eulitz, A. (2018). Influence of ceramic media composition on material removal in vibratory finishing. *Procedia CIRP, 72*, 1445–1450.

48. Nebiolo, W. P., & Chemicals, R. E. M. (2010). Increasing vibratory efficiency by optimizing the choice of media shape relative to part geometry. *Cell, 860,* 985–3758.
49. Cariapa, V., Park, H., Kim, J., Cheng, C., & Evaristo, A. (2008). Development of a metal removal model using spherical ceramic media in a centrifugal disk mass finishing machine. *The International Journal of Advanced Manufacturing Technology, 39*(1), 92–106.
50. Malkorra, I., Souli, H., Salvatore, F., Arrazola, P., Rech, J., Cici, M., ... Rolet, J. (2021). Modeling of drag finishing—Influence of abrasive media shape. *Journal of Manufacturing and Materials Processing, 5*(2), 41.
51. Uhlmann, E., Dethlefs, A., & Eulitz, A. (2014). Investigation of material removal and surface topography formation in vibratory finishing. *Procedia CIRP, 14,* 25–30.
52. Srivastava, S., Qin, C. Z., & Castagne, S. (2015). Effect of workpiece orientation, lubrication and media geometry on the effectiveness of vibratory finishing of Al6061. In *MATEC Web of Conferences* (Vol. 30, p. 4001). EDP Sciences.
53. Malkorra, I., Souli, H., Claudin, C., Salvatore, F., Arrazola, P., Rech, J., ... Rolet, J. (2021). Identification of interaction mechanisms during drag finishing by means of an original macroscopic numerical model. *International Journal of Machine Tools and Manufacture, 168,* 103779. https://doi.org/10.1016/j.ijmachtools.2021.103779
54. Salvatore, F., Grange, F., Kaminski, R., Claudin, C., Kermouche, G., Rech, J., & Texier, A. (2017). Experimental and numerical study of media action during tribofinishing in the case of SLM titanium parts. *Procedia CIRP, 58,* 451–456. https://doi.org/10.1016/j.procir.2017.03.251
55. Malkorra, I., Salvatore, F., Rech, J., Arrazola, P., Tardelli, J., & Mathis, A. (2020). Influence of lubrication condition on the surface integrity induced during drag finishing. *Procedia CIRP, 87,* 245–250. https://doi.org/10.1016/j.procir.2020.02.087
56. Hashemnia, K., & Pourandi, S. (2018). Study the effect of vibration frequency and amplitude on the quality of fluidization of a vibrated granular flow using discrete element method. *Powder Technology, 327,* 335–345. https://doi.org/10.1016/j.powtec.2017.12.097
57. Song, X., Chaudhari, R., & Hashimoto, F. (2014). Experimental investigation of vibratory finishing process. In *International Manufacturing Science and Engineering Conference* (Vol. 45813, p. V002T02A013). American Society of Mechanical Engineers.
58. Wong, B. J., Majumdar, K., Ahluwalia, K., & Yeo, S. H. (2019). Effects of high frequency vibratory finishing of aerospace components. *Journal of Mechanical Science and Technology, 33*(4), 1809–1815.
59. Kundrák, J., Mitsyk, A. V., Fedorovich, V. A., Markopoulos, A. P., & Grabchenko, A. I. (2021). Simulation of the circulating motion of the working medium and metal removal during multi-energy processing under the action of vibration and centrifugal forces. *Machines.* https://doi.org/10.3390/machines9060118
60. Nalli, F., Bottini, L., Boschetto, A., Cortese, L., & Veniali, F. (2020). Effect of industrial heat treatment and barrel finishing on the mechanical performance of Ti6AL4V processed by selective laser melting. *Applied Sciences, 10*(7), 2280.

61. Kacaras, A., Gibmeier, J., Zanger, F., & Schulze, V. (2018). Influence of rotational speed on surface states after stream finishing. *Procedia CIRP, 71*, 221–226. https://doi.org/10.1016/j.procir.2018.05.067

62. da Silva Maciel, L., & Spelt, J. K. (2020). Measurements of wall-media contact forces and work in a vibratory finisher. *Powder Technology, 360*, 911–920. https://doi.org/10.1016/j.powtec.2019.10.066

63. Ahluwalia, K., Mediratta, R., & Yeo, S. H. (2017). A novel approach to vibratory finishing: Double vibro-polishing. *Materials and Manufacturing Processes, 32*(9), 998–1003. https://doi.org/10.1080/10426914.2016.1232812

64. Boschetto, A., Bottini, L., Macera, L., & Veniali, F. (2020). Post-processing of complex SLM parts by barrel finishing. *Applied Sciences, 10*(4), 1382.

65. Na, W., Tingting, Z., Sheng-Qiang, Y., Wenhui, L., & Kai, Z. (2020). Experiment and simulation analysis on the mechanism of the spindle barrel finishing. *The International Journal of Advanced Manufacturing Technology, 109*(1), 57–74.

66. Malkorra, I., Salvatore, F., Arrazola, P., & Rech, J. (2020). The influence of the process parameters of drag finishing on the surface topography of aluminium samples. *CIRP Journal of Manufacturing Science and Technology, 31*, 200–209. https://doi.org/10.1016/j.cirpj.2020.05.010

67. Baghbanan, M. R., Yabuki, A., Timsit, R. S., & Spelt, J. K. (2003). Tribological behavior of aluminum alloys in a vibratory finishing process. *Wear, 255*(7), 1369–1379. https://doi.org/10.1016/S0043-1648(03)00124-8

68. Sofronas, A., & Taraman, S. (1979). Model development and optimization of vibratory finishing process. *International Journal of Production Research, 17*(1), 23–31.

69. Naeini, S. E., & Spelt, J. K. (2011). Development of single-cell bulk circulation in granular media in a vibrating bed. *Powder Technology, 211*(1), 176–186.

70. Uhlmann, E., & Kopp, M. (2021). Measurement and modeling of contact forces during robot-guided drag finishing. *Procedia CIRP, 102*, 518–523. https://doi.org/10.1016/j.procir.2021.09.088

71. Kang, Y. S., Hashimoto, F., Johnson, S. P., & Rhodes, J. P. (2017). Discrete element modeling of 3D media motion in vibratory finishing process. *CIRP Annals, 66*(1), 313–316. https://doi.org/10.1016/j.cirp.2017.04.092

72. Beigmoradi, S., & Vahdati, M. (2021). Investigation of vibratory bed effect on abrasive drag finishing: A DEM study. *World Journal of Engineering, ahead-of-p*(ahead-of-print). https://doi.org/10.1108/WJE-03-2021-0171

73. Thornton, C., Cummins, S. J., & Cleary, P. W. (2013). An investigation of the comparative behaviour of alternative contact force models during inelastic collisions. *Powder Technology, 233*, 30–46.

74. Makiuchi, Y., Hashimoto, F., & Beaucamp, A. (2019). Model of material removal in vibratory finishing, based on Preston's law and discrete element method. *CIRP Annals, 68*(1), 365–368. https://doi.org/10.1016/j.cirp.2019.04.082

75. Sood, A., & Mullany, B. (2021). Advanced surface analysis to identify media-workpiece contact modes in a vibratory finishing process. *Procedia Manufacturing, 53*, 155–161. https://doi.org/10.1016/j.promfg.2021.06.077

76. Uhlmann, E., Eulitz, A., & Dethlefs, A. (2015). Discrete element modelling of drag finishing. In *Procedia CIRP* (Vol. 31, pp. 369–374). Elsevier B.V. https://doi.org/10.1016/j.procir.2015.03.021

77. Vijayaraghavan, V., & Castagne, S. (2016). Sustainable manufacturing models for mass finishing process. *The International Journal of Advanced Manufacturing Technology, 86*(1), 49–57.

78. Wang, X., Yang, S., Li, W., & Wang, Y. (2018). Vibratory finishing co-simulation based on ADAMS-EDEM with experimental validation. *The International Journal of Advanced Manufacturing Technology, 96*(1), 1175–1185. https://doi.org/10.1007/s00170-018-1639-0

79. Toh, B. L., Gopasetty, S. K., Nagalingam, A. P., Alcaraz, J. Y. I. I., Jing, Z., Yeo, S. H., & Gopinath, A. (2021). Vibratory finishing of laser powder bed fused stainless steel 316 internal cooling channels. In *International Conference on Advanced Surface Enhancement* (pp. 13–16). Springer.

80. Naeini, S. E., & Spelt, J. K. (2009). Two-dimensional discrete element modeling of a spherical steel media in a vibrating bed. *Powder Technology, 195*(2), 83–90.

81. Uhlmann, E., Dethlefs, A., & Eulitz, A. (2014). Investigation into a geometry-based model for surface roughness prediction in vibratory finishing processes. *International Journal of Advanced Manufacturing Technology, 75*(5–8), 815–823. https://doi.org/10.1007/s00170-014-6194-8

Chapter 4

Ball burnishing of additively manufactured parts

Aref Azami
University of Strathclyde, Glasgow, UK

*Ramon Jerez-Mesa, Jordi Lluma-Fuentes and
Jose Antonio Travieso-Rodríguez*
Barcelona East School of Engineering (EEBE), Universitat Politècnica de
Catalunya (UPC), Barcelona, Spain

CONTENTS

4.1 INTRODUCTION

In recent years, additive manufacturing (AM) technologies have been developed significantly and can be considered an important revolution in the field of advanced manufacturing processes. Due to the remarkable ability of this family of technologies in the manufacture of parts with complex geometry, their demand in various industries, especially advanced sectors such as aerospace, medicine, and automobile, amongst others, is increasing. However, AM faces significant challenges in order to be used as a mass production method in various industries [1]. One of the most important challenges that AM technologies face is the surface quality of the parts produced by them. In fact, the high surface roughness and internal defects of the parts manufactured by AM are known as their main disadvantages. Surface roughness

DOI: 10.1201/9781003288619-4

is recognized as one of the most important factors of surface quality of any part that plays an important role in the life and performance of the part [2]. In fact, surface roughness is known as stress focal points that can lead to cracking and growth, reduced fatigue life, and ultimately failure of the part. Therefore, in order to reduce surface asperities in particular and enhance the surface integrity, it is necessary to perform post-processing operations on the parts, such as heat treatments or polishing [3]. The burnishing process is known as one of the finishing methods in which using plastic pressure applied by a hard ball or roller on the surface of the part, plastic deformation is done under hard working conditions, which increases the surface integrity. The burnishing process as a finishing operation can reduce roughness, increase hardness, and improve wear and corrosion resistance. It also causes compressive residual stresses in the part, which helps to improve the fatigue life of the part [4]. Due to its cheapness, availability, and also positive effects of this process on surface integrity, it can be used as a potential post-processing process for AM components.

4.2 BURNISHING PROCESS

Burnishing is known as one of the surface finishing processes which is used to improve the surface integrity of different parts. Unlike other abrasive finishing processes methods, such as grinding, in which material removal operations are performed through abrasive particles, in this process no material removal operation is performed, and the surface finishing is carried out only through the cold forming process [5]. In this process, cylindrical or spherical tools are used to apply pressure to the surface of the part. The pressure of the tool and its movement on the surface of the part cause the tool to come into direct contact with the surface asperities. Initially, this contact causes the elastic deformation of the surface roughness, but due to the long and continuous contact of the tool with the surface, eventually the local plastic deformation takes place and the surface is finished. No material removal operation is performed during the process and the surface is finished by filling the valleys of the surface with peaks (asperities) given by the pressure and movement of the plastic deformation tool. In the burnishing process, because the shear force is applied locally and its value is rarely greater than 10 micrometres, the overall geometry of the part does not change. Therefore, this process is not able to correct shape and form errors of the part significantly. It is used mostly to finish parts that require high surface quality, such as mould manufacturing, filleted bearing components, or curved surfaces like additively manufactured parts.

Surface roughness is one of the fundamental features of the workpiece that results from other previous production processes, especially machining operations, and is in fact the same peaks and valleys of the workpiece surface [6]. Surface roughness has a great impact on the final performance and life of the part. The burnishing process is able to remove surface roughness and create a smooth and finished surface. Also, due to the pressure and movement of the tool during the process, the surface of workpiece is subjected to cold work operation, which causes the displacement of the grain boundaries and thus produces compressive residual stresses in the part. Therefore, using the burnishing process not only reduces the surface roughness and smoothing it, but also increases the hardness, wear resistance, residual compressive stresses and fatigue life of the part [7]. Figure 4.1 shows a schematic of the branding process.

The burnishing process can be divided into two categories based on its various tools, namely Deep Rolling (DR) and Low Plasticity Burnishing (LPB). In the DR method, the aim is to increase the cold work and create high compressive residues on the surface of the part, while in the LPB method, the main goal is to reduce the surface roughness. One of the most important defining features of a burnishing tool is the shape and geometry of the indenter [9]. The geometry of the burnishing tool is divided into two types: roller and ball. Because roller tools are widely used because they are simpler to prepare and it is easier to adjust the force. Roller tools have various dimensions and profiles, and each has a specific feature and application. But in general, rollers with larger cylindrical lands provide a higher contact surface between the tool and the part, which allows a higher feed rate. However, compared to ball tools, roller tools are less adaptable to the part geometry, and are therefore inefficient for finishing the parts with complex geometries used widely in the advanced manufacturing industries. In addition, roller tools are less efficient at improving the hardness of the part, therefore compared to ball tools, they also have more force to achieve the same results as ball tools. As a result, in order to achieve better surface integrity, the use of ball tools is more efficient and more widely used. Figure 4.2 shows the fishbone diagram influencing the parameters in the burnishing process.

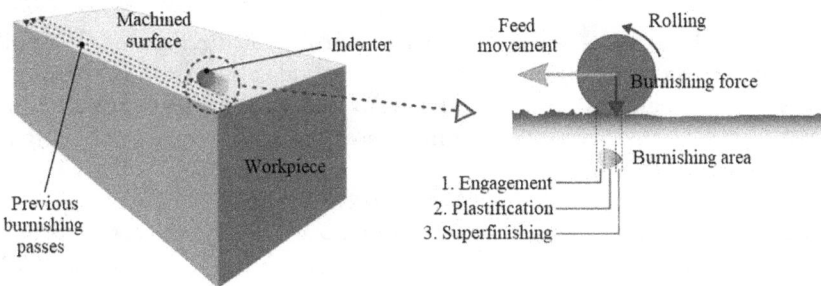

Figure 4.1 Schematic of the ball burnishing process [8].

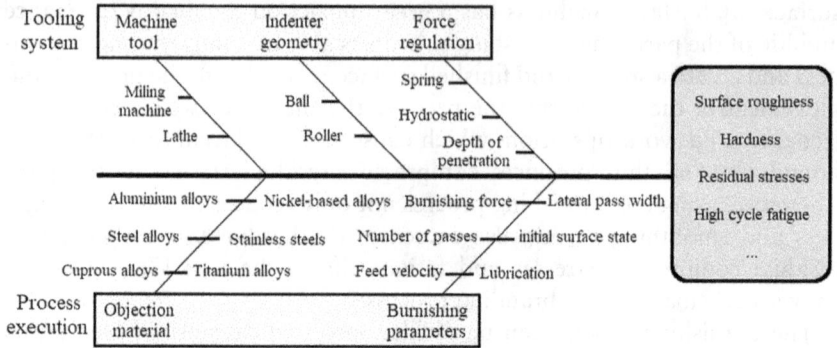

Figure 4.2 Fishbone diagram of the influential parameters of the burnishing process.

4.3 BURNISHING OF ADDITIVELY MANUFACTURED PARTS

Additive manufacturing (AM) technology is known as one of the advanced manufacturing methods that compared to traditional methods, has a high ability to manufacture parts with complex dimensions and geometries. Due to the remarkable ability of this technology in the manufacture of metal parts with complex geometry, this technology has been considered increasingly by various industries, especially advanced industries such as aerospace, medicine, and military. Despite all the features and advantages of AM, this technology also has disadvantages and limitations. One of the most important limitations of metal parts produced by AM is the high surface roughness as well as their internal defects such as voids. These defects reduce surface performance as well as fatigue strength. Therefore, to address these disadvantages, post-processing methods such as heat treatment, polishing/finishing, etc., are necessary to improve surface integrity [1–3].

4.3.1 Ball burnishing (BB)

Figure 4.3 illustrates a schematic of ball burnishing mechanism. The ball burnishing process is one of the finishing methods able not only to reduce the surface roughness, but to also improve the surface integrity, such as wear and corrosion resistance, compressive residual stresses, and fatigue life [4–10]. In this process, a hard ball is used as the deforming element (indenter). The materials used for balls are generally alumina, carbide ceramics, cemented carbide, silicon nitride ceramics, silicon carbide ceramics and bearing steels. During the ball burnishing, there is a spot contact between indenter (ball) and the surface of workpiece. Here, the indenter acts as a tool for surface layer deformation. For the specified normal force, Compared to roller burnishing, the ball burnishing provides higher pressure, microhardness, compressive residual stress, and fatigue strength [11].

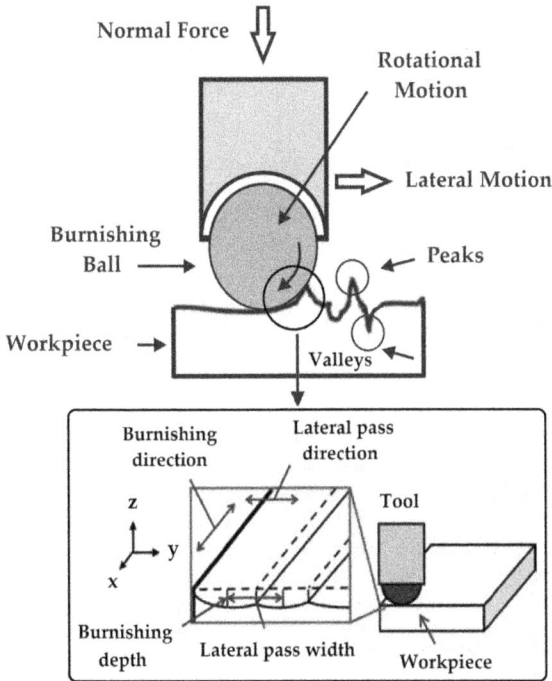

Figure 4.3 Ball burnishing mechanism.

4.3.1.1 Analytical models of BB

To understand comprehensively the ball burnishing mechanism, it is necessary to investigate the forces involved in the process. The literature shows that researchers have tackled with the issue through different perspectives and proposing models that have an increasing complexity because they include hypotheses that are more and more representative of the actual process.

4.3.1.1.1 Normal and tangential force

Figure 4.4 (a) shows a schematic diagram of the forces involved in the ball burnishing process. Teimouri et al. [12], proposed an analytical model to calculate the active forces and compare them with the experimental results. Of the various components of active forces, the ones that can be controlled by the experimental device are the vertical force (F_y) and the horizontal force (F_x) of the resultant force (F_R).

According to the Figure 4.4, the normal and tangential contact forces can be defined as follows:

$$F_n = F_R \cos(\varphi) \tag{4.1}$$

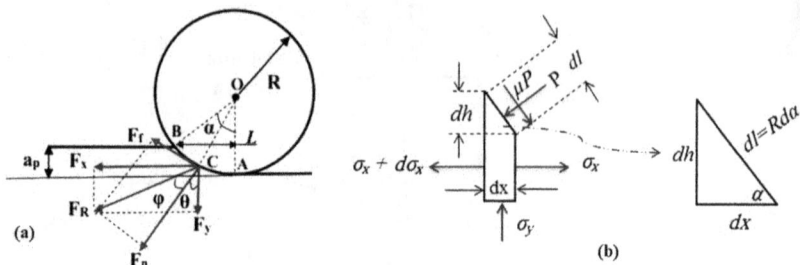

Figure 4.4 (a) Schematic diagram of forces involved in the burnishing process (b) Slab representation of the model [13].

$$F_f = F_R \sin(\varphi) \tag{4.2}$$

where F_n, F_f, F_R are the normal, tangential, and resultant forces respectively. In addition, φ is the angle of the contact plane that can be presented as follows:

$$\tan(\varphi) = \frac{F_f}{F_n} = \frac{\mu F_n}{F_n} = \mu \tag{4.3}$$

Based on Equation (4.3), the angle of the contact plane is approximately equal to the coefficient of friction. To calculate the forming force, the angle of contact (that is, BOA or α) can be presented as follows:

$$\cos\alpha_{max} = \frac{R - a_p}{R} = 1 - \frac{a_p}{R} \Rightarrow \alpha_{max} = \cos^{-1}\left(1 - \frac{a_p}{R}\right) \tag{4.4}$$

where α_{max}, a_p, and R are the angle of attack, the depth of burnishing, the radius of ball, respectively.

In the BOA triangle, the segment OC is tangent to the circle (ball) and is the bisector of the angle of attack (α_{maxs}). On the other hand, the normal force of contact is also tangent to the basic circle, which is consistent with OC. Hence, it can be said that the angle θ is half the angle of attack (that is, $\theta = \alpha_{max}/2$). Therefore, the horizontal force and and vertical force of burnishing can be calculated as follows:

$$F_x = F_R \sin(\varphi + \theta) \tag{4.5}$$

$$F_y = F_R \cos(\varphi + \theta) \tag{4.6}$$

Substituting Equation (4.1) into Equations (4.5) and (4.6), the horizontal and vertical components of the polishing force can be rewritten as:

By substituting Equation (4.1) in Equations (4.5) and (4.6), the horizontal and vertical components of forces can be presented as follows:

$$F_x = \frac{F_n}{\cos(\varphi)}\left(\sin(\varphi + \theta)\right) \tag{4.7}$$

$$F_y = \frac{F_n}{\cos(\varphi)}\left(\cos(\varphi + \theta)\right) \tag{4.8}$$

4.3.1.1.2 Slab method analysis

According to Equations (4.7) and (4.8), it is seen that the vertical and horizontal forces depend on the normal force of contact. Therefore, Teimouri et al. [12] proposed the slab technique analysis for calculating the normal force of contact. Figure 4.4(b) depicts a slab model used for the burnishing analysis.

During the rolling action of the tool, there is a neutral point where the direction of friction is changed [13]. The direction of friction is reversed once passing through the neutral point (exit side step). In other words, in the balanced equation, the sign of the friction at the entrance step is positive, and then the sign of the friction at the exit side step is negative. To calculate the normal force of contact, it is necessary to determine the normal contact pressure. The contact pressure can be presented as follows:

$$P = Y\left(\frac{h}{h_0}\right)\exp\left(\mu\left(B_{max} - B\right)\right) \tag{4.9}$$

Where B is a coefficient that can be determined as follows:

$$B = \sqrt{\frac{R}{h_1}}\tan^{-1}\sqrt{\frac{R}{h_1}}\alpha \text{ and } B_0 = \sqrt{\frac{R}{h_1}}\tan^{-1}\sqrt{\frac{R}{h_1}}\alpha_{max} \tag{4.10}$$

where Y is the yield stress, α is the angle of contact varying from 0 to α_{max}, and h is the instantaneous thickness which changes with the variation of the contact from h_0 to h_1.

Equation (4.9) defines the equation of contact pressure on the entrance side. Also, The contact pressure on the exit side can be calculated as follows:

$$P = Y\left(\frac{h}{h_1}\right)\exp\left(\mu B\right) \tag{4.11}$$

The presence of inlet and outlet sides in the rolling operation of the polishing tool creates a neutral point (that is, α_N) and the change in contact pressure changes from upward to downward direction. Also, at this point, the contact pressure on the two sides is equal.

To obtain the position of the neutral point, the contact pressure on the entrance side (that is, Equation (4.9)) must be equal to the contact pressure on the exist side (that is, Equation (4.11)). Hence, the angle of contact at the neutral point is calculated from following equation:

$$\alpha_N = \sqrt{\frac{R}{h_1}} \tan\left(\frac{\sqrt{\frac{R}{h_1}}}{2\mu B}\left[\mu B_{max} + \ln\left(\frac{h_1}{h_0}\right)\right]\right) \tag{4.12}$$

According to the above explanation, it can be inferred that in the entrance side, when the contact angle is between α_0 and α_N, the contact pressure is calculated by Equation (4.9). On the other hand, when the rolling reaches the existence side (i.e. the contact angle is between and zero), the contact pressure is determined by Equation (4.11).

From the above description, it can be concluded that the contact pressure is represented by Equation (4.9) when the angle of contact is between α_0 and α_N on the entrance side. On the other hand, by reaching roll to the existence side (that is, when the angle of contact is between α_N and zero), the contact pressure is represented by Equation (4.11).

4.3.1.1.3 Consideration of the time perspective

In order to fully understand the changes in force during the burnishing process, it is necessary to determine those changes over the processing time. The time-domain analysis is performed to compare the force–time curve provided by the analytical model with the results obtained by the dynamometer. During the burnishing process, the angular motion of indenter (ball) is caused by contact with the surface that moves in the direction of the feed with the feed velocity (v_f). The angular velocity (ω) of the indenter (ball) is determined as follows:

$$\omega = \frac{v_f}{R} \tag{4.13}$$

By finding the angular velocity, the time required for the angle of contact to reach the neutral point can be found as follows:

$$\omega \int_0^{t_N} dt = \int_0^{\alpha_N} d\alpha \Rightarrow t_N = \frac{\alpha_N}{\left(\dfrac{v_f}{R}\right)} \tag{4.14}$$

On the other hand, once the angle of contact angle reaches α_{max}, one load cycle is finished; therefore, the corresponding time (that is. t_f for this angle) is:

$$\omega \int_0^{t_f} dt = \int_0^{\alpha_{max}} d\alpha = \Rightarrow t_f = \frac{\alpha_{max}}{\left(\dfrac{v_f}{R}\right)} \tag{4.15}$$

The contact pressure from the angular displacement region (that is, α) can be converted to the time-domain variation (that is, t) by the following equation:

$$\omega \int_0^t dt = \int_0^\alpha d\alpha = \Rightarrow t_f = \frac{\alpha}{\left(\dfrac{v_f}{R}\right)} \tag{4.16}$$

The strain hardening behaviour of the workpiece material can be considered as another parameter that affects the contact pressure. This effect changes the yield strength during the process. Based on this effect, the mean flow stress value (Y_m) can be calculated as follows [12]:

$$Y_m = \frac{K\left(\ln \dfrac{h_0}{h_0 - a_p}\right)^n}{n+1} \tag{4.17}$$

where K is the coefficient of material strength and n is the strain hardening exponent.

To calculate the burnishing force, the contact pressure should be multiplied to the contact area. The contact area is a curved surface where the radius depends on the depth of burnishing. According to the Hertz theory of contact mechanics, the radius of the contact is calculated as follows [14]:

$$r_c = \sqrt{2Ra_p} \tag{4.18}$$

Therefore, the contact area can be calculated by Equation (4.19):

$$A_c = \pi r_c^2 = 2\pi Ra_p \tag{4.19}$$

Considering the contact pressure multiplied by the area of contact and the time-domain analysis (Equations (4.13)–(4.16)) and the strain hardening effect on the flow stress (Equation (4.17)), the normal force of contact is presented by:

Once the entrance side is $0 < \alpha < a_N$ and the relevant is $0 < t < t_N$, the normal force of contact is defined as follows:

$$
F_n^{en} = \frac{K\left(\ln\dfrac{h_0}{h_0 - a_p}\right)^n}{n+1}\left(\frac{R\left(1 - \cos\left(\dfrac{v_f}{R}t\right)\right) + h_1}{h_0}\right)\exp\left(\mu\left(B_0 - B\right)\right)2\pi R a_p
$$

(4.20)

On the other hand, Once the existence side is $a_N < \alpha < 0$ and the relevant is $t_N < t < 0$ the normal force of contact is defiend as follows [12]:

$$
F_n^{en} = \frac{K\left(\ln\dfrac{h_0}{h_0 - a_p}\right)^n}{n+1}\left(\frac{R\left(1 - \cos\left(\dfrac{v_f}{R}t\right)\right) + h_1}{h_0}\right)\exp\left(\mu B\right)2\pi R a_p
$$

(4.21)

By determining the F_n, the horizontal and vertical forces can be represented by substituting Equations (4.20) and (4.21) into Equations (4.7) and (4.8):

$$
F_x = \begin{cases} F_n^{en}\dfrac{\sin(\varphi + \theta)}{\cos\varphi} & 0 < t < t_N \\[2mm] F_n^{ex}\dfrac{\sin(\varphi + \theta)}{\cos\varphi} & t_N < t < t_f \end{cases}
$$

(4.22)

$$
F_y = \begin{cases} F_n^{en}\dfrac{\cos(\varphi + \theta)}{\cos\varphi} & 0 < t < t_N \\[2mm] F_n^{ex}\dfrac{\cos(\varphi + \theta)}{\cos\varphi} & t_N < t < t_f \end{cases}
$$

(4.23)

In all cases, the models explained above do not consider the original topology of the surface whatsoever, although evidence shows that it is a significantly determinant factor to predict the results and efficiency of the process [15].

4.3.1.2 Ball burnishing of AM parts

Karthick et al. [16], developed Low plasticity burnishing (LPB) process as cost-effective post-processing treatment for finishing of as-built electron beam additive manufacturing (EBAM) parts. They proposed a sequential method to investigate the surface integrity of as-built electron beam additive manufacturing nickel-based superalloy 718 (AB) versus grinding and LPB. In this method, they investigated the effects of using ball burnishing, without and with grinding process, on the surface integrity such as microhardness, surface roughness, microporosity, and residual stresses of EBAM components. Their research was divided into several stages. They used ball burnishing and grinding processes, separately and in combination, to finish the as-built EBAM components. Initially, they finished the as-built (AB) EBAM components separately using grinding (G) and ball burnishing (LPB) methods and then, investigated the effects of these processes on surface integrity. Then in the next step, the parts were finished using the combined method consisting of grinding + LPB. In the combined method, the as-built EBAM parts were firstly polished by the grinding process and in the next step, the ground parts were finished by the ball burning process. Figure 4.5 presents the pre- and post-finished images of the parts.

The results obtained for surface roughness, microhardness and microporosity parameters indicated that the use of the combined method AB + G + LPB, respectively, were much more efficient compared to the separate use of G and LPB processes. In addition, the improvement of microhardness and reduction of surface porosity due to plastic deformation resulting from the burnishing process also increased the residual stresses of the parts. They found that, compared to other arrangements, the use of AB + G + LPB processes, respectively, increases the penetration depth of the microhardness below the machined surface, which is especially true when applying more burnishing pressure from 10 to 40 N reduces microporosity and increases microhardness and residual stresses. Figure 4.6 shows the optical micrographs of as-built (AB) + Grinding (G) + Low plasticity burnishing (LPB) processed Inconel 718 parts under various burnishing pressures.

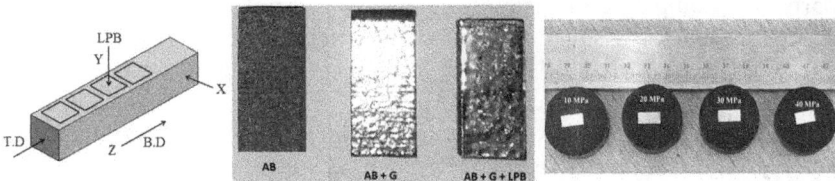

Figure 4.5 Pre-finished and post-finished parts of EBAM-built Inconel 718 displayed by CAD [16].

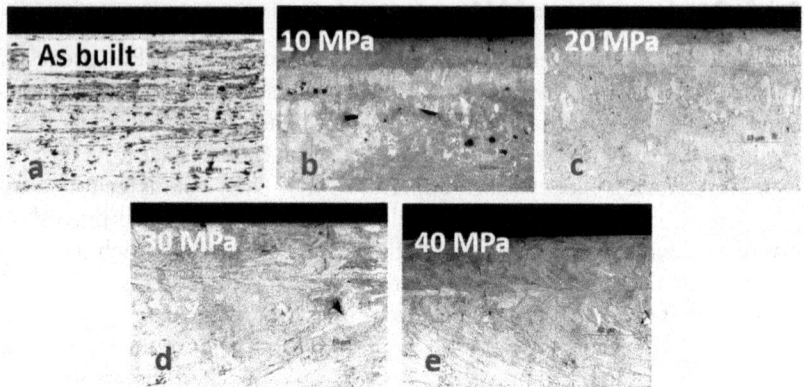

Figure 4.6 Optical micrographs of as-built (AB) + Grinding (G) + Low plasticity burnishing (LPB) processed Inconel 718 parts under various burnishing pressures [16].

Besides the metal parts made by AM, the parts made by other techniques, especially the Fused Filament Fabrication (FFF) method, also require precision and quality in accordance with specified standards. The surface finish of parts made by FFF method is strongly influenced by the parameters of printing and the round from of the filaments, which result in the so-called stair-stepping effect. To address this issue, various optimization methods such as strategy of slicing, raster angle, part orientation, air gap, thickness of layer, and orientation of building have been developed. Another way to enhance the surface finish of Fused Filament Fabrication (FFF) parts is to utilize chemical and mechanical polishing methods. The main advantage of chemical polishing methods such as vapour smoothing, painting, electroplating or metallization, and laser finishing, is their ability to uniformly finish parts with complex geometry. However, the hazardous nature of some of the solvents used and the lack of mechanical properties of the components because of chemical adsorption of vapours are usually identified as major disadvantages of chemical polishing methods. Although the main problem with mechanical polishing/finishing processes is their inability to polish parts with complex geometries using an automatized routine, they can still be used to finish parts made by AM due to their high efficiency, availability and cheapness, especially, the FFF components that have simpler shapes with practical needs. Chueca de Bruijn et al. [17], developed the BB process as a new post-processing technique for finishing of the FFF components to achieve higher surface integrity. They used the engineering-grade thermoplastic PEI Ultem (Ultem) to model the materials and to study the influences of various BB parameters on the surface roughness and mechanical properties of the parts. The purpose of this study was to evaluate the feasibility of using the BB process as a new post-processing method for the FFF components. They printed the parts in two different directions. The first batch of parts were printed horizontally (X-Flat orientation) in order to maximize the upper and lower surfaces of the components. By contrast, the second batch parts were printed at a 90° rotation angle (X-edge direction) with respect to the X axis of the first batch parts, resulting in a wide surface with a typical lateral profile. They also used Taguchi design method to optimize process parameters. Comparison of microscopic images of the parts showed a smoother surface and less porosity of the parts finished by ball burnishing. Compared to the surface left over from the printing, the ball-burnished surface showed a marked decrease in the size and shape of the voids for both the X-Edge and X-Flat samples (Figure 4.7). Regardless of the print direction of the part (X-Flat or X-Edge), the surface roughness values of Ra and Rz also decreased significantly. Ra and Rz improved by 73% and 57% for the X-Flat sample, and more than 80% for the X-Edge sample, respectively. However, the greatest reduction in surface roughness was when the orientation of the BB process was perpendicular to the printing orientation. They proved that the value of two parameters of BB force and the passes number have the greatest effect on reducing the surface roughness. Table 4.1 shows the optimal values obtained for the ball burnishing process parameters. In terms of mechanical performance, the results of the static test did not show many changes. Also, the modulus, yield point, and flexural strength showed small changes in the flexural and tensile tests, which could be due to the deformation of the original plastic and the thinning of the external layer of the sample due to the BB process. The results of the S-N R-1 curves indicated that the ball burnishing of the X-Edge and X-Flat models increased their fatigue life performance by at least 2 times. Increasing the residual

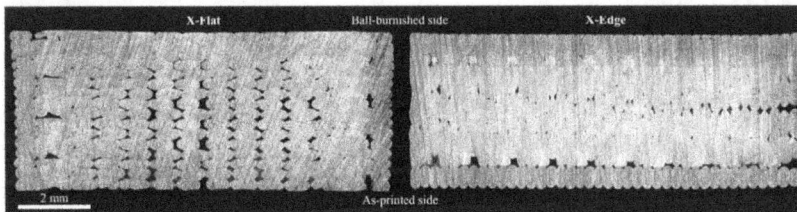

Figure 4.7 Cross-section of the X-Flat and X-Edge BB parts (Only the top faces of the samples were mechanically processed, and the bottom faces were kept in their as-printed state. The white scale bar is 2 mm) [17].

Table 4.1 Optimal values of BB parameters obtained from design of experiments (DOE) analysis [17]

Applied force [N]	Tool passes	Width of lateral path [mm]	Feed speed [mm min^{-1}]
400	10	0.32	2000

compressive stress on the surface of the BB component and increasing the hardness of the component reduces generation and growth voids and cracks, which leads to a significant increase in fatigue life. Figure 4.8 depicts the scanning electron microscopy (SEM) micrographs and 3D surface images of X-Flat, X-Edge, pristine, and BB parts.

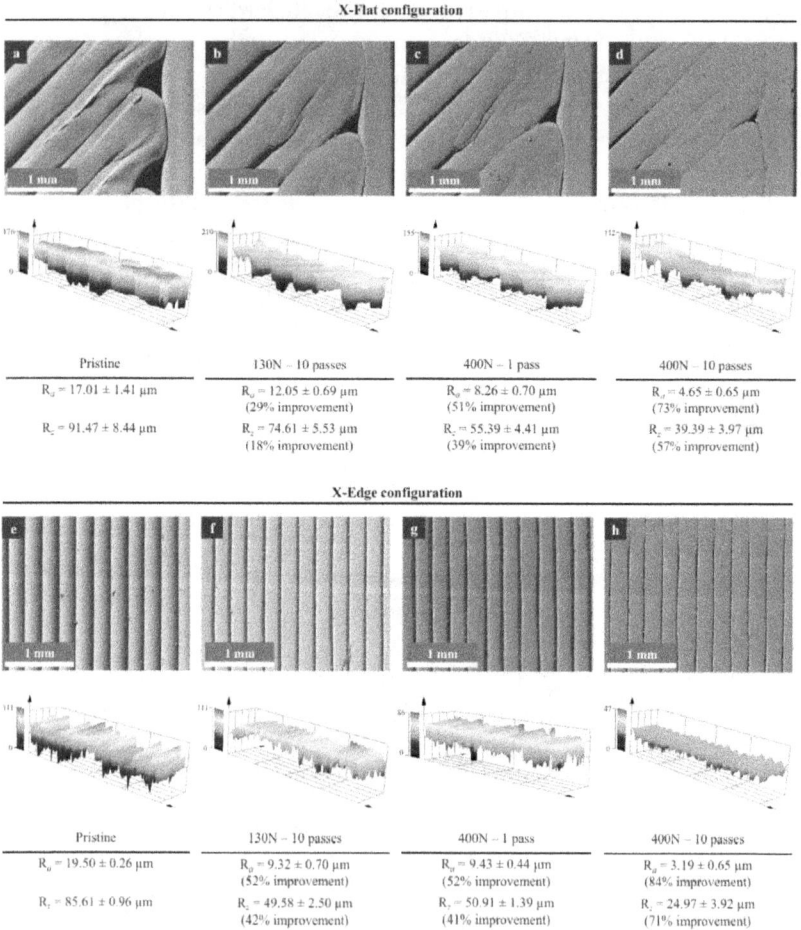

X-Flat configuration

Pristine	130N – 10 passes	400N – 1 pass	400N – 10 passes
$R_a = 17.01 \pm 1.41$ µm	$R_a = 12.05 \pm 0.69$ µm (29% improvement)	$R_a = 8.26 \pm 0.70$ µm (51% improvement)	$R_a = 4.65 \pm 0.65$ µm (73% improvement)
$R_z = 91.47 \pm 8.44$ µm	$R_z = 74.61 \pm 5.53$ µm (18% improvement)	$R_z = 55.39 \pm 4.41$ µm (39% improvement)	$R_z = 39.39 \pm 3.97$ µm (57% improvement)

X-Edge configuration

Pristine	130N – 10 passes	400N – 1 pass	400N – 10 passes
$R_a = 19.50 \pm 0.26$ µm	$R_a = 9.32 \pm 0.70$ µm (52% improvement)	$R_a = 9.43 \pm 0.44$ µm (52% improvement)	$R_a = 3.19 \pm 0.65$ µm (84% improvement)
$R_z = 85.61 \pm 0.96$ µm	$R_z = 49.58 \pm 2.50$ µm (42% improvement)	$R_z = 50.91 \pm 1.39$ µm (41% improvement)	$R_z = 24.97 \pm 3.92$ µm (71% improvement)

Figure 4.8 Scanning electron microscopy (SEM) micrographs and 3D surface images of X-Edge and X-Flat, pristine and BB parts using various parameters (The white scale bar is 1 mm. The size of Surface images is 4,350 × 925 µm) [17].

4.3.2 Ultrasonic vibration-assisted ball burnishing (UVABB)

In recent years, research has been conducted on the use of ultrasonic vibration and its effects on the performance of various manufacturing processes [18]. Due to the significant advantage of using ultrasonic vibration in machining processes and the positive effect of tool vibration on the output characteristics of the process, the ball burnishing process was developed with the help of ultrasonic vibration (UV) [19–20]. In this process, in addition to the forward motion of the tool, an additional vibratory and/or ultrasonic motion is applied to it, which causes the tool to have a reciprocating motion in the forward direction. Compared to the traditional burnishing operations, the use of vibratory and/or ultrasonic motion results in a more homogeneous surface integrity, good contact stiffness, better lubrication holding capacity, and greater resistance to abrasion and friction [8]. It should be noted that the difference between the vibration and ultrasonic ball burnishing process is in the frequency of their vibration. So that the amount of vibration in the vibration method is in the range of 20 to 50 Hz and the amount of vibration in the ultrasonic method is in the range of kHz. Figure 4.9 shows the structure of the ultrasonic vibration wing tool.

4.3.2.1 Analytical models of UVABB

Once the BB process is accompanied by ultrasonic longitudinal vibration in the vertical direction (the vibration direction corresponds to the normal component of static force), the total force resulted from both the dynamic and static force can be expressed as follows:

$$F_y = F_{ys} + F_{yd}$$

where F_y is the total vertical force, F_{ys} is the value of average normal static force, and F_{yd} is the average normal dynamic force due to UV.

4.3.2.1.1 Static force

To calculate the static force, the analytical model developed by Teimouri et al. [12] are used (see Equations (4.1) to (4.23)). They also proposed an analytical model for Ultrasonic vibration-assisted ball burnishing [21].

4.3.2.1.2 Dynamic force

The UV of the BB applies vibration force to the surface of workpiece. To calculate the force, it is necessary to identify the kinematics of motion. The

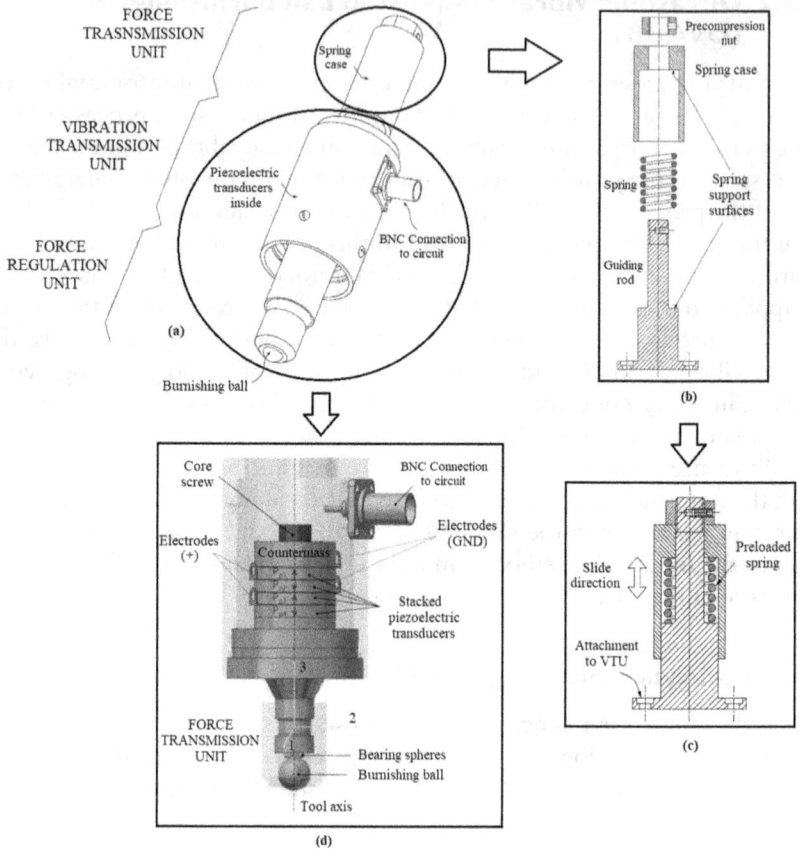

Figure 4.9 (a) Schematic of the UVABB tool, (b) Force transmission unit (FTU) components, (c) Vibration transmission unit (VTU), (d) and Force transmission unit (FTU) components [20].

displacement, velocity, and acceleration equations can be represented as follows:

$$\begin{cases} U = A\sin(\omega t) \\ \dot{U} = A\omega\cos(\omega t) \\ \ddot{U} = -A\omega^2\sin(\omega t) \end{cases} \tag{4.24}$$

Based on the motion equation (i.e. Newton's second law), the dynamic force is defined as:

$$F_{dy} = m\ddot{U} = -mA\omega^2\sin(\omega t) \tag{4.25}$$

By considering the $\omega = 2\pi f$ and $m = \frac{4}{3}\rho\pi R^3$, the Equation (4.25) can be rewritten as:

$$F_{dy} = -\frac{16}{3}\rho\pi^3 R^3 A f \sin(2\pi ft) \tag{4.26}$$

To calculate the total force, it is necessary to integrate the average static force from Equation (4.23) with dynamic forces (i.e. Equation (4.26)). As outlined in Figure 4.10, the average static force is constant and the dynamic force has a vibrational characteristic, therefore, the sum of them also has a sinusoidal characteristic of the ultrasonic frequency. According to Figure 4.11, the average value of the amplitude of the total impact force can be obtained by applying the Fourier summation as proposed by Tabatabaei et al. [22]:

$$F_a = \frac{t_c}{T}F_{sy} + \frac{2F_{sy}}{\pi}\sum_{n=1}^{\infty}\frac{1}{n}\sin\left(\frac{t_c}{T}n\pi\right)\cdot\cos(2\pi fnt) \tag{4.27}$$

Where the value of $\frac{t_c}{T}$ is called the tool-work contact rate and can be adjusted according to the finishing conditions. In addition, F_{sy} is the static force value given by Equation (4.23); n is the number of Fourier sum terms that should be obtained by trial and error. T is the vibration period and can be obtained by inverting the frequency, and t_C is the contact time during ultrasonic irradiation and can be calculated according to nonlinear elasto-plastic contact mechanics [23]:

$$t_C = 5\cdot087\left(\frac{\rho^2 R^5}{E_{eq}^2 V^2}\right) \tag{4.28}$$

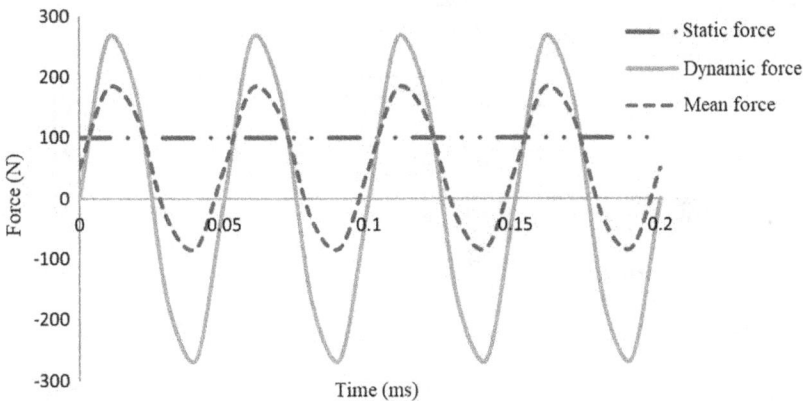

Figure 4.10 Schematic diagram of the static and dynamic forces and their average values during UVBB process [13].

In the Equation (4.28), V is the impact velocity; E_{eq} is the equivalent elasticity modulus of contact represented as follows:

$$E_{eq} = \frac{E}{2(1-v)} \qquad (4.29)$$

4.3.2.2 Ultrasonic vibration-assisted ball burnishing (UVBB) of AM parts

Due to the significant effect that the use of ultrasonic vibration has on the improvement of surface roughness, hardness, residual stresses, and surface integrity, it can be used as a new method for polishing/finishing AM parts. Salmi et al. [24], developed the Ultrasonic-Assisted Ball burnishing (UABB) method as a new process for finishing the parts made by AM. To this end, they investigated the effects of process parameters such as feed, change of direction, spring compression, and ultrasonic vibration on the surface roughness, hardness and surface quality of cobalt-chrome cylindrical parts, and hexagonal 316L stainless steel parts made by AM. Figure 4.11 shows

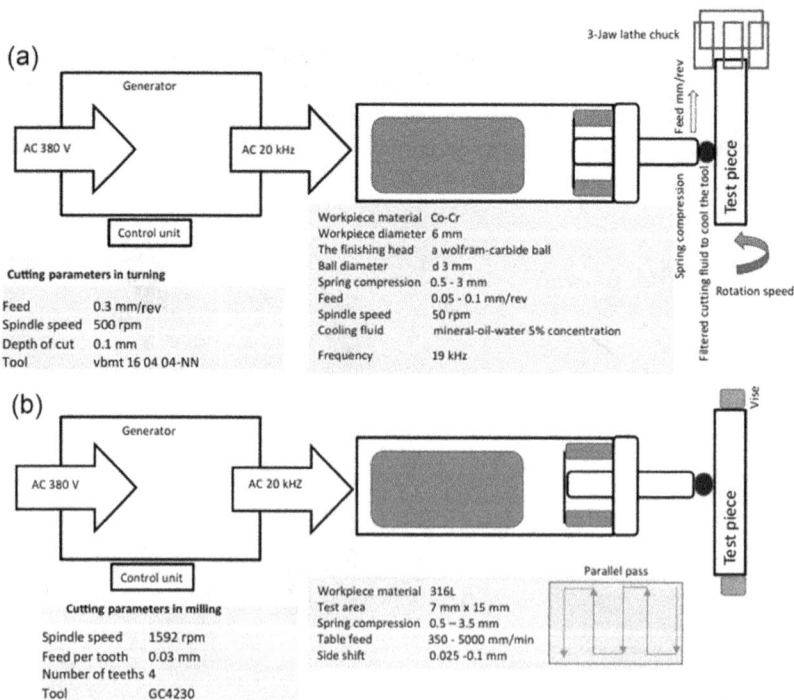

Figure 4.11 Schematic of the UABB system-the measurements and data acquisition process. (a) Co-Cr lathe operation and (b) 316L milling operation [24].

the schematic and details of the process parameters. To better understand the effects of the operation, they compared the results of the UABB process with the results of the initial printing. The surface quality of the Co-Cr components obtained by the UABB method was approximately 32 times that of the original print (i.e. from Ra = 5.66 μm to Ra = 0.18 μm). Experimental outcomes showed that the most effective parameter of the process in finishing Co-Cr parts was the feed velocity; however, Pareto analysis showed that the relative displacements between the tool and the surface of workpiece play an important role in achieving the optimal state. They also found that the interaction between feed velocity and spring compression together had a greater impact on process performance than spring compression alone. The use of UABB process for 316L stainless steel parts reduced the surface roughness from the initial value of Ra = 7.39 to the final value of 0.55 μm, which was approximately 13 times less than the surface roughness of the original print. Experimental results showed that the interaction between feed velocity and directional change played an important role in polishing 316L stainless steel parts, however, Pareto analysis results showed that the milling path plays an important role in achieving optimal condition. They found that displacements between the tool and the workpiece, as well as spring compaction, are the most influential process parameters on surface quality. Also, after the UABB process, the hardness increased by an average of 47.4% for Co-Cr parts and 70.7% for 316L stainless steel parts. Figure 4.12 shows the components after the UABB process.

Compared to traditional methods, the additive manufacturing technology has a high capability in manufacturing parts with complex geometry and internal structures. The manufacture of parts with complex geometry, such as 3D inner structures in metal moulds, by additive manufacturing (AM) not only increases their performance, but also reduces production costs and time. However, metal parts made by AM have disadvantages such as high surface roughness and inner defects such as voids, which reduce the fatigue life and ultimately fracture the part. In order to eliminate these disadvantages, various post-processing processes such as milling, finishing, and heat treatment must be performed on AM components. Because post-processing

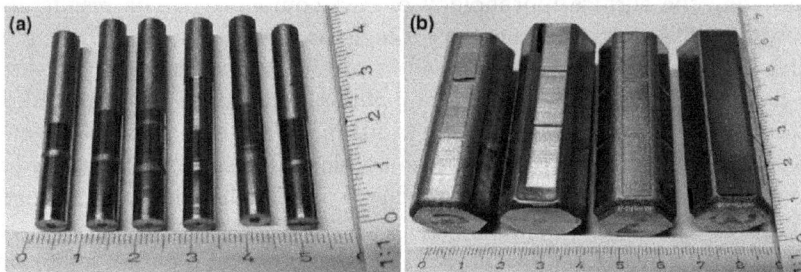

Figure 4.12 (a) cylindrical Co-Cr parts. (b) 316L Stainless steel prims. Images of the parts after UABB [24].

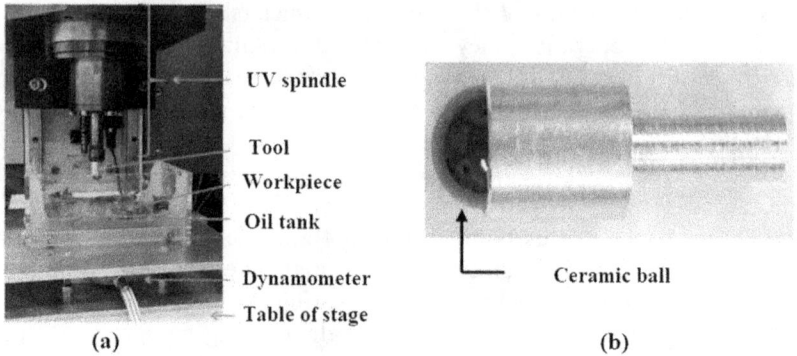

Figure 4.13 (a) Experimental set-up, and (b) burnishing tool [25].

processes are not able to eliminate these defects alone, the development of new processes is essential. Teramachi and Ian [25] developed the ultrasonic-assisted ball burnishing (UABB)) process to enhance surface quality and eliminate surface and subsurface defects of the workpieces made by AM. Figure 4.13 shows the experimental set-up and tool. They investigated the influence of UV and lateral path width of tool on the surface and subsurface integrity of AlSI10 Mg AM components.

For this purpose, they used a silicon nitride ceramic ball as a burnishing tool and considered the direction of burnishing in the direction of laser scanning on the surface of the part, in parallel, and the direction of lateral pass in the direction of laser scanning perpendicular. As can be seen in Figure 4.14, unburnished surfaces had non-level areas and voids which caused these surfaces to become rough. However, these non-level areas were dramatically smoothed out after the BB process. It was also found that the voids were either eliminated or reduced after the burnishing operation. This is due to the pressure exerted by the tool on the surface and the subsequent deformation of the plastic and the flow of material due to the movement of the ball on the surface of the part. The use of UV played a significant role in eliminating surface and subsurface defects, especially voids, and generally improving the surface morphology of AM components. The results also showed a 98% reduction in surface roughness as well as a 24% increase in hardness for AlSI10 Mg AM. The effect of UV assistance was investigated by comparing the results without and with UV (experiments A and B). In addition, the effect of lateral pass width in UVAB was also investigated (experiment C). The burnishing depth and lateral pass width were fixed at 0.5 mm during the process in experiments A and B, while they were changed in experiment C.

The metal powder bed fusion (PBF) method is one of the most important methods of AM. This method due to its high capabilities such as production of flexible mechanical parts, parts with multiple complexities, as well as

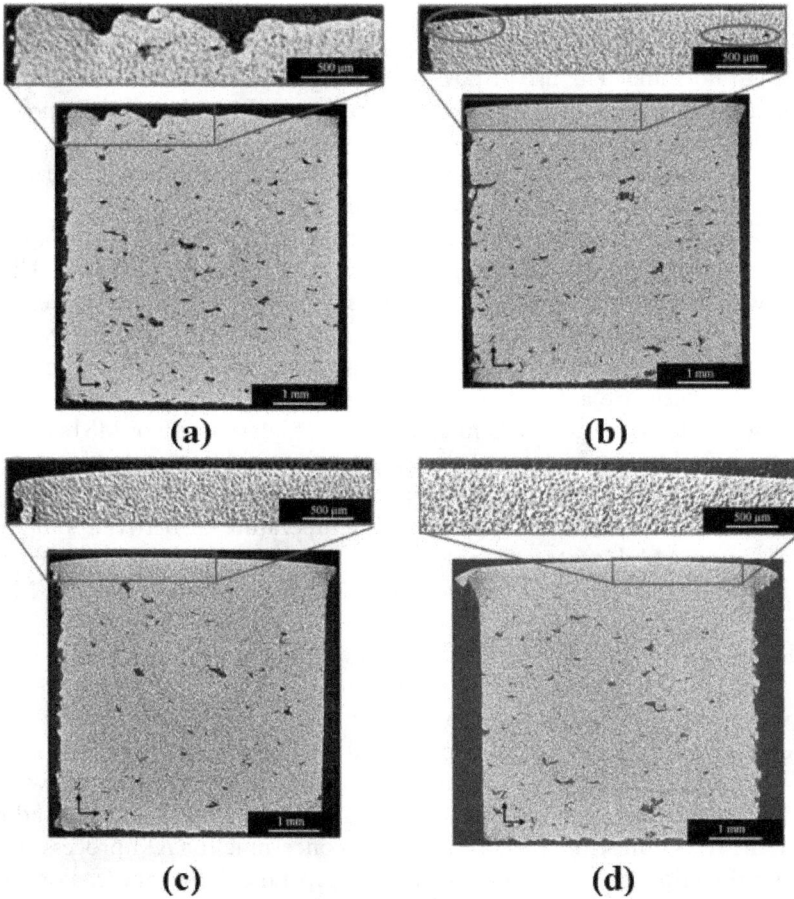

Figure 4.14 Cross-sectional images of the unburnished and burnished samples: (a) unburnished, (b) experiment A, (c) experiment B, and (d) experiment C [25].

parts with complex geometry of small to medium size can be a good alternative to the subtractive (like machining) or forming production. For instance, this method can be used to produce metal insert tools with cooling channels used in cutting applications and mould-injection applications. Precise fabrication of uniformly cooling channels by PBF technology improves the cooling process and dimensional accuracy in injection moulds, which increases their efficiency and longevity. However, the PBF method also has disadvantages, such as high surface roughness, poor near-shape properties, and surface and subsurface defects (such as voids). Another disadvantage of the PBF method when producing precision mechanical parts is the difference in surface roughness of different areas of a part. In other words, a part may not have consistent surface roughness, and each area may have a different value.

This has made the parts produced by PBF require post-processing methods such as hot isostatic pressing, heat treatment, shot-blasting, chemical engraving, and finishing. The use of post-processing methods reduces the surface roughness, increases the stiffness and compressive residual stresses, improves the microstructure and ultimately increases the fatigue life of the parts. Due to the high performance of the UABB process in improving the surface quality (surface roughness, hardness, residual stresses, etc.) of various parts, especially metal parts, Ituarte et al. [26], tried to use this process as a new post-processing method for the parts made by the PBF AM method. Their goal was to develop the UABB process as a new one-step post-processing method by optimizing process parameters and increasing productivity. They used Maraging steel MS1 powder as a raw material to produce the part. Maraging steels are a group of high-strength and low-carbon nickel-based alloy steels that have good machinability. Also, high strength of MS1s makes them a suitable material for tooling applications. Maraging steel requires heat treatment by age hardening for precipitation of nanometre-sized intermetallic compounds, at an approximate temperature of between 480 and 490° C after AM. Heat treatment helps to enhance the hardness and strength of these steels. Table 4.2 describes the material composition of the MS1 powder material.

Regarding the ultrasonic vibration-assisted ball burnishing process, Figure 4.15(a) shows the illustration of the prism or test workpiece. During the AM process, each workpiece was oriented vertically in the Z-axis, and the X–Z side surfaces or faces were post-processed by ultrasonic vibration-assisted ball burnishing. Selection of this orientation aimed at processing the worst surface quality achievable by AM, located in the lateral X–Z planes. In addition, the figure illustrate the crucial details concerning the AM process such as the direction of the layers, which are perpendicular to the Z-axis; the table feed direction, which is parallel to the Z-axis; and the side-shift direction, which is perpendicular to the Z-axis.

Figure 4.15(b) shows the schematics of the post-processing of the workpiece. In this study, a spring-compression type of ultrasonic burnishing tool was used as the burnishing tool. They used Taguchi design method and analysis of variance (ANOVA) to design the experiment as well as optimize the

Table 4.2 Maraging MS1 properties [26]

Material physical and chemical composition (Wt%)	Density $\left(\dfrac{g}{cm^3}\right)$
Fe (Balance), Ni (17 – 19 wt%), Co (8.5 – 9.5 wt%), Mo (4.5 – 5.2 wt%), Ti (0.6 – 0.8 wt%), Al (0.05 – 0.15 wt%), Cr, Cu (each ≤ 0.1 wt%), C (≤ 0.03 wt%), Mn, Si (each ≤ 0.1 wt%), P, S (each ≤ 0.01 wt%)	8.0 – 8.1

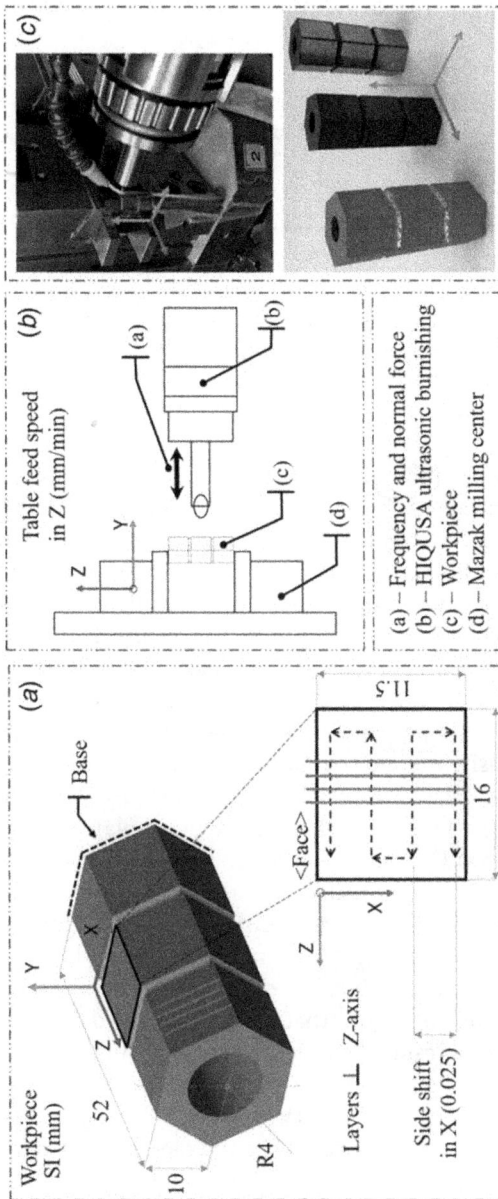

Figure 4.15 (a) CAD of the test prism and ultrasonic burnishing tool path; (b) schematic of the ultrasonic burnishing and components; and (c) pictures of the manufacturing process and workpieces [26].

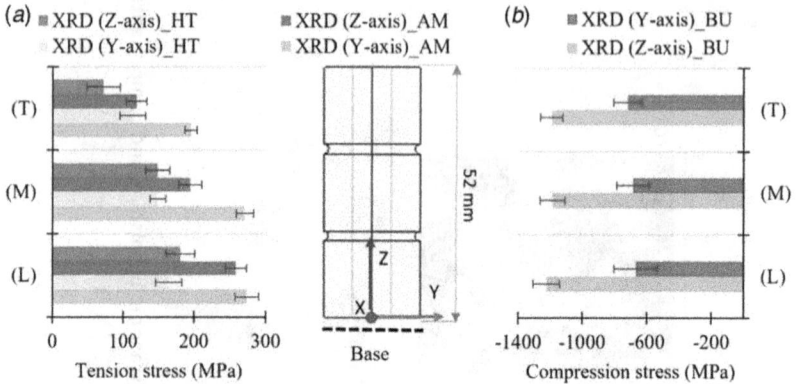

Figure 4.16 Average XRD residual stress state of the workpiece as a function of the build height. (a) As-built after additive manufacturing (AM) and heat treatment (HT) process. (b) After ultrasonic vibration-assisted ball burnishing (BU) [26].

process parameters. The process parameters that were examined were spring compression (C), table feed (FS) speed and burnishing ball diameter (D).

Measurements were divided into 3 steps of the discrete manufacturing process, including as-built condition, after age hardening, and after vibration-assisted ball burnishing. Figure 4.16(a) shows the initial residual stress state of the workpiece surfaces, comparing the results as-built, that is, after both AM and the age-hardening heat treatment. The displayed XRD stress results are in the direction of the Z- and Y-axis (i.e. residual stresses in the building orientation and perpendicular to the building orientation, respectively). Simultaneously, the chart represents the XRD residual stress as a function of the build height, thus dividing the workpiece in three areas: low region (L) (i.e. closer to the base plate of the build platform or base of the workpiece), mid region (M), and top region (T).

Figure 4.16(b) illustrates the mean values of the XRD residual stress state after ultrasonic burnishing. The result displayed in the graph includes the results from all the differently treated surfaces, corresponding to three height areas along the build direction (i.e. Z-axis). The results of the XRD analysis after ultrasonic burnishing show that the process had a homogenization effect, as the gradient of residual stress has been minimized along the Z-axis.

They proved that the use of UVABB process reduces the surface roughness and increases the micro hardness (HIV) of the 18% Nickel Maraging tool steel made by AM in a PBF process. The average surface roughness (Ra) values decreased from an initial 6.24 µm to values lower or equal to 1.31 µm, with the lowest value of 0.14 µm, depending on process parameters. The surface macro-hardness increased from an initial 50 HRC (i.e. approximately 505 HV1) after age hardening to 53 HRC (i.e. approximately 567 HV1) after ultrasonic burnishing. The direct measurements of surface microhardness (HV1) of the burnished faces varied between 503 and 630 HV1, depending on process parameters.

The advantage of implementing a surface modification by single-pass ultrasonic vibration-assisted ball burnishing is that the overall post-processing of AM Maraging steel for tooling applications can be simplified into a three-step process that includes (i) AM of the tool insert; (ii) age hardening of the material; and (iii) CNC surface modification by single-pass ultrasonic vibration-assisted ball burnishing to fulfil engineering requirements in terms of surface hardness (HV1) and surface roughness (Ra).

4.4 CONCLUSION

In this chapter, the application of ball burnishing (both in conventional method and assisted with ultrasonic vibrations) as a new post-processing approach for finishing of AM parts has been discussed. Using plastic pressure applied by a hard ball on the surface of the part, plastic deformation is performed under hard working conditions, which increases the surface integrity. The interest in finishing AM components in the burnishing process is that pressure-induced deformation fills the voids and thereby reduces porosity when reinforced by plastically deformed materials. The revised bibliography taken into account proves that using burnishing process for finishing of AM parts is a relatively new issue with a long path to go in the future. Results are promising so far, but due to lack of sufficient references, it is necessary that further research be done to investigate the effects of various process parameters on different AM parts comprehensively.

REFERENCES

1. Altenberger, I., Nalla, R. K., Sano, Y., Wagner, L., & Ritchie, R. O. (2012). On the effect of deep-rolling and laser-peening on the stress-controlled low- and high-cycle fatigue behavior of Ti–6Al–4V at elevated temperatures up to 550° C. *International Journal of Fatigue*, 44, 292–302. https://doi.org/10.1016/J. IJFATIGUE.2012.03.008
2. Amini, C., Jerez-Mesa, R., Travieso-Rodríguez, J. A., Llumà, J., & Estevez-Urra, A. (2020). Finite element analysis of ball burnishing on ball-end milled surfaces considering their original topology and residual stress. *Metals 2020*, 10(5), 638. https://doi.org/10.3390/MET10050638
3. Azami, A., & Azizi, A. (2017). Rotational abrasive finishing (RAF); novel design for micro/nanofinishing. *The International Journal of Advanced Manufacturing Technology*, 91(9), 3159–3167. https://doi.org/10.1007/S00170-017-0016-8
4. Azami, A., Azizi, A., & Khoshanjam, A. (2018). Nanofinishing of stainless-steel tubes using Rotational Abrasive Finishing (RAF) process. *Journal of Manufacturing Processes*, 34, 281–291. https://doi.org/10.1016/J.JMAPRO. 2018.06.027

5. Azami, A., Azizi, A., Khoshanjam, A., & Hadad, M. (2020). A new approach for nanofinishing of complicated-surfaces using rotational abrasive finishing process. *Materials and Manufacturing Processes*, *35*(8), 940–950. https://doi. org/10.1080/10426914.2020.1750631

6. Chueca de Bruijn, A., Gómez-Gras, G., & Pérez, M. A. (2021). On the effect upon the surface finish and mechanical performance of ball burnishing process on fused filament fabricated parts. *Additive Manufacturing*, *46*, 102133. https://doi.org/10.1016/J.ADDMA.2021.102133

7. Gómez-Gras, G., Travieso-Rodríguez, J. A., Jerez-Mesa, R., Llumà-Fuentes, J., & Gomis de la Calle, B. (2016). Experimental study of lateral pass width in conventional and vibrations-assisted ball burnishing. *The International Journal of Advanced Manufacturing Technology*, *87*(1), 363–371. https://doi. org/10.1007/S00170-016-8490-Y

8. Ituarte, I. F., Salmi, M., Papula, S., Huuki, J., Hemming, B., Coatanea, E., Nurmi, S., & Virkkunen, I. (2020). Surface modification of additively manufactured 18% nickel maraging steel by ultrasonic vibration-assisted ball burnishing. *Journal of Manufacturing Science and Engineering*, *142*(7). https://doi. org/10.1115/1.4046903

9. Jerez-Mesa, R., Landon, Y., Travieso-Rodríguez, J. A., Dessein, G., Llumà-Fuentes, J., & Wagner, V. (2018a). Topological surface integrity modification of AISI 1038 alloy after vibration-assisted ball burnishing. *Surface and Coatings Technology*, *349*, 364–377. https://doi.org/10.1016/J.SURFCOAT.2018.05.061

10. Jerez-Mesa, R., Travieso-Rodríguez, J. A., Gómez-Gras, G., & Llumà-Fuentes, J. (2018b). Development, characterization and test of an ultrasonic vibration-assisted ball burnishing tool. *Journal of Materials Processing Technology*, *257*, 203–212. https://doi.org/10.1016/J.JMATPROTEC.2018.02.036

11. Jerez-Mesa, R., Travieso-Rodríguez, J. A., Landon, Y., Dessein, G., Llumà-Fuentes, J., & Wagner, V. (2019). Comprehensive analysis of surface integrity modification of ball-end milled Ti-6Al-4V surfaces through vibration-assisted ball burnishing. *Journal of Materials Processing Technology*, *267*, 230–240. https://doi.org/10.1016/J.JMATPROTEC.2018.12.022

12. Karthick Raaj, R., Vijay Anirudh, P., Karunakaran, C., Kannan, C., Jahagirdar, A., Joshi, S., & Balan, A. S. S. (2020). Exploring grinding and burnishing as surface post-treatment options for electron beam additive manufactured Alloy 718. *Surface and Coatings Technology*, *397*, 126063. https://doi.org/10.1016/ J.SURFCOAT.2020.126063

13. Lee, J. Y., Nagalingam, A. P., & Yeo, S. H. (2021). A review on the state-of-the-art of surface finishing processes and related ISO/ASTM standards for metal additive manufactured components. *Virtual and Physical Prototyping*, *16*(1), 68–96. https://doi.org/10.1080/17452759.2020.1830346

14. Liao, Z., la Monaca, A., Murray, J., Speidel, A., Ushmaev, D., Clare, A., Axinte, D., & M'Saoubi, R. (2021). Surface integrity in metal machining – Part I: Fundamentals of surface characteristics and formation mechanisms. *International Journal of Machine Tools and Manufacture*, *162*, 103687. https:// doi.org/10.1016/J.IJMACHTOOLS.2020.103687

15. Maleki, E., Bagherifard, S., Bandini, M., & Guagliano, M. (2021). Surface post-treatments for metal additive manufacturing: Progress, challenges, and opportunities. *Additive Manufacturing*, *37*, 101619. https://doi.org/10.1016/ J.ADDMA.2020.101619

16. Salmi, M., Huuki, J., & Ituarte, I. F. (2017). The ultrasonic burnishing of cobalt-chrome and stainless-steel surface made by additive manufacturing. *Progress in Additive Manufacturing*, 2(1), 31–41. https://doi.org/10.1007/S40964-017-0017-Z

17. Sherif, H. A., & Almufadi, F. A. (2016). Identification of contact parameters from elastic-plastic impact of hard sphere and elastic half space. *Wear, 368–369*, 358–367. https://doi.org/10.1016/J.WEAR.2016.10.006

18. Tabatabaei, S. M. K., Behbahani, S., & Mirian, S. M. (2013). Analysis of ultrasonic-assisted machining (UAM) on regenerative chatter in turning. *Journal of Materials Processing Technology*, 213(3), 418–425. https://doi.org/10.1016/J.JMATPROTEC.2012.09.018

19. Teimouri, R., & Amini, S. (2019). Analytical modeling of ultrasonic burnishing process: Evaluation of active forces. *Measurement, 131*, 654–663. https://doi.org/10.1016/J.MEASUREMENT.2018.09.023

20. Teimouri, R., Amini, S., & Ashrafi, H. (2018). An analytical model of burnishing forces using slab method: https://doi.org/10.1177/0954408918781481, *233*(3), 630–642.

21. Teramachi, A., & Yan, J. (2019). Improving the Surface Integrity of Additive-Manufactured Metal Parts by Ultrasonic Vibration-Assisted Burnishing. *Journal of Micro and Nano-Manufacturing*, 7(2). https://doi.org/10.1115/1.4043344

22. Travieso-Rodríguez, J. A., Gómez-Gras, G., Dessein, G., Carrillo, F., Alexis, J., Jorba-Peiro, J., & Aubazac, N. (2015). Effects of a ball-burnishing process assisted by vibrations in G10380 steel specimens. *The International Journal of Advanced Manufacturing Technology*, 81(9), 1757–1765. https://doi.org/10.1007/S00170-015-7255-3

23. Wagoner, R. H., & Chenot, J.-L. (1997). *Fundamentals of Forming*. https://www.wiley.com/en-us/Fundamentals+of+Metal+Forming+-p-9780471570042

24. Wang, J., Geng, Y., Li, Z., Yan, Y., Luo, X., & Fan, P. (2022). Study on the vertical ultrasonic vibration-assisted nanomachining process on single-crystal silicon. *Journal of Manufacturing Science and Engineering*, 144(4). https://doi.org/10.1115/1.4052356

25. Wang, Q. Jane and Zhu, D. (2013). Hertz theory: Contact of spherical surfaces. In Y.-W. Wang Q. Jane and Chung (Ed.), *Encyclopedia of Tribology* (pp. 1654–1662). Springer US. https://doi.org/10.1007/978-0-387-92897-5_492

26. Wang, Z., Luo, X., Liu, H., Ding, F., Chang, W., Yang, L., Zhang, J., & Cox, A. (2021). A high-frequency non-resonant elliptical vibration-assisted cutting device for diamond turning microstructured surfaces. *International Journal of Advanced Manufacturing Technology*, 112(11–12), 3247–3261. https://doi.org/10.1007/S00170-021-06608-3/TABLES/5

Chapter 5

A critical review of mechanical-based post-processing techniques for additively manufactured parts

Abdul Wahab Hashmi, Harlal Singh Mali and Anoj Meena
Malaviya National Institute of Technology, Jaipur, India

Shadab Ahmad
Shandong University of Technology, Zibo, China

Ana Pilar Valerga Puerta
Universidad de Cádiz, Puerto Real, Cádiz, Spain

Maria Elizete Kunkel
Universidade Federal de São Paulo, São Paulo, Brazil

CONTENTS

DOI: 10.1201/9781003288619-5

5.1 INTRODUCTION

Rapid prototyping techniques that join materials one layer at a time to create objects gave birth to the additive manufacturing (AM) techniques, and this allows designers to create detailed physical prototypes in a matter of few hours direct output from the computer-aided design (CAD) model in three dimensions (3D).

The processes are best suited for parts with freely shaped curves and features and are generally complex in design with a limited percentage of plane surface area. But as seen in Figure 5.1d, a significant problem for commercial AM applications is the "Stair Casing Effect", which causes poor surface finish [1].

AM manufacturing steps ideas are shown in the following steps:

1. Using any commercial CAD software, the 3D CAD model generated in Figure 5.1a.
2. 3D CAD model slicing of two layers Figure 5.1b.
3. By stacking one by one of those 2D layers, the physical output is generated in Figure 5.1c.

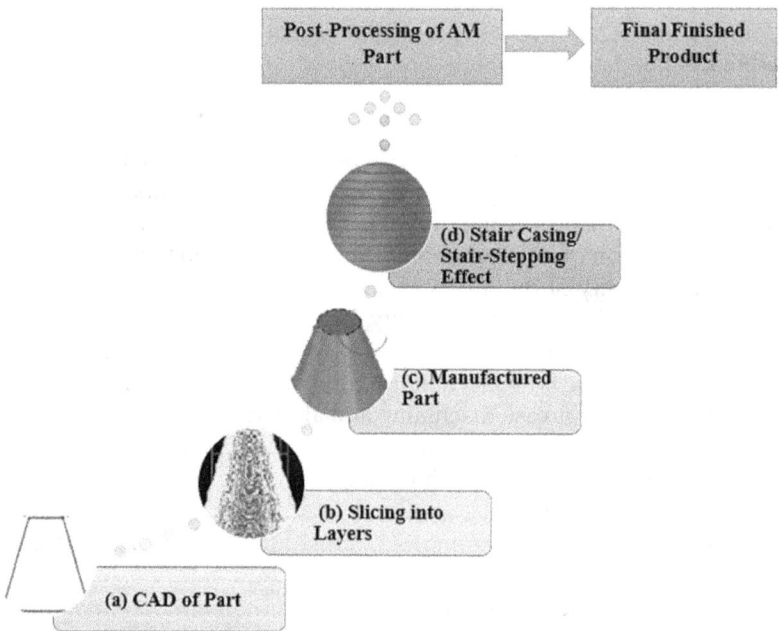

Figure 5.1 The fourth industry revolution (Industry 4.0) [1–3] is to illustrate the process of AM layer-based and the effect of stair-stepping – (a) CAD of part; (b) slicing; (c) AM actual output; and (d) effect of the staircase.

The poor surface finish model was generated due to the stacking of 2D layers due to the effect of stair casing seen in Figure 5.1d.

Numerous academics have investigated different factors of processes associated to AM in an effort to lessen the impact of the staircase, such as the orientation, the thickness of the layer, and material deposition orientation. Various post-processing methods depend on the application and the model material applied to enhance the quality of the surface.

However, AM creates components and structures on every layer one by one concerning the CAD-3D models, dissimilar conservative subtractive technology used to form solid material, subtract or cut and for final part fabricating, it is blocked. Comparing AM techniques to traditional manufacturing techniques, there are many advantages. It provides more flexibility for CAD designers to plan and optimize the topological structure, thus making it possible to produce metal parts in close shape with complex details and reduce the need for costly moulds. Simultaneously, it provides lead time decrease, reduces waste from those materials, and reduces costs. The UK Technology Strategy Board stressed AM's strategic position as a gamer in future manufacturing. In the next 25 years, buildings and complete bodies have been expected to be available for printing. AM processes have recently been introduced in plans of production and research in the automotive, aerospace, and biomedical industries [4–8]. The global AM market stood at approximately $3.7 billion in 2015, and its rapid growth is estimated to continue to exceed $6.5 billion by 2019 [9]. The materials range, including plastic, ceramics, and polymers, can be printed by AM. Polymers are the most common 3D printing materials because they are readily available and less expensive. Metals AM, however, has higher strength for material strategy and select choice. AM materials like stainless steels, titanium, nickel super-alloys, and tungsten have alloys and many metals, which are attractive particularly where time-consuming and expensive exhibited from subtractive machining. Nevertheless, AM has faced many technological problems, including raw materials, production speed, CAD software, and reliability, since its establishment in the 1980s. The next key issue that reduces use of AM products which shows the poor quality of AM surfaces is surface finishing. It is regarded as regular mechanical product properties of quality. In general, the strength of exhaustion is strongly exaggerated on the final surface because of cracks initiation and poor surface quality. However, surface quality was demonstrated to be sensitive to material strength and corrosion resistance. Material consistency can be improved on rough surfaces. Meanwhile, AM techniques or 3D printing (3DP), Rapid Prototyping (RP), or fabrication of solid form is an emerging technology in the domain of creating various three-dimensional structures, particularly for the automotive and aerospace industries, manufacturing industry, bio-implants, i.e. orthopaedic/dental implants, artificial human organ 3D printing, e.g. heart valves, ear, nose, tissue engineering, drug delivery, orthoses, and prostheses. Each mentioned area's major involvement is shown in Figure 5.2.

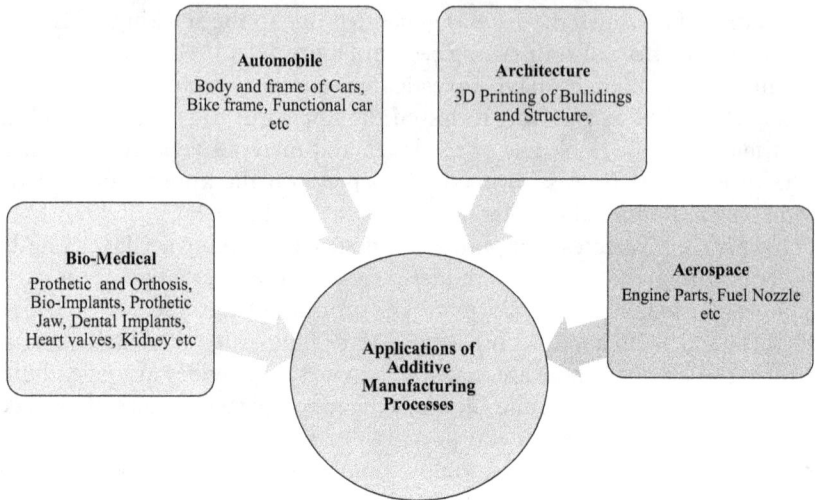

Figure 5.2 Different applications of AM processes.

The distinctive process denoted basic and economical method which produces different shapes given various materials like composites, metals, polymers, ceramics, etc. [10].

This review aims to undertake a detailed critical evaluation of the existing literature on the various mechanical-based post-processing methods for enhancing surface quality in AM parts.

5.2 CLASSIFICATION OF AM TECHNOLOGIES BASED ON USAGE OF STANDARD AND CLASSIC METHODS

As a basic premise of AM technology, By adding layers upon layers of material and joining them together, Directly from a CAD-generated 3D model, a three-dimensional object is built. Many firms have developed and launched new techniques. Because the technology is still relatively new, the companies developing and launching new technologies have devised their own marketing labels for the operation, even though the underlying procedures may be the same. However, companies that use certain AM technology have their own names. The classification of AM technologies based on the usage of standard and classic methods is shown in Figure 5.3.

Table 5.1 shows the classification of AM technologies based on standard and classic methods usage.

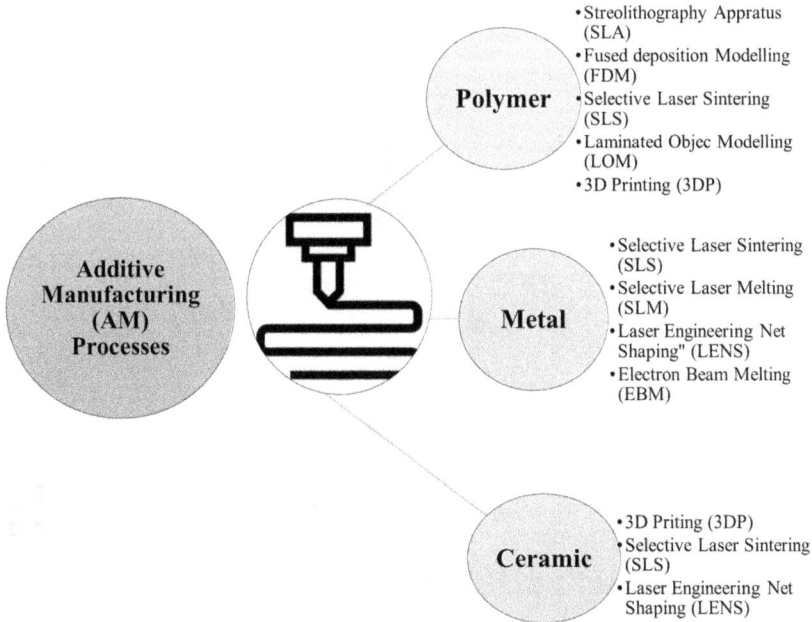

Figure 5.3 Classification of AM processes.

5.3 NEED FOR POST-PROCESSING OF AM PARTS

Layer-wise characterization of AM, tool-free production part. The laser sintering and melting process of laser beams are referred to as technologies of AM composed of higher strength for broad usage of industry. This study, therefore, analyzes the typical surface characteristics of these two AM techniques. An overview of commercially available systems is an AM process alternative. The metal AM (PBF) parts surface is distinguished by periodic due to the distribution of random powder. In addition, the production process in the layer ends up strong depending on the property surface and angle of inclination defining manufactured parts orientation within the powder bed system. Knowing the surface properties dependent on the angle could help assess and, when process preparation, consider the surface's specific quality in every AM part area. The loose powder cannot be used again and cannot be melted by the laser beam process. Because the process of laser beam fusion takes place at the ambient air temperature, a limited recycling effort of the material, i.e. the removal of agglomerates, is needed. AM parts had the roughest surfaces, making them poorly finished. Figure 5.4 illustrates the numerous surface flaws that develop throughout the AM operations.

AM part surface quality can be improved by controlling the process parameters before the actual 3D printing of the product. This control of process parameters comes under the pre-processing technique phase, shown

Table 5.1 Classification of AM technologies based on standard and classic methods usage

AM method	Type of process	Typical materials	Standard	Resolution of feature	Surface Roughness (μm)	Ref.
Vat photo polymerization	Stereo lithography (SLA)	Acrylics and epoxies, PPF ceramics hydrogels, cells	Energy source as UV laser beam used for repeat photo solidifies resin layers on the surface.	50–100 μm	2–40 μm	[11, 12]
	Digital Light Processing (DLP)	Polymers, composites, zirconia, elastomers	Below are the resin light projectors used.	35–120 μm	10–30 μm	[13, 14]
Powder Bed Fusion	Selective Laser Sintering (SLS)	Nylon, PEEK, PLGA, PCL, PLA, PVA, HA	By atomic diffusion, sinter powder particles are utilized by the CO_2 laser beam.	50–100 μm	5–35 μm	[15]
	3D Printing (3DP)	Shape memory alloys, starch, cellulose, PLGA, PCL PLA Al2O3	Binding of particles.	100 μm	12–27 μm	[16]
	Selective laser melting (SLM)	Cp-Ti, alloys of cobalt, steel, Ti-6Al-4V, aluminium, ceramics	Until fully melted and together, a fused bed of metallic powder has melted particles in definite locations generated by the laser beam.	30–150 μm	10–20 μm	[17, 18]
	Electron Beam Melting (EBM)	Alloys of Ti-6Al-4V, Co-Cr-Mo, Cp-Ti, β-Ti al, Inconel 718	In a high vacuum, the electron is generated in which the metal powders melt.	Min size of feature: 0.1mm 20.3–25.4 μm	10–50 μm	[19]
	Direct Metal Laser Sintering (DMLS)	Ti-6Al-4V, Co-Cr	Instead of plastic powders, metal alloy powders are used.	89–97 μm	15–60 μm	[20]

Material Jetting	Continuous Printing (CIJ)	PCL, ABS, polyamide, PLA and their composites	Stream liquid continuous breaking forming droplets.	Accuracy: 0.1–0.3 mm: 40 μm	3–30 μm	[21]
Binder Jetting	Binder Jetting	Magnetic materials, intermetallics, solid oxide fuel cells, biodegradable alloys, alloys of shape memory, steel, polymers, and BaTiO3	Raw material powder bed having print binder feed.	25 μm	4–12 μm	[22]
Material Extrusion	Fused deposition modelling (FDM)	ABS, polyamide, PLA, and their composites	Through fine nozzle thermoplastic extrusion through a fine nozzle.	100–150 μm	9–40 μm	[23]
Sheet Lamination	Laminated Object Manufacturing (LOM)	Ceramics, aluminium, titanium, copper, stainless steel, fabrics, plastics, composites, and synthetic materials	By bonding, sequentially laminating, and cutting, the 3D object is formed.	200–300 μm	6–27 μm	[24]
Deposition of directed energy	LENS–Laser engineer net shape	Magnetic alloys, metals, permalloys, Ti–tungsten alloys	The object is fused with a Laser.	10–120 μm	45–200 μm	[25, 26]
	Electron Beam Additive Manufacturing (EBAM)	SS, cobalt alloys, nickel alloys, copper-nickel alloys, tantalum, titanium alloys	The electron beam is used to create an object.	90 m	20–54 m	

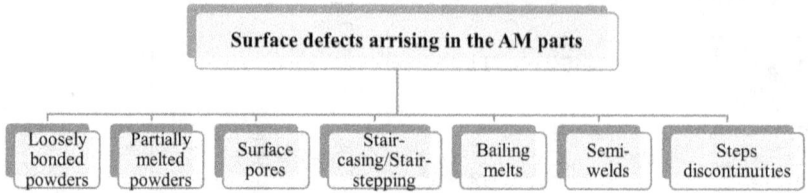

Figure 5.4 Surface flaws arise during the AM process.

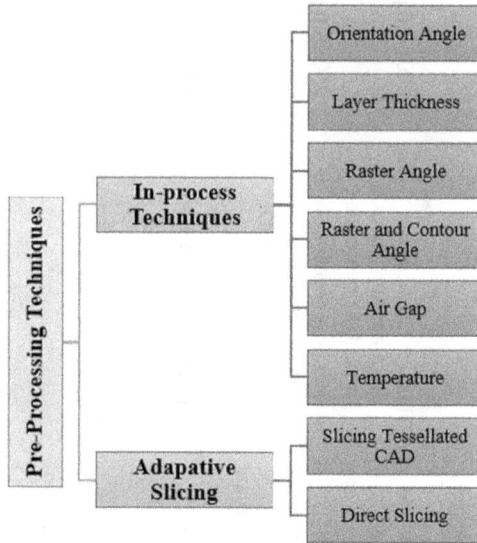

Figure 5.5 Surface flaws arise during the AM process.

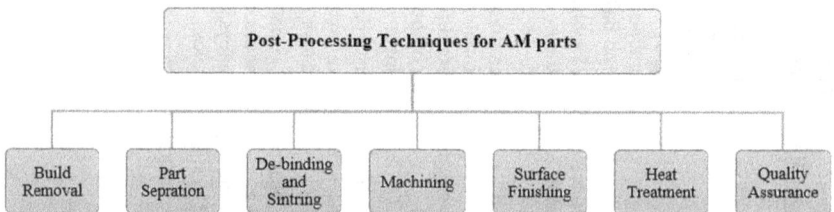

Figure 5.6 Post-processing techniques for AM parts.

in Figure 5.5. However, pre-processing procedures are restricted in their ability to improve the surface quality of AM parts. Additional post-processing techniques are required [27–30]. The post-processing techniques are discussed in detail in the below sections of this chapter.

Based on the kind of AM processes and their requirements for post-processing, the post-processing techniques for AM components include the following operations, as illustrated in Figure 5.6.

5.4 CLASSIFICATION OF POST-PROCESSING TECHNIQUES TO ENHANCED THE SURFACE QUALITY OF THE AM PART

After the production phase, the present context leads to different post-processing techniques and are categorized into four categories: mechanical, chemical, thermal, and hybrid methods, as shown in Figure 5.7. Post-processing methods that are covered in the section below include these four methods.

5.5 MECHANICAL-BASED SURFACE QUALITY IMPROVEMENT METHODS

The methods are based on cutting mechanically or peak surface profile stress. The strategies are usually followed by traditional finishing of metal methods, whereas plastics of acrylonitrile butadiene styrene (ABS) reaction

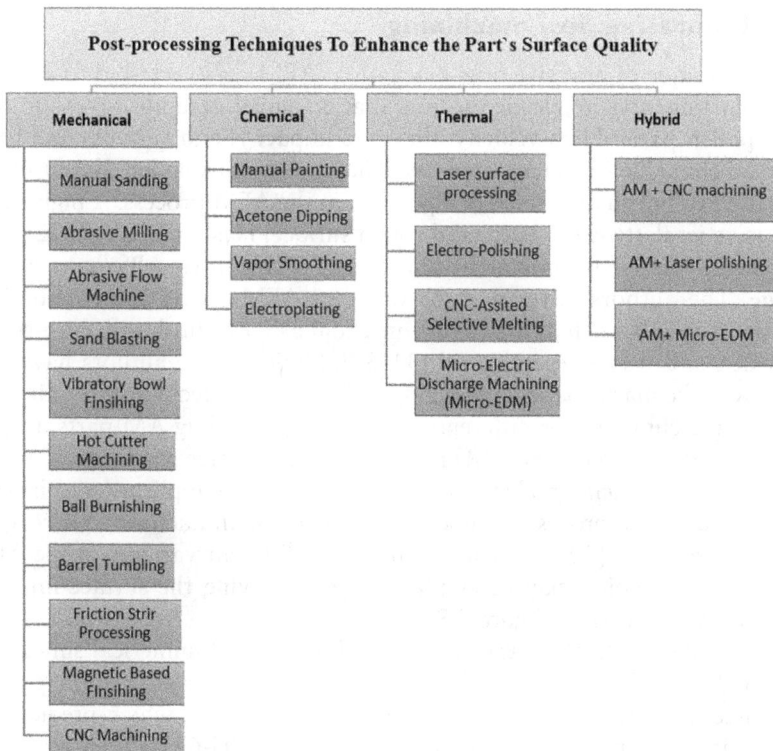

Figure 5.7 Classification of techniques for enhancing the surface quality of AM parts.

vary from metals. Spencer et al. (1993a; 1993b) successfully tested various techniques for the abrasive finishing of stereolithography pieces using abrasive blast, tumbling barrel, centrifuge tumbling, ultrasonic vibratory bowl, and abrasion. Abrasive media impact in such finishing of unnecessary mass materials were removed from the edges and corners. Abrasive finishing was achieved [31, 32].

5.5.1 Manual sanding

In the case of sandpapers (120 and 320 grit), steel wool, and fillings, Stratasys recommended different handheld methods of eliminating small bits of support, and strands resemble hairs and husks. Hot knives manually eliminate seams and use them to fill raw material gaps (Stratasys, 1997) [33].

The manual methods are simple and economical, but they cannot be checked, measured, consistent, and precise because they are based on the operator's skills. In addition, the manual finish of the methods makes it reliable for fast tools application and moulding when dealing with components with shapes of complicate.

5.5.2 Abrasive flow machining

The technique of abrasive flow machining (AFM), also referred to as abrasive blasting, uses an elastic medium that is loaded with abrasives to finish and polish parts. High-velocity abrasives impact rough surfaces and leave burrs behind before being smoothed down. Abrasive-laden elastic media are used to polish and finish components in the AFM process. A high-speed abrasive jet flattens the burrs and rough surface. Hashmi et al. (2022) have employed the AFM process to finish the FDM-printed ABS parts at this stage. The authors have used the different AFM media from natural or waste materials, such as waste polymer, coal ash, rice husk ash, corn-starch powder, and waste vegetable oil [34–37]. Similarly, the authors have also reviewed the mathematical modelling and simulation technique for abrasive finishing techniques for different applications, including AM parts, to predict the material removal (MR) mechanism and surface roughness [38–39], and studied artificial intelligence and machine vision application to investigate the different process parameters during any manufacturing process, i.e. AM applications [40–41]. The schematic of different variants of the AFM process [42], which may be employed for improving the surface finish of AM parts, is shown in Figure 5.8.

Bouland et al. (2019) performed experimental and numerical simulation for AFM-based finishing of the laser powder bed fused (L-PBF) part to enhance the surface quality. The numerical simulation, which predicts the indentation of abrasive particles to the surface of a Ti-6Al-4 V alloy, utilizes a computation fluid dynamic (CFD) technique [43]. Abrasive flow finishing (AFF) was utilized by Mali et al. (2018) to smooth FDM-printed ABS

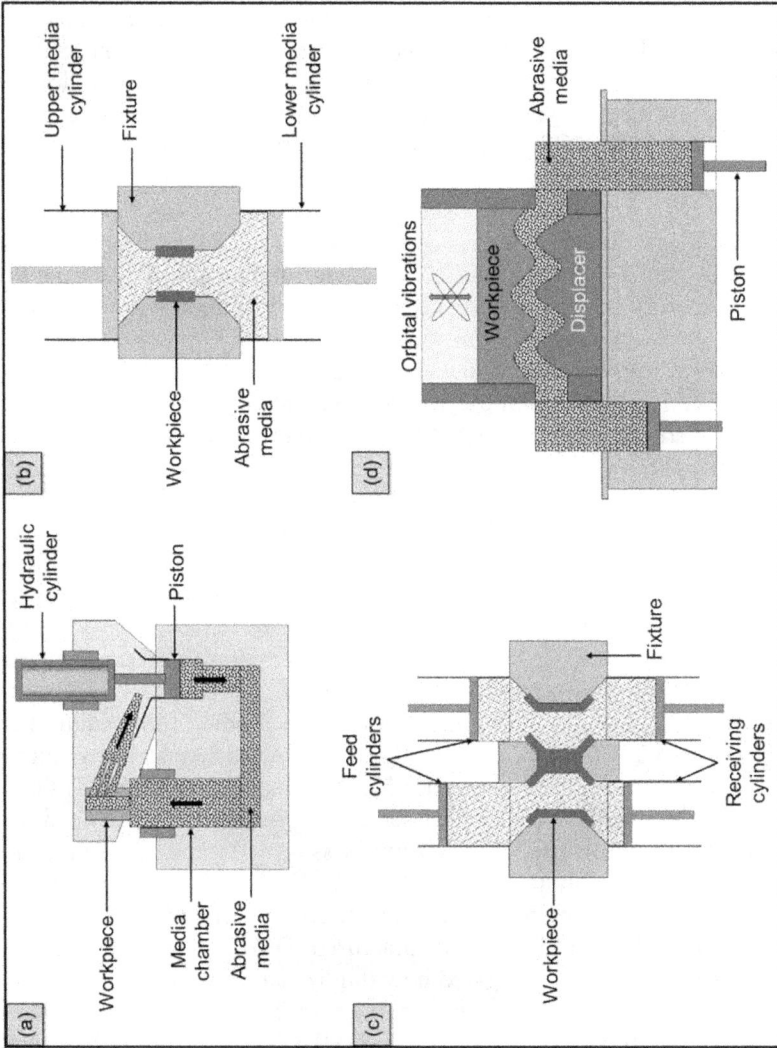

Figure 5.8 The different variants of AFM process (a) One-way AFM, (b) Two-way AFM, (c) Multi-way AFM, and (d) Orbital AFM [42].

surface quality. The results showed that on the internal and external surface, a maximum improvement of ΔRa 21.37 µm and a maximum improvement of ΔRa 6.27 µm were obtained, respectively [44]. On additively built Inconel 625 samples, Neda et al. (2018) employed mixed chemical abrasive flow polishing. The authors discovered that the surface roughness and texture of the inferior surface of IN625 samples might be greatly improved by removing semi-welded particles using a combination of chemical and abrasive flow polishing [45]. The AFM process was used by Duval-Chaneac et al. (2018) to improve the surface finish of the internal surface of additively manufactured (SLM) tube. The authors used the AFM procedure to finish the non-heat-treated and heat-treated maraging steel 300 tube shape parts. Surface roughness was observed to be in the 12–14 µm range at first, then reduced to 2–10 µm. The results showed that ploughing mechanisms predominated more during finishing these SLM surfaces that had not been heated than those that had [46]. Nagalingam et al. (2018) proposed a method to modify the internal surface of additively manufactured items by managing hydrodynamic cavitation erosion with abrasive particles. Hydrodynamic cavitation abrasive finishing was the name given to this method (HCAF). The finishing cylindrical specimens of aluminium alloy AlSi10Mg were fabricated using AM technique, i.e. Direct Metal Laser Sintering (DMLS). The results showed that up to 40% of the initial surface roughness (Ra) was reduced, and hydrodynamic cavitation erosion has been identified as a key wear process in this HCAF process [47]. Tan et al. (2017) showed the utilization of ultrasonic cavitation abrasive finishing of additively manufactured (DMLS) Inconel 625 parts. The smallest partially melted particles were removed from the workpiece by heterogeneous cavitation nucleation, according to scans made using a scanning electron microscope (SEM). The workpieces' initial average surface roughness Ra values ranged from 6.5–7 µm to 3.65 µm in the finished product [48]. Anilli et al. (2017) employed the combined method of electrochemical machining (ECM) and AFM process to finish the additively manufactured (SLM) single and double chamber nozzles of the laser cutting machine. A combined finishing process sequence can increase roundness by 40% and reduce average surface roughness by about 85% [49]. E. Atzeni et al. (2016) employed a newly developed finishing technique, namely Abrasive Fluidized Bed (AFB) polishing, to finish the additively manufactured (DMLS) flat AlSi10Mg substrates. The researchers investigated how fluidized abrasives interacted with AlSi10Mg substrates. In a fluidized bed, the rotating substrate was polished more quickly with a range of abrasives. DMLS produced AlSi10Mg substrates with a 16.72 µm starting surface roughness. With improved surface roughness following AFB, AlSi10Mg flat substrates can have final average roughness Ra of around 1.5 µm [50]. Williams et al. (2007) employed AFM process to finish the profile edge laminae of sealing conformal channels [51]. Leong et al. (1997) employed an abrasive jet deburring procedure to finish the jewellery models fabricated using the stereolithography (SLA) process.

The authors investigated several process variables such as abrasive medium, nozzle distance, air pressure, and blasting time. The results revealed that the duration of abrasive blasting is significantly relevant compared to the distance of abrasive blasting. The orientation of the jewellery models has an impact on the effectiveness of the deburring process as well. Surface roughness was reduced by abrasive jet deburring by 70% while using dry air as the carrier and glass beads as the abrasive media [52]. AFM investigated SLA components surface finishing by Williams et al. (1998). The parameters of AFM include the grit size of abrasive and pressure of AFM media. The surface roughness of SLA parts was improved by multiple AFM cycles. The results showed that media grit size, build orientation, pressure extrusion, and grit size interaction all have a major impact on MRR results [53]. Linchao et al. (2021) have employed the combined method of electrochemical and mechanical polishing (EMCP) for additively manufactured (L-PBF) internal channels [54]. The schematic of EMCP set-up is shown in Figure 5.9 below.

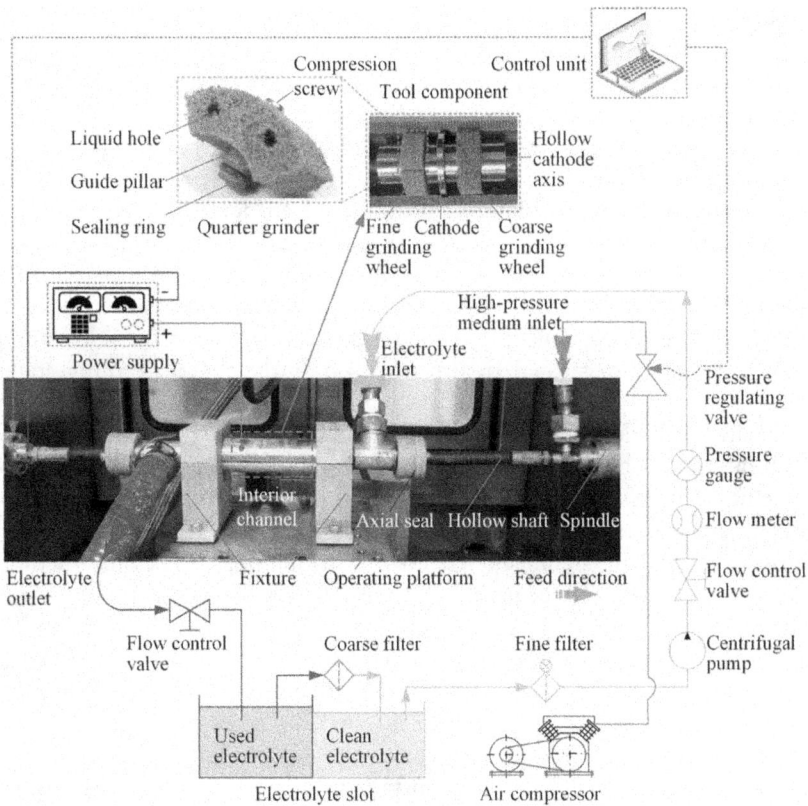

Figure 5.9 The schematic of ECMP to enhance the surface quality of AM parts [54].

Cong Ni et al. (2022) employed the combined AFM and ultrasonically assisted AFM (UAAFM) technique to finish an additively manufactured (SLM) micro-channel using 316 L SS powder. The authors have also simulated the behaviour of media flow using CFD simulation [55]. Dixit et al. (2022) have also employed the AFM process to finish the FDM-printed ABS and PLA parts. The authors have developed and characterized the new AFM media of xanthan gum and compared the results for improvement of surface roughness (Ra) and material removal rate (MRR) [56]. Jalui et al. (2021) have used a AFM process to finish the additively manufactured (L-PBF) internal channels and thin wall of titanium. The results showed 40% and 38% of surface roughness improvement inside the internal channel and outside the thin wall of the titanium workpiece, respectively [57].

5.5.3 CNC staircase machining and abrasive milling

Vispute et al. (2018) have employed a new hybrid additive, subtractive manufacturing (HASM) technique to improve the surface finish of FDM-printed ABS parts. HASM set-up developed by using three axes CNC milling machine. A novel indigenous material deposition tool (MDT) was developed for deposition of material in the manner of layer-by-layer, which has acted as additive manufacturing of any complex shaped part by using ABS material in the form of the filament of the pallet. A CNC milling machine was used to remove the material to enhance the surface finish of the part manufactured by MDT, which acted as subtractive manufacturing. A tool path for additive, subtractive manufacturing has been developed by using customized software modules in MATLAB software. The initial surface roughness of additive manufactured parts was reduced from 19.495 µm up to 2.068 µm after CNC machining [58]. Different post-processing techniques effects such as hand finishing, shot peening, tumbling, CNC machining, painting in spray, or surface treatment based on the chemical to investigate the significance of roughness in surface and their effects on accuracy in the dimension of Nylon and Alumide and ABS thermoplastic materials workpiece fabricated by the laser sintering (LS) method and fused deposition modelling (FDM) respectively. They performed hypothesis testing such as the Chi-square test ($x2$) and statistic Z-observed (Zobs) methods of testing to compare the significant effects of various post-processing methods [59]. Oyelola et al. (2016) investigated the surface integrity and machining behaviour of additively manufactured (DMD) Ti-6Al-4V workpiece. A wire of Ti-6Al-4V having a diameter of 1.2mm is used in the DMD process as the feedstock material. A hollow cylindrical-shaped workpiece having an external diameter of 73 ±0.5mm and an internal diameter of 70 ±0.5mm and with a height of 61.7±0.5mm and wall thickness of 2.7±0.5mm is fabricated for turning operation. The initial surface roughness 250 µm was reduced up-to 0.822 µm and 2.140 µm in the coated and uncoated regions after the machining [60]. Löber et al. (2016) compared different post-processing techniques to enhance the surface finish of additively manufactured (SLM) parts. The post-processing techniques such as sandblasting, grinding, and plasma and electrolytic polishing are used to improve surface roughness. The workpiece sample was fabricated by stainless steel X2CrNiMo17-12-2 (or 316l) powder 45 μm diameter by SLM process, which has a cuboidal shape with dimensions of 10×10×10 mm³. The combination of a mechanical polishing method followed by plasma or electropolishing decreased the surface roughness up-to-the minimum value of 0.12 μm. This combination was only possible for flat structures [61]. Galantucci et al. (2014) used lamellar paper of abrasive bulk with very little distinction in dimension. From micro sandblasting and electroplating, the abrasive friction following physical vapour deposition showed small dimensional changes (Galantucci et al., 2014). CNC milling was proposed by Kulkarni and Dutta (2000), where the G-code program was produced by adaptive slicing in line with the part geometry. Nevertheless, the subtractive machining method could not reach the features of parts with intricate shapes [62, 63]. Blair (1998). Used as a material removal process by chip forming, sandpapers action by abrasive machining of various rotated sizes in grit. The smaller size of the media improved the material's removal level and increased the layer's toughness. In order to achieve a 90% increase in surface finishing [64]. Pragana et al. (2022) have reviewed the various hybrid AM methods. The authors have explained the hybrid CNC machining and AM techniques, and it is shown in Figures 5.10 and 5.11 [65].

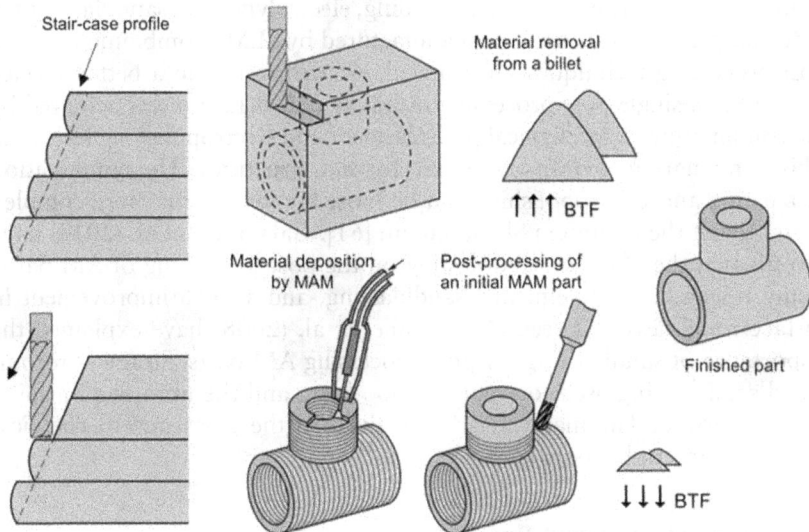

Figure 5.10 The schematic of CNC machining of AM parts [65].

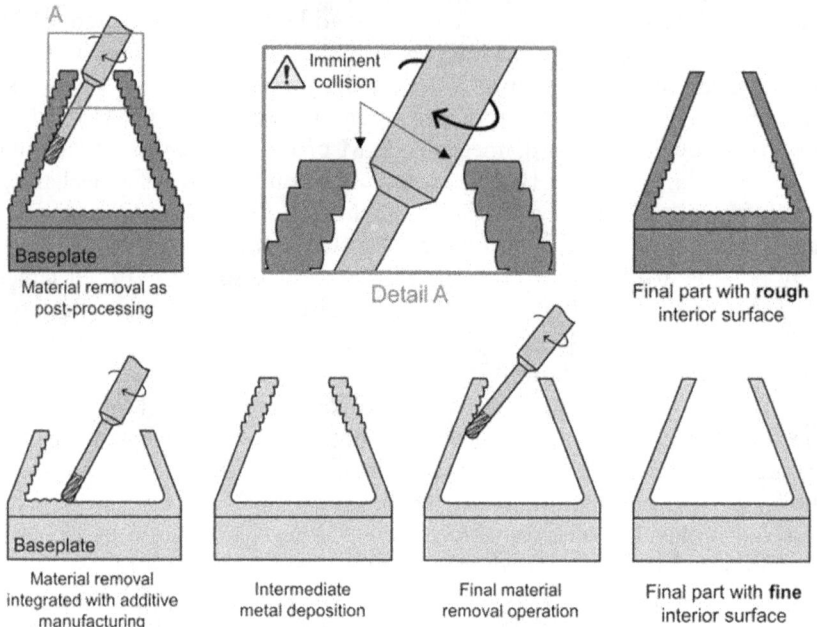

Material removal as post-processing

Detail A

Final part with **rough** interior surface

Material removal integrated with additive manufacturing

Intermediate metal deposition

Final material removal operation

Final part with **fine** interior surface

Figure 5.11 The schematic of hybridization of CNC machining with AM process [65].

5.5.4 Sandblasting

Lukas Löber et al. (2011) have made a comparison of different post-processing techniques like plasma polishing, grinding, electrolytic, and sandblasting for reducing parts of SR roughness manufactured by SLM. Combining multiple post-processing techniques (mechanical, electro) results in a better surface finish than a single post-processing technique. SR 0.12 μm was achieved by the combination of mechanical pre-treatment and electropolishing followed. This combination is only favourable for flat structures. The combination of blasting and electropolishing can be most favourable for more complex parts, giving the minimum SR of 3.6 μm [61]. Galantucci et al. (2014) have investigated the effect of sandblasting on the post-processing of AM parts. Many researchers implemented sandblasting, and a 96% improvement in surface roughness was seen [62]. Zinniel et al. (2008) have explained the importance of sandblasting for post-processing AM parts. Stratasys recommend sandblasting used for vapour smoothing, and the ultra-fine finishing process proved. The matte finish was given to the parts due to chemical glossy surface and vapour [66].

5.5.5 Vibratory bowl finishing

Into vibration finishing, parts of mass finishing are completed using a spring-mass system by means of media of abrasive action and finely compounds of

Figure 5.12 The schematic of vibratory bowl finishing of AM parts [69].

abrasive and water combined. The shaft of the water driven by the belt is connected to two opposite sets of eccentric weight. Different material removal levels are achieved depending on time, medium size, media composition, media weight, and compounds. Schmid et al. (2009) examined the greater role in the forming and finishing of abrasive media and the type of surface via the form of media in the pyramid of times (3 to 4 hours) of long machining [67], Trivedi (2014) estimates a rise of 31.67% and 4.59% in surface rugging and hardness. It improved dimensional stability for ABS parts by working longer hours with a lower medium weight [68]. Atzenia et al. (2022) have also employed the vibratory bowl finishing for additively manufactured (L-PBF) parts. The schematic of the machine set-up for vibratory bowl finishing of AM parts is shown in Figure 5.12 [69].

5.5.6 Barrel tumbling

Tumbling barrels are mass finish processes used for fine finishing, deburring, and parts conditioning of surface morphology. The pieces are placed in a compound of abrasive, water, and media loaded in a closed rotating tube. This process needs fewer starting, execution, and maintenance costs, and can machine various geometries without any attachments. For the estimation of the removal of material rate during the process of barrel finishing, Boschettoet al. developed an FDM thickness layer, orientation, and time-based mathematical model [70, 71]. In various geometrics of ceramic media used by Fisher and Schöppner (2013), 52% reported triangular media of surface roughness decrease. Frequency of rotation, media and time type size were used to test the content removal level for the main parameters [72]. Boschetto and Bottini (2015) recognize a different orientation angle for barrel finishing of FDM printed parts as the most important parameters for surface roughness and dimensional accuracy. The topical peak removal rate occurs at an angle of 90° and a minimum of 18° [73]. The schematic of barrel finishing of AM parts is shown in Figure 5.13 [74].

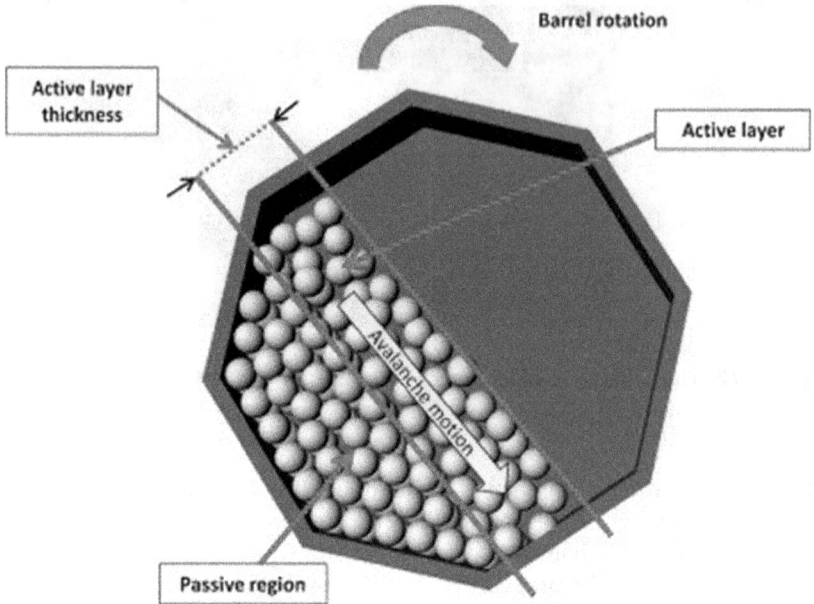

Figure 5.13 The schematic of vibratory bowl finishing of AM parts [74].

5.5.7 Hot cutter machining

A designed empirical model of a flat machine straight-edge hot cutter for predicting the end of roughness in the surface by adaptive slicing. This method proved effective for decreasing surface roughness and cutting direction, and rake angle were major parameters by incorporating hot cutter machines with adaptive technologies [75]. The virtual hybrid system was proposed (Pandey et al. 2006). Numerical controlled machines traversing in X–Y directions can be attached to the hot cutters, with different shapes, along the periphery of the slice. Although the simulated process indicates significant improvements in surface finish, the complex mechanism has expensive and time-consuming realistic manufacturing complications [76]. The schematic of hot cutter machining of FDM printed parts is shown in Figure 5.14.

Taufik et al. (2020) have developed a thermal-assisted novel finishing technique using CNC assisted selective melting tool to improve the surface finish of FDM printed parts. The schematic of this newly designed CNC selective melting is shown in Figure 5.15 [77].

5.5.8 Ball burnishing

Vinitha et al. (2012) have used the ball burning process to press and fill the valleys with the peaks of the surface profile. The workpiece was held in

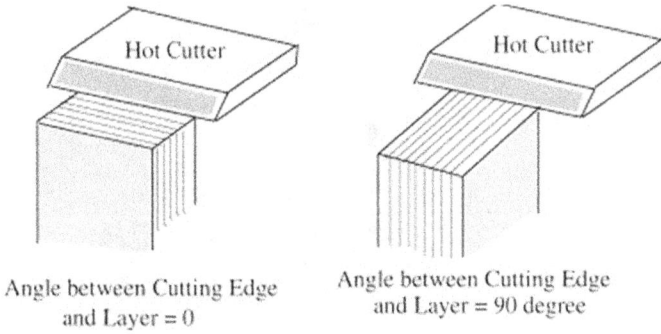

Angle between Cutting Edge and Layer = 0

Angle between Cutting Edge and Layer = 90 degree

Figure 5.14 The schematic of hot cutter machining of FDM parts [75].

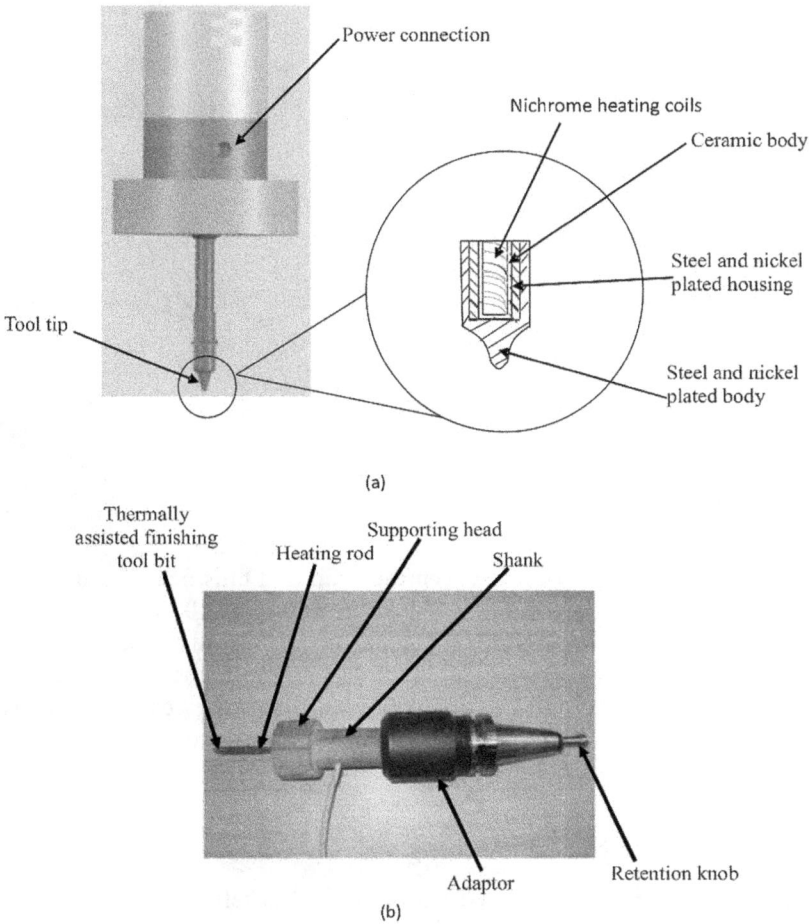

Power connection

Nichrome heating coils

Ceramic body

Steel and nickel plated housing

Tool tip

Steel and nickel plated body

(a)

Thermally assisted finishing tool bit

Heating rod

Supporting head

Shank

Adaptor

Retention knob

(b)

Figure 5.15 The schematic of thermally assisted finishing tool of FDM parts [77].

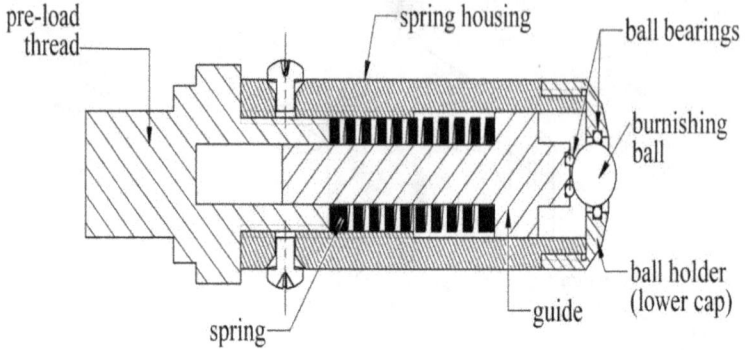

Figure 5.16 The schematic of designed ball burnishing tool [79].

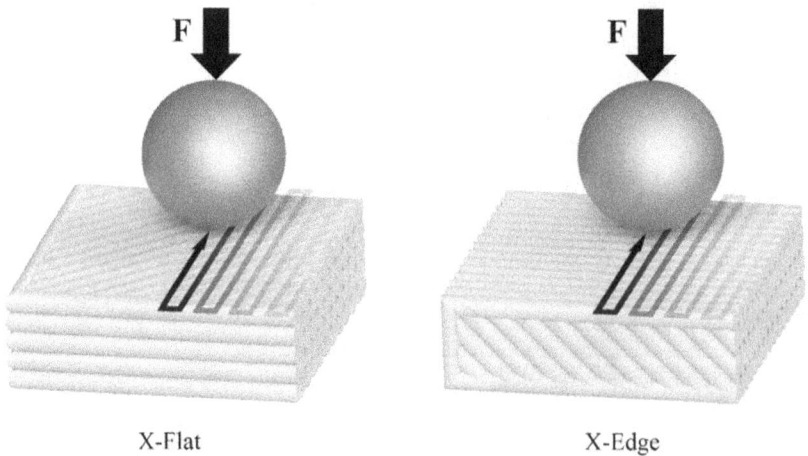

X-Flat   X-Edge

Figure 5.17 The schematic of ball burnishing of AM parts [79].

the lathe machine tool post between the head and tailstock. The increased speed of spindle and penetration improved finishing, ABS parts wear, and the strength applied by the burnishing machine increased the hardness [78]. Bruijn et al. (2021) have investigated the surface finish and mechanical performance of FDM printed parts, which were post-processed using the ball burnishing technique. The designed burnishing tool and schematic of the finishing set-up are shown in Figures 5.16 and 5.17 [79].

5.5.9 Magnetic based finishing

Sometimes the finishing assisted magnetic field is referred to as magnetic abrasives finishing, also known as the technique of surface finishing of an abrasive particle forced against the target surface by means of a magnetic field. This makes it possible to finish conventionally inaccessible surfaces (e.g. an internal layer of a long, curved pipe). A range of applications has been developed to include the manufacture of medical components, fluid systems, optics, dies and moulds, electronic elements, micro-electromechanical systems (MEMS), and mechanical components. Jiang Guo et al. (2019) have developed the novel rotating vibrating magnetic abrasive polishing method to improve the surface finish of the complex internal surface of a double-layered tube structure manufactured by Inconel 718 material using the SLM process. The uneven surface due to partially melted powders during the SLM process was finished, and the surface roughness was improved from Ra of around 7 µm to less than Ra 1 µm [80]. The schematic of rotating-vibrating magnetic abrasive polishing for the finishing of the AM parts is shown in Figure 5.18.

Kumar et al. (2018) have used a ball end magnetorheological finishing process (BEMRF) for nano finishing of FDM printed parts. The primary finishing techniques (facing and lapping) and secondary finishing techniques (BEMRF) have been used for finishing FDM printed PLA. The surface roughness improvement was recorded from Ra = 20 µm the surface roughness to Ra = 81 nm final value by combining the finishing process of primary

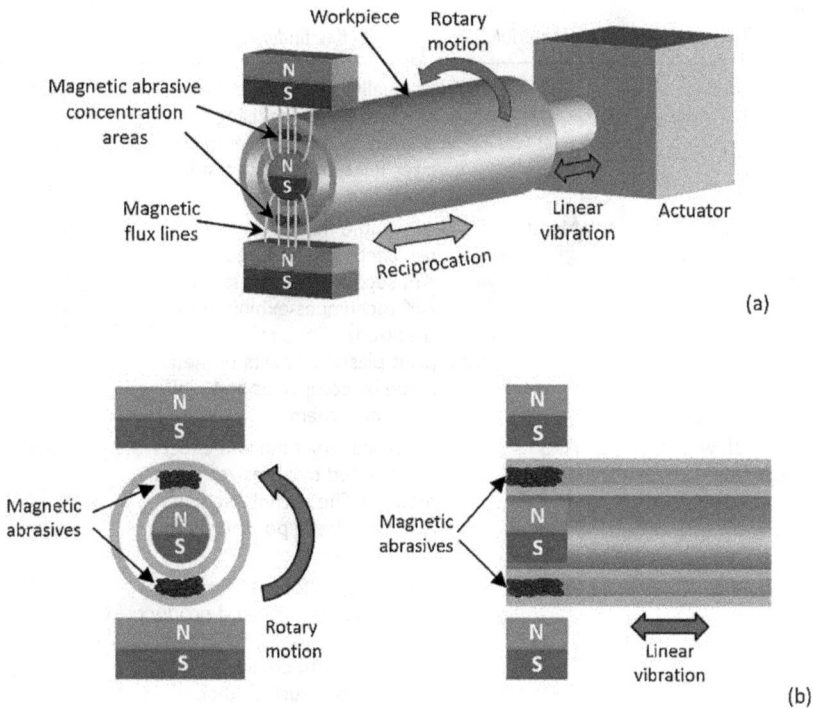

Figure 5.18 The schematic of rotating-vibrating magnetic abrasive polishing of AM parts [80].

and secondary [81]. Zhang et al. (2018) have used a magnetic-based finishing (MAF) method to finish additively manufactured (SLM) 316L stainless steel parts. The permanent magnet grade N50 Nd-Fe-B magnet-based tool was designed and fabricated for polishing AM parts. The improvement of surface roughness was found from ~12 to 3 µm (Ra) after a 75-minute polishing [82]. Yamaguchi ET al. (2017) have used a magnetic field-assisted post-processing method for surface roughness improvement and residual stress additively manufactured (SLM) 316L steel parts. A combination of post-processing methods like sanding, magnetic assisted polishing (MAP), and magnetically assisted burnishing has been applied. The roughness of the SLM-processed surfaces reduces in the wide range from initial roughness 100 µm up to final roughness of 0.1 µm. The combined post-processing methods MAP and MAB impart compressive residual stress at the surface of SLM manufactured parts [83].

Table 5.2 shows the summary of past research on mechanical-based surface quality improvement methods.

Table 5.2 Summary of past research on mechanical based surface quality improvement methods

Surface quality improvement methods	Material	Key findings	References
Manual sanding	ABS	High-quality 3D-printed parts can be processing using manual sanding by hand to remove layer lines and create a smooth finish.	[42]
CNC staircase machining and abrasive milling	Metal	Low volume manufacture, end-use parts, and adopted as AM secondary process for AM techniques exhibited by traditional CNC machining. 3D print plastic or parts of metals made by companies and CNC machining them.	[59]
Abrasive flow machining	ABS	AFM processing time and effects are affected by abrasive medium. The cutting medium is a viscoelastic polymer with abrasive particles.	[44]
Sandblasting	ABS	Sandblasting uniformly abrades the surface of a printed product to remove smooth, reflecting extrusions and create a matte, homogenous surface that scatters light equally.	[66]

(Continued)

Table 5.2 (Continued)

Surface quality improvement methods	Material	Key findings	References
Vibratory bowl finishing	Polymer	Abrasive action media effects mass finishing of parts and through vibrating spring-mass system fine abrasive compounds mixed.	[67]
Hot cutter machining	Polymer	The numerically controlled machine attaches with different shapes of hot cutters along the periphery of slices traversing in X-Y directions.	[75]
Ball burnishing	ABS	While burnishing the tool, the workpiece lay between the head and tailstock in the post tool machine. The surface finish and ABS parts wear enhanced by the depth of penetration increase and spindle speed. The burnishing tool increases the hardness while applying force.	[78]
Magnetic based finishing	Metal	Magnetic field-assisted finishing, also called magnetic abrasive finishing, forces abrasive particles against a surface using a magnetic field.	[80]

5.6 CONCLUSIONS AND FUTURE DIRECTION

In this chapter, the authors have critically reviewed the mechanical-based post-processing techniques to enhance the surface quality of the AM parts. The literature relating to different mechanical-based post-processing techniques to improve the surface finishing of the AM parts has been analyzed in this work. There are numerous mechanical-based techniques that can be used to improve the surface polish of components made using AM processes. Choosing a suitable method according to the surface finish specification can improve the surface finish. The analysis of numerous pieces of work demonstrates that systemic post-processing can reduce the Ra value of AM parts. While performing post-processing procedures, most of the authors have worked to improve the overall performance of AM parts. This relation has many assumptions, including abrasive volume, HCM feed speed, and abrasive flow rate. In this study, it has been found that mechanical-based post-processing techniques can also enhance the surface finish of the polymer and metal AM parts. The current review shows that abrasive-based finishing techniques, such as the AFM process, and magnetic-based finishing techniques, such as MAF, MRF, MRAFF, or BEMRF, are the most favourable and promising techniques to achieve the surface finish at a nano-level of surface roughness.

5.6.1 Challenges and future direction

AM techniques may take the form of service activity, and therefore must be adapted to ensure companies that wish to offer this service achieve greater growth and profitability, considering the eventual recipient of the additive manufacturing service, including sales to consumers and sales to other businesses and organizations.

There is currently no vertical platform for additive manufacturing-based advanced manufacturing (3DP) and customized manufacturing technologies. The problem of the contract for 3DP solution companies is made partly possible by present service providers. Customers managing materials, weight, delivery time, performance, cost, location, and catalogue (direct e-commerce choices) in real-time will ensure that additive manufacturing is on the cloud as a trustworthy service.

REFERENCES

[1] X.-Y. Zhang, G. Fang, and J. Zhou, "Additively manufactured scaffolds for bone tissue engineering and the prediction of their mechanical behavior: A review", *Materials*, vol. 10, no. 1, p. 50, 2017.

[2] N. Kumbhar and A. Mulay, "Post processing methods used to improve surface finish of products which are manufactured by additive manufacturing technologies: A review", *Journal of The Institution of Engineers (India): Series C*, vol. 99, no. 4, pp. 481–487, 2018.

[3] N. Bhattacharjee, A. Urrios, S. Kang, and A. Folch, "The upcoming 3D-printing revolution in microfluidics", *Lab on a Chip*, vol. 16, no. 10, pp. 1720–1742, 2016.

[4] M. Wirth and F. Thiesse, 2014. "Shapeways and the 3D printing revolution", *Proceedings of the European Conference on Information Systems (ECIS) 2014*, Tel Aviv, Israel, June 9–11, 2014, ISBN 978-0-9915567-0-0, http://aisel. aisnet.org/ecis2014/proceedings/track18/3

[5] H.-J. Chang, A. Andreoni, and M. L. Kuan, "International industrial policy experiences and the lessons for the UK", 2013.

[6] S. C. Cox et al., "Adding functionality with additive manufacturing: Fabrication of titanium-based antibiotic eluting implants", *Materials Science and Engineering: C*, vol. 64, pp. 407–415, 2016.

[7] S. Li, H. Hassanin, M. M. Attallah, N. J. Adkins, and K. Essa, "The development of TiNi-based negative Poisson's ratio structure using selective laser melting", *Acta Materialia*, vol. 105, pp. 75–83, 2016.

[8] C. Qiu, N. J. Adkins, H. Hassanin, M. M. Attallah, and K. Essa, "In-situ shelling via selective laser melting: Modelling and microstructural characterisation", *Materials & Design*, vol. 87, pp. 845–853, 2015.

[9] C. Qiu, Sheng Yue Chunlei, Nicholas JE Adkins, Mark Ward, Hany Hassanin, Peter D. Lee, Philip J. Withers, and Moataz M. Attallah, "Influence of processing conditions on strut structure and compressive properties of cellular lattice structures fabricated by selective laser melting", *Materials Science and Engineering: A*, vol. 628, pp. 188–197, 2015.

[10] C.-C. Yeh, "Trend analysis for the market and application development of 3D printing", *International Journal of Automation and Smart Technology*, vol. 4, no. 1, pp. 1–3, 2014.

[11] M. Seifi, A. Salem, J. Beuth, O. Harrysson, and J. J. Lewandowski, "Overview of materials qualification needs for metal additive manufacturing", *Jom*, vol. 68, no. 3, pp. 747–764, 2016.

[12] W. E. Frazier, "Metal additive manufacturing: a review", *Journal of Materials Engineering and Performance*, vol. 23, no. 6, pp. 1917–1928, 2014.

[13] F. P. Melchels, J. Feijen, and D. W. Grijpma, "A review on stereolithography and its applications in biomedical engineering", *Biomaterials*, vol. 31, no. 24, pp. 6121–6130, 2010.

[14] T. Chartier, C. Chaput, F. Doreau, & M. Loiseau (2002). "Stereolithography of structural complex ceramic parts", *Journal of Materials Science*, vol. 37, pp. 3141–3147.

[15] V. Chan, P. Zorlutuna, J. H. Jeong, H. Kong, and R. Bashir, "Three-dimensional photopatterning of hydrogels using stereolithography for long-term cell encapsulation", *Lab on a Chip*, vol. 10, no. 16, pp. 2062–2070, 2010.

[16] C. Provin, S. Monneret, H. Le Gall, and S. Corbel, "Three-Dimensional Ceramic Microcomponents Made Using Microstereolithography", *Advanced Materials*, vol. 15, no. 12, pp. 994–997, 2003.

[17] Y.-L. Cheng and F. Chen, "Preparation and characterization of photocured poly (ε-caprolactone) diacrylate/poly (ethylene glycol) diacrylate/chitosan for photopolymerization-type 3D printing tissue engineering scaffold application", *Materials Science and Engineering: C*, vol. 81, pp. 66–73, 2017.

[18] R. B. Osman, A. J. van der Veen, D. Huiberts, D. Wismeijer, and N. Alharbi, "3D-printing zirconia implants; a dream or a reality? An in-vitro study evaluating the dimensional accuracy, surface topography and mechanical properties of printed zirconia implant and discs", *Journal of the mechanical behavior of biomedical materials*, vol. 75, pp. 521–528, 2017.

[19] K. Shahzad, J. Deckers, J.-P. Kruth, and J. Vleugels, "Additive manufacturing of alumina parts by indirect selective laser sintering and post processing", *Journal of Materials Processing Technology*, vol. 213, no. 9, pp. 1484–1494, 2013.

[20] G. V. Salmoria, C. H. Ahrens, P. Klauss, R. A. Paggi, R. G. Oliveira, and A. Lago, "Rapid manufacturing of polyethylene parts with controlled pore size gradients using selective laser sintering", *Materials Research*, vol. 10, no. 2, pp. 211–214, 2007.

[21] G. Salmoria, J. Leite, R. Paggi, A. Lago, and A. Pires, "Selective laser sintering of PA12/HDPE blends: Effect of components on elastic/plastic behavior", *Polymer Testing*, vol. 27, no. 6, pp. 654–659, 2008.

[22] M. Schmidt, D. Pohle, and T. Rechtenwald, "Selective laser sintering of PEEK", *CIRP annals*, vol. 56, no. 1, pp. 205–208, 2007.

[23] K. Yasuda et al., "Biomechanical properties of high-toughness double network hydrogels", *Biomaterials*, vol. 26, no. 21, pp. 4468–4475, 2005.

[24] K. Lu and W. T. Reynolds, "3DP process for fine mesh structure printing", *Powder technology*, vol. 187, no. 1, pp. 11–18, 2008.

[25] H. Seitz, W. Rieder, S. Irsen, B. Leukers, and C. Tille, "Biomed Mater Res", *B Appl Biomater*, vol. 74, p. 782, 2005.

[26] E. Vorndran et al., "3D powder printing of β-tricalcium phosphate ceramics using different strategies", *Advanced Engineering Materials*, vol. 10, no. 12, pp. B67–B71, 2008.

[27] A. W. Hashmi, H. S. Mali, A. Meena, V. Puerta, and M. E. Kunkel, "Surface characteristics improvement methods for metal additively manufactured parts: A review", *Advances in Materials and Processing Technologies*, pp. 1–40, 2022. https://doi.org/10.1080/2374068x.2022.2077535

[28] A. W. Hashmi, H. S. Mali, and A. Meena, "Improving the surface characteristics of additively manufactured parts: A review", *Materials Today: Proceedings*, 2021, in press. https://doi.org/10.1016/j.matpr.2021.04.223

[29] A. W. Hashmi, H. S. Mali, and A. Meena, "The surface quality improvement methods for FDM printed parts: A review", *Materials Forming, Machining and Tribology*, pp. 167–194, 2021. https://doi.org/10.1007/978-3-030-68024-4_9

[30] A. W. Hashmi, H. S. Mali, and A. Meena, "Surface quality improvement methods of additively manufactured parts: A review", *Solid State Technology*, vol. 63, pp. 23477–23517, 2020.

[31] J. D. Spencer, "Vibratory finishing of stereolithography parts", in *1993 International Solid Freeform Fabrication Symposium*, 1993.

[32] J. D. Spencer, R. Cobb, and P. Dickens, "Surface finishing techniques for rapid prototyping", *Technical Papers-Society of Manufacturing Engineers-All Series*, 1993.

[33] S. H. Ahn, M. Montero, D. Odell, S. Roundy, and P. K. Wright, "Anisotropic material properties of fused deposition modeling ABS", *Rapid Prototyping Journal*, vol. 8, no. 4, pp. 248–257, 2002. https://doi.org/10.1108/13552540210441166

[34] A. W. Hashmi, "An experimental investigation of viscosity of a newly developed natural polymer-based media for abrasive flow machining (AFM) of 3D printed ABS parts", *Journal of Engineering Research*, 2021 December 22. https://doi.org/10.36909/jer.13643

[35] A. W. Hashmi, H. S. Mali, A. Meena, K. K. Saxena, A. P. V. Puerta, and D. Buddhi, "A newly developed coal-ash-based AFM media characterization for abrasive flow finishing of FDM printed hemispherical ball shape", *The International Journal on Interactive Design and Manufacturing*, pp. 1–16, 2022. https://doi.org/10.1007/s12008-022-00982-2

[36] A. W. Hashmi, H. S. Mali, and A. Meena, "Experimental investigation on abrasive flow Machining (AFM) of FDM printed hollow truncated cone parts", *Materials Today: Proceedings*, vol. 56, pp. 1369–1375, 2022. https://doi.org/10.1016/j.matpr.2021.11.428

[37] A. W. Hashmi, H. S. Mali, and A. Meena, "Design and fabrication of a low-cost one-way abrasive flow finishing set-up using 3D printed parts", *Materials Today: Proceedings*, vol. 62, pp. 7554–7563, 2022. https://doi.org/10.1016/j.matpr.2022.04.647

[38] A. W. Hashmi, H. S. Mali, A. Meena, K. K. Saxena, A. P. Puerta, C. Prakash, D. Buddhi, J. P. Davim, and D. S. Abdul-Zahra, "Understanding the mechanism of abrasive-based finishing processes using mathematical modeling and numerical simulation", *Metals*, vol. 12, no. 8, p. 1328, 2022 August.

[39] A. W. Hashmi, H. S. Mali, and A. Meena, "A critical review of modeling and simulation techniques for loose abrasive based machining processes", *Materials Today: Proceedings*, vol. 56, Part 4, pp. 2016–2024, 2022.

[40] A. W. Hashmi, H. S. Mali, A. Meena, I. A. Khilji, M. F. Hashmi, and S. N. B. M. Saffe, "Artificial intelligence techniques for implementation of intelligent machining", *Materials Today: Proceedings*, vol. 56, pp. 1947–1955, 2022. https://doi.org/10.1016/j.matpr.2021.11.277

[41] A. W. Hashmi, H. S. Mali, A. Meena, I. A. Khilji, M. F. Hashmi, and S. N. B. M. Saffe "Machine vision for the measurement of machining parameters: A review", *Materials Today: Proceedings*, vol. 56, pp. 1939–1946, 2021. https://doi.org/10.1016/j.matpr.2021.11.271

[42] N. Dixit, V. Sharma, and P. Kumar. "Research trends in abrasive flow machining: A systematic review", *Journal of Manufacturing Processes*, vol. 64, pp. 1434–1461, 2021 April 1.

[43] C. Bouland, V. Urlea, K. Beaubier, M. Samoilenko, and V. Brailovski, "Abrasive flow machining of laser powder bed-fused parts: Numerical modeling and experimental validation", *Journal of Materials Processing Technology*, vol. 273, p. 116262, 2019.

[44] H. S. Mali, B. Prajwal, D. Gupta, and J. Kishan, "Abrasive flow finishing of FDM printed parts using a sustainable media", *Rapid Prototyping Journal*, vol. 24, no. 3, pp. 593–606, 2018. https://doi.org/10.1108/RPJ-10-2017-0199

[45] N. Mohammadian, S. Turenne, and V. Brailovski, "Surface finish control of additively-manufactured Inconel 625 components using combined chemical-abrasive flow polishing", *Journal of Materials Processing Technology*, vol. 252, pp. 728–738, 2018.

[46] M. Duval-Chaneac, S. Han, C. Claudin, F. Salvatore, J. Bajolet, and J. Rech, "Experimental study on finishing of internal laser melting (SLM) surface with abrasive flow machining (AFM)", *Precision Engineering*, vol. 54, pp. 1–6, 2018.

[47] A. P. Nagalingam and S. Yeo, "Controlled hydrodynamic cavitation erosion with abrasive particles for internal surface modification of additive manufactured components", *Wear*, vol. 414, pp. 89–100, 2018.

[48] K. L. Tan and S. H. Yeo, "Surface modification of additive manufactured components by ultrasonic cavitation abrasive finishing", *Wear*, vol. 378, pp. 90–95, 2017.

[49] M. Anilli, A. G. Demir, and B. Previtali, "Additive manufacturing of laser cutting nozzles by SLM: processing, finishing and functional characterization", *Rapid Prototyping Journal*, vol. 24, no. 3, pp. 562–583, 2018. https://doi.org/10.1108/RPJ-05-2017-0106

[50] E. Atzeni, M. Barletta, F. Calignano, L. Iuliano, G. Rubino, and V. Tagliaferri, "Abrasive Fluidized Bed (AFB) finishing of AlSi10Mg substrates manufactured by direct metal laser sintering (DMLS)", *Additive Manufacturing*, vol. 10, pp. 15–23, 2016.

[51] R. E. Williams, D. F. Walczyk, and H. T. Dang, "Using abrasive flow machining to seal and finish conformal channels in laminated tooling", *Rapid Prototyping Journal*, vol. 13, no. 2, pp. 64–75, 2007. https://doi.org/10.1108/13552540710736740

[52] K. F. Leong, C. K. Chua, G. S. Chua, and C. H. Tan, "Abrasive jet deburring of jewellery models built by stereolithography apparatus (SLA)," *Journal of Materials Processing Technology*, vol. 83, no. 1–3, pp. 36–47, 1998.

[53] R. E. Williams and V. L. Melton, "Abrasive flow finishing of stereolithography prototypes," *Rapid Prototyping Journal*, vol. 4, no. 2, pp. 56–67, 1998. https://doi.org/10.1108/13552549810207279

[54] L. An, D. Wang, and D. Zhu. "Combined electrochemical and mechanical polishing of interior channels in parts made by additive manufacturing," *Additive Manufacturing*, vol. 51, p. 102638, 2022 March 1.

[55] C. Ni, and Y. Shi. "Abrasive flow finishing of micro-channel produced by selective laser melting", *Materials and Manufacturing Processes*, pp. 1–5, 2022 August 6.

[56] N. Dixit, V. Sharma, and P. Kumar, "Experimental investigations into abrasive flow machining (AFM) of 3D printed ABS and PLA parts", *Rapid Prototyping Journal*, vol. 28, no. 1, pp. 161–174, 2021 October 3. https://doi.org/10.1108/RPJ-01-2021-0013

[57] S. S. Jalui, T. J. Spurgeon, E. R. Jacobs, A. Chatterjee, T. Stecko, and G. P. Manogharan. "Abrasive flow machining of additively manufactured titanium: Thin walls and internal channels", In *2021 International Solid Freeform Fabrication Symposium 2021*. University of Texas at Austin.

[58] M. Vispute, N. Kumar, P. K. Jain, P. Tandon, and P. M. Pandey, "On the surface finish improvement in hybrid additive subtractive manufacturing process", in *Innovative Design, Analysis and Development Practices in Aerospace and Automotive Engineering (I-DAD 2018)*: Springer, 2019, pp. 443–449.

[59] J. Nsengimana, J. Van der Walt, E. Pei, and M. Miah, "Effect of post-processing on the dimensional accuracy of small plastic additive manufactured parts", *Rapid Prototyping Journal*, vol. 25, no. 1, pp. 1–12, 2019. https://doi.org/10.1108/RPJ-09-2016-0153

[60] O. Oyelola, P. Crawforth, R. M'Saoubi, and A. T. Clare, "Machining of additively manufactured parts: implications for surface integrity", in *Procedia Cirp*, 2016, vol. 45: Elsevier, pp. 119–122.

[61] L. Löber, C. Flache, R. Petters, U. Kühn, and J. Eckert, "Comparison of different post processing technologies for SLM generated 316l steel parts", *Rapid Prototyping Journal*, vol. 19, no. 3, pp. 173–179, 2013. https://doi.org/10.1108/13552541311312166

[62] L. Galantucci, M. Dassisti, F. Lavecchia, and G. Percoco, "Improvement of fused deposition modelled surfaces through milling and physical vapor deposition", in *1st Workshop on the State-of-the-Art and Challenges of Research Efforts at POLIBA*, 2014, vol. 1, pp. 87–92.

[63] P. Kulkarni and D. Dutta, "On the integration of layered manufacturing and material removal processes", *The Journal of Manufacturing Science and Engineering*, vol. 122, no. 1, pp. 100–108, 2000.

[64] B. M. Blair, "Post-build processing of stereolithography molds", *School of Mechanical Engineering, Georgia Institute of Technology Atlanta …*, 1998.

[65] J. P. Pragana, R. F. Sampaio, I. M. Bragança, C. M. Silva, and P. A. Martins, "Hybrid metal additive manufacturing: A state-of-the-art review", *Advances in Industrial and Manufacturing Engineering*, vol. 2, p. 100032, 2021 May 1.

[66] R. L. Zinniel, "Surface-treatment method for rapid-manufactured three-dimensional objects", ed: Google Patents, 2014.

[67] M. Schmid, C. Simon, and G. Levy, "Finishing of SLS-parts for rapid manufacturing (RM)–a comprehensive approach", *Proceedings SFF*, pp. 1–10, 2009.

[68] J. S. Chohan and R. Singh, "Pre and post processing techniques to improve surface characteristics of FDM parts: a state of art review and future applications", *Rapid Prototyping Journal*, vol. 23, no. 3, pp. 495–513, 2017. https://doi.org/10.1108/RPJ-05-2015-0059

[69] E. Atzeni, A. Balestrucci, A. R. Catalano, L. Iuliano, P. C. Priarone, A. Salmi, and L. Settineri, "Performance assessment of a vibro-finishing technology for

additively manufactured components", *Procedia CIRP*, 88, pp. 427–432, 2020 Jan 1.

[70] A. Boschetto, L. Bottini, and F. Veniali, "Microremoval modeling of surface roughness in barrel finishing", *The International Journal of Advanced Manufacturing Technology*, vol. 69, no. 9–12, pp. 2343–2354, 2013.

[71] A. Boschetto, V. Giordano, and F. Veniali, "3D roughness profile model in fused deposition modelling", *Rapid Prototyping Journal*, vol. 19, no. 4, pp. 240–252, 2013. https://doi.org/10.1108/13552541311323254

[72] M. Fischer and V. Schöppner, "Some investigations regarding the surface treatment of Ultem* 9085 parts manufactured with fused deposition modeling", in *24th Annual International Solid Freeform Fabrication Symposium, Austin*, 2013, pp. 12–14.

[73] A. Boschetto and L. Bottini, "Surface improvement of fused deposition modeling parts by barrel finishing", *Rapid Prototyping Journal*, vol. 21, no. 6, pp. 686–696, 2015. https://doi.org/10.1108/RPJ-10-2013-0105

[74] A. Boschetto, and L. Bottini, "Roughness prediction in coupled operations of fused deposition modeling and barrel finishing", *Journal of Materials Processing Technology*, vol. 219, pp. 181–192, 2015 May 1.

[75] P. M. Pandey, N. V. Reddy, and S. G. Dhande, "Improvement of surface finish by staircase machining in fused deposition modeling", *Journal of materials processing technology*, vol. 132, no. 1–3, pp. 323–331, 2003.

[76] P. M. Pandey, N. Venkata Reddy, and S. G. Dhande, "Virtual hybrid-FDM system to enhance surface finish", *Virtual and Physical Prototyping*, vol. 1, no. 2, pp. 101–116, 2006.

[77] M. Taufik, and P. K. Jain. "Thermally assisted finishing of fused deposition modelling build part using a novel CNC tool", *Journal of Manufacturing Processes*, 59, pp. 266–278, 2020 Nov 1.

[78] M. Vanitha, A. N. Rao, and M. KedarMallik, "Optimization of Speed parameters in Burnishing of Samples Fabricated by FDM", *International Journal of Mechanical and Industrial Engineering (IJMIE)*, vol. 2, no. 2, pp. 10–12, 2012.

[79] A. C. de Bruijn, G. Gómez-Gras, and M. A. Pérez. "On the effect upon the surface finish and mechanical performance of ball burnishing process on fused filament fabricated parts", *Additive Manufacturing*, vol. 46, p. 102133, 2021 October 1.

[80] J. Guo et al., "Novel rotating-vibrating magnetic abrasive polishing method for double-layered internal surface finishing", *Journal of Materials Processing Technology*, vol. 264, pp. 422–437, 2019.

[81] A. Kumar, Z. Alam, D. A. Khan, and S. Jha, "Nanofinishing of FDM-fabricated components using ball end magnetorheological finishing process", *Materials and Manufacturing Processes*, vol. 34, no. 2, pp. 232–242, 2019.

[82] J. Zhang, W. G. Tai, H. Wang, A. S. Kumar, W. F. Lu, and J. Y. H. Fuh, "Magnetic abrasive polishing of additively manufactured 316L stainless steel parts", in *Proceedings of euspen's 18th International Conference*, 2018, pp. 4–8.

[83] H. Yamaguchi, O. Fergani, and P.-Y. Wu, "Modification using magnetic field-assisted finishing of the surface roughness and residual stress of additively manufactured components", *CIRP Annals*, vol. 66, no. 1, pp. 305–308, 2017.

Chapter 6

Chemical post-processing for fused deposition modelling

Ana Pilar Valerga
University of Cádiz Av. Universidad de Cádiz, Puerto Real, Cádiz, Spain

Mir Irfan Ul Haq
Shri Mata Vaishno Devi University, Katra, India

Severo R. Fernandez-Vidal
University of Cádiz Av. Universidad de Cádiz, Puerto Real, Cádiz, Spain

CONTENTS

DOI: 10.1201/9781003288619-6

6.1 INTRODUCTION

Today, more so than ever, manufacturing processes play a crucial role in design and product development. It is especially common for a good idea to fail because of the technical complexity that its fabrication involves, or the adjustments it would require in an already stablished production line. These problems are evidence of the growing necessity for more versatile and flexible manufacturing techniques, that could be adapted to the individual requirements of each product, and not the other way around. With this purpose in mind, additive manufacturing is increasingly becoming an alternative to the more traditional production methods.

Additive manufacturing allows not only the creation of complex geometries, but also a reduction in waste material. When used in combination with topology optimization techniques, it could even minimize the energy required to produce a part. The adaptability of this technology is excellent, opening the possibility of quickly introducing new revisions for a product. Initially, applications were limited to prototyping, and the acquisition cost was high. However, while the number applications are rapidly growing over the past few years, the entry cost is also decreasing. The average price has dropped significantly recently, spreading the use of additive manufacturing both for professional and receptive purposes.

Among the different additive manufacturing variants, the adoption of Fused Deposition Modelling (FDM) or Fused Filament Fabrication (FFF) is the most extended. This technique places molten polymeric filament into a printing bed, giving shape to the final part. Additionally, it permits the introduction of different reinforcements within the polymeric resin. Subsequently, the properties of the final part could be enhanced, closing the gap with metals and traditional composites when it comes to specific strength.

However, one of the main shortcomings of this technology, as in other additive manufacturing technologies, is the poor surface finish. Many authors have studied the different possibilities of improving this finish by means of pre-processing and post-processing techniques. The former consists of modifying the manufacturing parameters and strategies, while the latter is carried out once the part is finished.

In this chapter, a review is made of all the manufacturing parameters of the FDM process and how they affect the surface quality of the manufactured parts. Many authors studied the influence of FDM process parameters on the surface roughness of manufactured parts. In general, layer thickness and part orientation were found to be the most significant factors that influence the surface finish of FDM parts.

Most authors agree that, although quality improvement is possible using the most correct manufacturing parameters, the staircase effect or the impact of layer stacking is impossible to eliminate. Therefore, post-processing techniques have traditionally been used. There are many post-processing techniques that are catalogued in this chapter, but special emphasis is placed on chemical finishing techniques.

There are numerous articles in the literature in which chemical techniques are applied for this purpose. However, there is also evidence that the nature of the material is altered, thus modifying other properties of the parts, such as their mechanical or thermal resistance. This chapter provides an overview of the most commonly used techniques and the impact they can have on the parts.

6.2 FUSED DEPOSITION MODELLING (FDM)/FUSED FILAMENT FABRICATION (FFF)

Additive manufacturing was born in the 1980s. Several developers worked on it in the same period of time, and it took shape as a result. Hideo Kodoma of the Nagoya Municipal Institute of Industrial Research developed two 3D printing technologies in 1981. In 1984, the Frenchmen Alain Le Méhauté, Olivier de Witte, and Jean Claude André developed another version of this technology; three weeks later, the American Chuck Hull filed his own patent for stereolithography (SLA). In 1987, the first SLA printer was commercialized, which was based on an ultraviolet laser to cure a photosensitive resin in localized spots, one layer at a time, layer by layer. Subsequently, DLP (digital light processing) technology appeared which, thanks to a projector, projected light onto a resin vat to cure a whole layer at once, unlike SLA technology which cured one spot at a time. And so, with the coming together of these scientists and many others, the technology was consolidated and the change in manufacturing began [1, 2].

In 2010, 3D printing took another big leap forward when desktop 3D printers became available on the market.

In the early 1990s, the inventor Stratasys developed a new method for 3D printing: Fused Deposition Modelling (FDM). This material extrusion material extrusion (MEX, material extrusion) is an additive manufacturing process whereby an additive process by which the chosen material is delivered through a nozzle or orifice. FDM technology is based on the creation of parts from the deposition of layers of thermoplastic material on a hot base. This technique requires three main elements: the base on which the part is to be on which the part will be printed, the material from which the part will be made in the form of filament or pellets, and an extrusion head or extruder [3] (Figure 6.1).

The machine is heated to a temperature of around 200° C (depending on the material), so that the filament, with a diameter between 1.75 or 2.85 mm, is melted and extruded through the nozzle and it is deposited on the hot bed with a mobility in the three axes x, y, and z. In this way, the three-dimensional shape of the workpiece is created layer by layer.

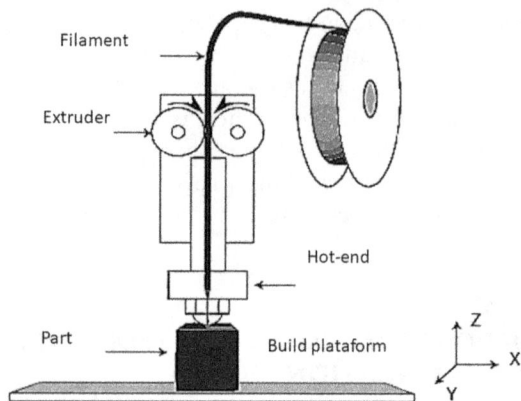

Figure 6.1 FDM process schematic.

Stratasys established a registered trademark with the name FDM. So, in 2006, the members of the RepRap Project coined the term Fused Filament Fabrication (FFF) as an alternative to provide an expression that was not legally restricted in its use [4]. However, it was not until 2009 when the main patent related to this process expired [5], and it was not until 2011 that the scientific community began to use this free term, with a current predilection for the acronym FDM even in open-source use.

The first step in obtaining a part using FDM technology is the creation of a three-dimensional virtual model of the part using CAD software. The parts must be saved in a format readable by specific manufacturing software, such as ".stl". However, these programs are becoming more and more flexible, and there are now other formats such as ".amf" or ".3mf".

Subsequently, using a CAM type software called 'slicer', the different printing parameters will be chosen and modified. Once the printer is at the optimum temperature and the file is loaded, the printing of the part begins, which sometimes requires supports for its manufacture. Once the printing is finished, the supports are removed, and the piece is finished as shown in Figure 6.2.

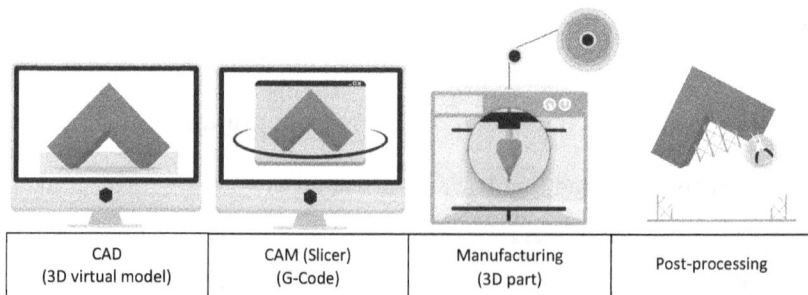

CAD (3D virtual model)	CAM (Slicer) (G-Code)	Manufacturing (3D part)	Post-processing

Figure 6.2 Schematic procedure for manufacturing with FDM.

The main advantage of this printing method is its versatility, since it permits to print with a wide variety of polymers, which can also be reinforced. The relative simplicity of the process, especially when compared with other additive manufacturing techniques, makes it an interesting option capable of fabricating parts under short production times. The main drawback is its limited accuracy and precision, which can lead to difficulties obtaining good part repeatability.

6.3 DEFECTOLOGY

A defect is a deviation of a part manufactured by any technology compared to the virtual or theoretical model of the part. At the same time, it is not possible to achieve a defect-free part, as depending on the evaluation scale; there will always be some kind of deviation.

Parts made using FDM technology tend to have some very characteristic defects (Table 6.1) that can be related to the inappropriate use of some of

Table 6.1 Characteristic defects of FDM technology

Defect and description	Image
Bubbles and porosities: Bubbles are intrinsic to the process. Working with inadequate parameters leads to the appearance of more bubbles, causing an increase in the weakness of the structure. 　　There are numerous studies to achieve a density close to 100% or not, but always controlled for many applications [6, 7][reference]. Porosity can be used as an advantage for example in medical implantology.	
Air gap or path spacing: It is a defect similar to the previous one, with the only difference being that it is produced by an insufficient overlap between deposited filaments, either due to an inadequate thickness or flow.	

(Continued)

Table 6.1 (Continued)

Defect and description	Image

Sewing:

It is produced at the beginning and end of each layer due to the change of trajectory that the head must make. It is necessary to take this seam to an area where it does not affect or can be concealed, for example in a corner. For cylindrical parts, this can be avoided by making a spiral trajectory, so that the height (z axis) increases as the x and y trajectories are made, without there being an actual change of layer.

Cracks and crevices:

A poor adhesion between layers is produced, which may be due to internal stresses that may appear in the part. If this defect appears only in the first layer, it is called Warping [8] and causes a bending of the part due to internal stresses. It can also be associated with a high coefficient of thermal expansion of the materials and is solved by reducing the temperature difference between the environment and the manufacturing process. If it only occurs in the first layer, it can be solved by using some kind of adhesive or by using higher platform temperatures.

Non self-supporting structures:

In this technology, it is essential to create support structures when the part consists of a cantilevered part. Otherwise, the action of gravity itself causes large deviations in these areas due to the detachment of the cantilevers. If the supports are used properly, defects can nevertheless occur due to the removal of the supports.

Deformations:

The largest deformations occur when the parts have a small cross-section. The material is extruded and, without having time for the layer to cool, the nozzle passes over it again and places another layer. This causes the layer to overflow, resulting in high dimensional and shape deviations.

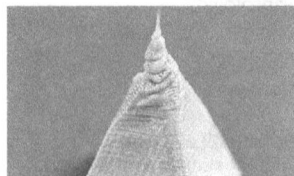

(Continued)

Table 6.1 (Continued)

Defect and description	Image
Residual swaths: These are residues of material that remain lying in areas of vacuum displacement. These swaths appear due to insufficient retraction of the wire, and because of the temperature of the head and the action of gravity, the material falls freely through the nozzle in areas between parts.	
Bumps: These are excess material deposited in specific areas. Normally this defect is associated with an incorrect extrusion or a defect in the material itself. It can also be associated with incorrect adhesion of material in some areas or poor shrinkage.	
Nozzle output: Occasionally, the nozzle exit leaves a mark on the top surface of the part. This is caused by an inadequate gap, too much flow versus low velocity resulting in material overflow, or even high temperature resulting in poor solidification.	
Staircase/stair-step effect: The layer-by-layer fabrication process generates a staircase effect (SE) on the surface of parts fabricated by AM impossible to be eliminated. SE results in a volumetric error, which is the difference between the volume of the material used for the fabricated part and the volume specified by the CAD model. SE adversely affects the surface accuracy of parts fabricated by AM [9, 10].	 designed model fabricated object

the parameters and/or manufacturing paths, mainly, as well as to the limitations of the technology itself. The control and minimization of these defects means that despite the great expansion of this technique, there is still significant scope for improvement.

As in any AM technology, the process of converting the model into STL and the subsequent slicing to obtain the CAM, simplifies the geometry, losing resolution in most cases, especially when processing circular or small parts [11, 12]. To this must be added the restriction of achieving small details due to the technology itself, and the characteristic defects of the technology. These defects can be reduced with the use of some manufacturing parameters, as well as other system variables. However, the surface finish of the parts will depend to a greater extent on the extrusion head and layer height used. Therefore, poor surface finish and low dimensional accuracy seem to be a major obstacle against the commercial production of parts, customized or not, by FDM [13].

Techniques used to improve surface finish fall into two categories, pre-processing and post-processing [14]. All surface refinement methods adopted prior to manufacturing FDM parts are classified into pre-processing and deal with modification in design and manufacturing parameters. On the other hand in post-processing, the parts are treated after the extrusion is completed under the nozzle [15].

Therefore, although there are still some doubts about its applicability in mass production, the use of FDM in the industry is increasing due to new technological advances that allow controlling these defects. As a developing technology for creating precision objects with high material possibilities, FDM may offer a way to replace conventional manufacturing techniques in some cases in the near future [16].

6.4 PRINTING PARAMETERS

During printing, it is very important to find the right set of parameters to minimize post-processing. The ultimate goal should be to ensure good adhesion and bonding between the deposited filaments and the polymer layers. In this way, loads can be transmitted efficiently, and mechanical properties could be maximized. In addition, the final percentage of voids present in the printed part should be as low as possible, avoiding the formation of early cracks. This is why the manufacturing parameters, or pre-processing, is very important in the manufacture of the parts.

An overview of the main FDM manufacturing parameters is given below.

6.4.1 Temperature

Printing temperature is a vital parameter for the printing process. The viscosity of a polymer has a strong dependence on temperature, affecting the printability of the material and how well it flows out of the printed head and into the previously deposited layer. It is safe to assume that with higher temperatures the final void content could be reduced [17], as a consequence of the decreased viscosity and longer cool down times that the polymer is going to experience. This effect promotes stronger and much better entangled links between polymer chains. The threshold that marks the upper limit for the appropriate temperature will be that which starts degrading the polymer, i.e. the temperature at which the chains present in the polymer start to break and unentangle.

In order to determine the correct printing temperature, analysis like the Differential Scanning Calorimetry (DSC) could be done. This technique studies the changes in heat flow that the polymer experiences during a heat up – cool down cycle. Therefore, it is possible to characterize the different temperature transitions, being glass transition and melting point the most important ones. It is always advised to print above the melting temperature, to ensure that the totality of the polymer matrix is molten [18].

Polymers are characterized by their degree of crystallinity. The chains that form the inner structure of the polymer could be arranged either in a chaotic or aligned way. When the chains are partially aligned the polymer contains a high degree of crystallinity. This aligned organization affects the final mechanical and thermal properties, providing a strong and brittle behaviour and slightly higher temperature resistance. Conversely, if the organization of the chains is chaotic, the polymer is predominantly amorphous. This results in a more elastic behaviour, with lower temperature resistance [19].

The glass transition temperature Tg, and the melting point or melting temperature Tm, are key in order to understand how a polymer is going to perform during its operational life. The glass transition temperature marks the beginning of an evolution from a glassy state to a rubbery one, i.e. from a high level of crystallinity to a more amorphous polymer. On the other hand, the melting point or melting temperature is the limit at which polymer chains can freely move without much resistance. Therefore, the viscosity greatly reduces, and the polymer can be easily extruded.

6.4.2 Bed temperature

While printing, it is also possible to set a specific temperature for the printer bed. This temperature is important whenever the printed part experiences warpage since a higher bed temperature could prevent this defect from happening. The different cooling states between layers could trigger the onset of tensions, caused by the different levels of shrinkage along the printed part. Once the polymer exits the printer head, it expands to some extent before shrinking again. This phenomenon is called warpage, and negatively affects the mechanical properties of a printed part [20].

By increasing the bed temperature, this differential in tension between the first layers may be reduced or even prevented. Typical bed temperature values available in most printers are between 60 and 120°, and used in conjunction with a heated chamber allow an easier printing process for polymers like polyamide, which tend to shrink during the printing process.

6.4.3 Chamber temperature

Most modern printers often come with an enclosure surrounding the printing bed and the printer head. In some cases, this chamber can be heated, offering many beneficial effects. The first of them is the possibility of having a slower and more constant cooling rate for the printed part. This foments the adhesion between layers, giving them more time to entangle their chains and form a strong bond. This event promotes higher mechanical properties, especially improving vertical or interlayer strength, known to be one of the weak points of 3D printed structures [21].

Also, as mentioned before, the diminution in cooling rate helps at reducing tensions caused by shrinkage and prevents warpage by keeping a more controlled temperature throughout the printed part. Therefore, due to the many benefits that this option brings, its use is highly recommended whenever it is possible.

6.4.4 Printing speed

Selecting the right print head speed is crucial for the end result, as this is a parameter that greatly affects the rest of the variables. A good balance should be achieved specially between printing speed and printing temperature, since they directly affect each other [22]. For instance, if a higher printing speed is designated, the filament will have a lower residence time inside the heat block of the printer head. Therefore, the actual temperature of the filament will be lower than what was initially chosen. This could eventually affect the adhesion between layers and tracks, as can be seen in Figure 6.3.

Furthermore, printing speed is also related with the shear rate that the filament experiences just before leaving the nozzle. There is an inverse proportionality between viscosity and shear rate. This means that the higher the speed the lower the viscosity, thus flowability during the printing operation could also be increased by printing speed.

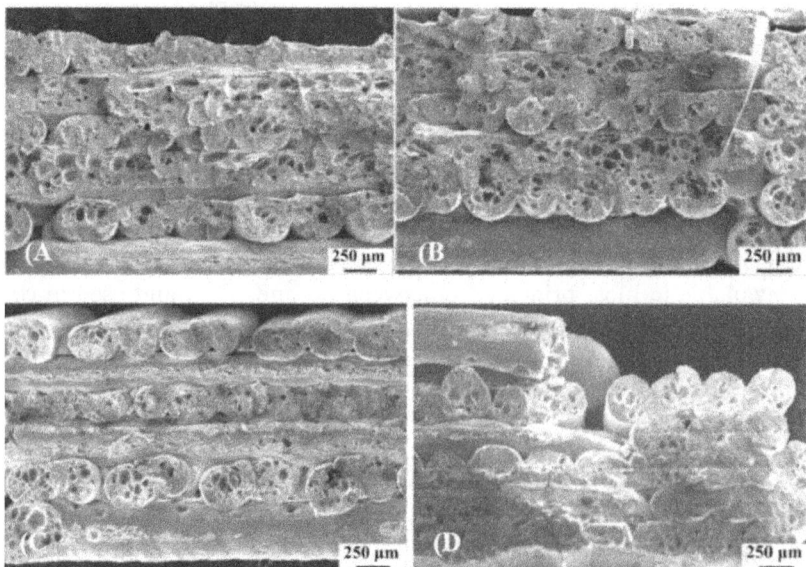

Figure 6.3 Cross-section, after fracture. Printing speeds of 15 mm/s (A), 20mm/s (B), 25mm/s (C), 30 mm/s (D) [18].

Another important aspect to consider when selecting the appropriate speed is surface finish and detailing. If the geometry of the printed part is very complex, selecting a low speed might be necessary in order to correctly define and capture all the details. Hence, a right balance between printing speed and printing temperature is of the utmost importance.

6.4.5 Layer height

The coalescence between layers and the level of void content could be improved through the use of a thinner layer height. Ideally, the dimensions of the layer height should always be lower than the nozzle diameter. Therefore, the printed track will be squeezed against the previously deposited layer and the contact areas between them will increase notably (Figure 6.4). Furthermore, voids formed between layers and tracks will also be reduced, contributing to achieve a lower level of defects, and increasing the mechanical properties [23]. Additionally, thinner layer heights will also translate into a longer printing time, although better surface quality could be achieved.

6.4.6 Extrusion width

The slicer software usually calculates the amount of material that should be extruded to meet a specific extrusion width and layer height. These two parameters, together with the printing speed, set the necessary flow rate to obtain a constant and seamlessly extrusion.

The extrusion width is a parameter that, in contrast to the layer height, should always be higher than the nozzle diameter. The reason behind this is that the polymer experiences an expansion after exiting the nozzle, the

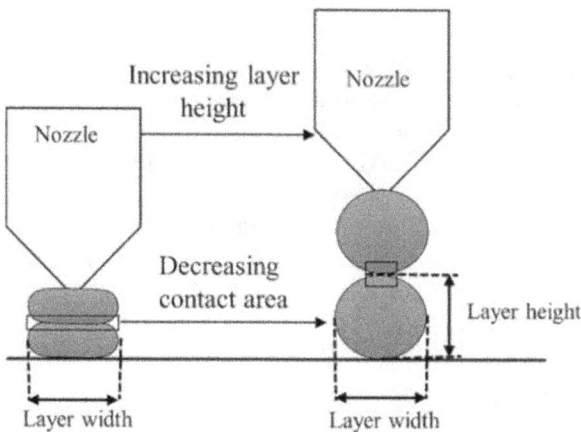

Figure 6.4 Effects of layer height.

so-called die swell. This phenomenon is motivated by the relaxation of the polymer chains after being compressed in order to pass through the nozzle.

Some slicers make it possible to select a negative air gap. The air gap is the space relative to a deposited track and the next one. Successive tracks within a same layer will contact between them through the raster's side walls. If each track or raster is placed with a small negative onset regarding the next one, their contact area will increase and the void content will be reduced, leading to improved mechanical properties [24].

6.4.7 Extrusion multiplier

This parameter is a useful resource in cases where, for some reason, the printer over- or under-extrudes. This tool is essential for printing with continuous fibres, since the behaviour of the duplet matrix-fibre could be different than what is expected. In these occasions, the extrusion multiplier could help tweaking the flow rate until the right level of extrusion is achieved. Over-extrusion could lead to accuracy defects and variations, while under-extrusion could prevent good filling between successive printed tracks.

6.4.8 Nozzle diameter

The diameter of the nozzle is directly related with achievable tolerances and final surface finish. A thinner nozzle allows lower layer heights, thinner extrusion widths, and a more detailed printing process. However, and especially when fibres are used, there is a size limitation for the nozzle to let pass through the fibres. Moreover, even if the fibre is applied from a secondary nozzle directly into the part, the layer height should be the adequate [25, 26].

6.4.9 Raster angle

Defined as the angle that forms the raster with the X axis of the printing bed. This is a parameter of the utmost importance, as a consequence of the anisotropic behaviour of these structures. It is very important to align the deposited filaments with the axis of the applied load for which the part was designed (Figure 6.5). By doing so, most of the filaments strength would be acting against the aforementioned load, leading to higher stresses withstood by the material. In an effort to obtain close-to isotropic materials, the raster angle could be easily tailored for each layer, allowing the use of any necessary lay-up sequence [27].

6.4.10 Cooling rate

Cooling rate is controlled mainly through fan speeds, and it is a useful parameter to successfully print complex geometries or low viscosity polymers.

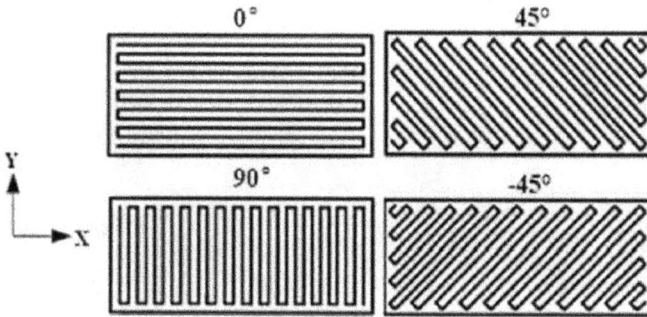

Figure 6.5 Different raster orientations relative to printer bed.

Printing at high temperatures greatly reduces the polymer's viscosity and its stability once it is deposited on the printed part. This could be beneficial to fill the gaps created between successive material depositions but could lead to inaccuracies due to the unstable behaviour of a low viscous polymer. In order to resolve this issue, an increased cooling rate may be applied, so that the polymer can solidify faster and maintain its intended geometry.

However, the effect of a reduction in cooling time could prevent the polymer from creating a strong bond with the previously deposited layer. As the entanglement of polymer chains is a time and temperature dependent process, reducing the available time for cooling down also reduces the level of bonding, resulting in a lower interlayer strength [28].

6.4.11 Build orientation

Due to the anisotropic behaviour of composites, the build orientation directly affects the final properties of the printed part. Ideally, most of the filaments should be aligned with the loading direction. This is something achievable through infill or raster orientation. However, the building orientation could also determine how the weakest strength axis is oriented. Traditionally, interlayer or Z direction has the lowest strength for a printed part. The way in which this problem is resolved could greatly affect the final behaviour of a printed part. Furthermore, choosing an efficient build orientation for complex geometries could lead to less support material needed and a more stable printing.

The degree of adhesion between layers influences the resulting mechanical properties of a printed part considerably. Phenomena like delamination could occur if a good interlayer bonding is not achieved [29]. The union of two subsequent layers has an important dependency on the viscosity of the molten polymer, which fluctuates according to the process temperature and cooling rate.

When two successive beads of polymer are deposited one upon the other, the upper bead re-melts the already cooled down layer. Due to this increase

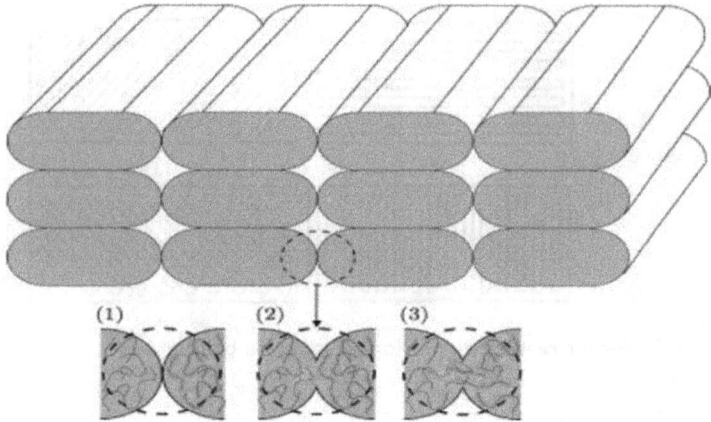

Figure 6.6 Bond formation between successive depositions.

in temperature, a neck starts to grow at the contact point, and interfacial interactions in shape of diffusion between beads occur. This is the basic principle that allows bond formation (Figure 6.6). The polymers internal chains entangle together in a randomized fashion and create a strong link between them. According to several publications [30], keeping a low cooling rate has a positive effect on bond formation, allowing more time for the diffusion process and randomization of polymer chains.

6.5 POST-PROCESSING TECHNIQUES

The FDM/FFF process generally has a poor surface finish and low dimensional accuracy. This seems to be the main obstacle to the commercial production of parts manufactured by this process. For this reason, numerous investigations have been developed to remedy the dimensional accuracy with an anticipation of possible deviations and others to control the surface finish and reduce some of the defects of the parts [31].

However, these techniques are used not only to improve the surface finish, but also can be used as a preliminary stage in bonding or coating operations. In addition, much other research shows that many of these techniques, especially chemical treatments, substantially modify the material, and thus the mechanical behaviour of the parts. Figure 6.7 shows the post-processing techniques for FDM currently being studied.

As can be seen in the diagram, the nature of these treatments is very different, and so will be the outcome. Depending on the results required, the material, the time, and the in-service constraints among other aspects, it has to be examined which of the existing techniques is most appropriate for the requests.

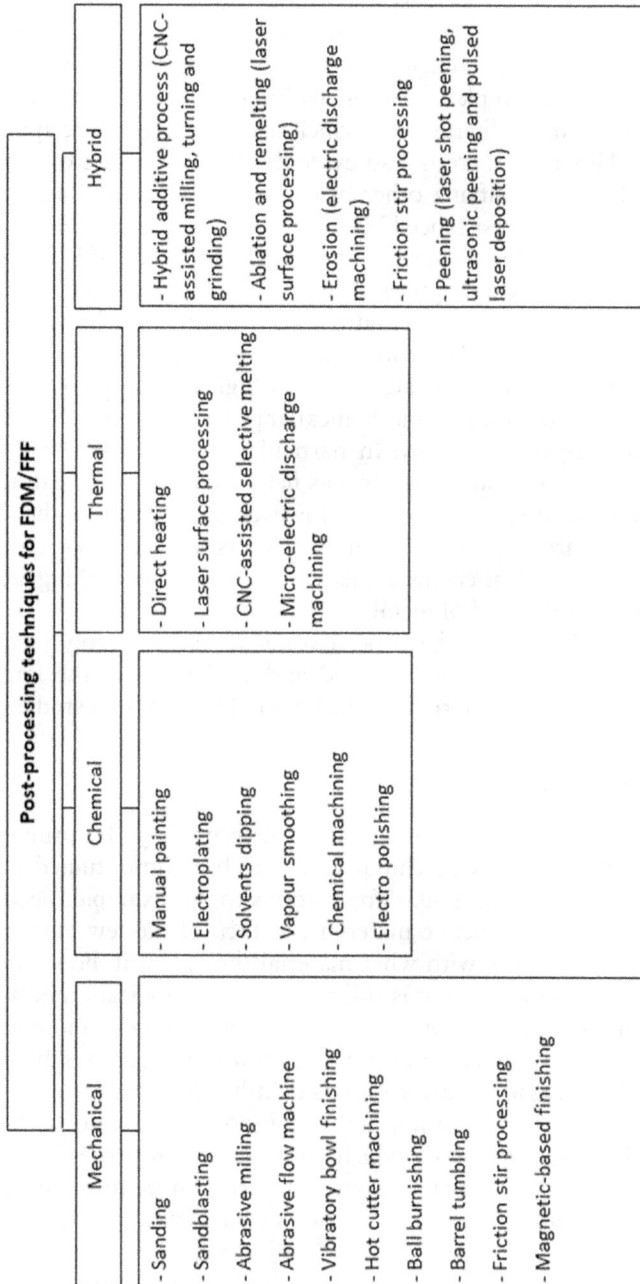

Figure 6.7 Classification of post-processing techniques used in FDM.

6.6 CHEMICAL POST-PROCESSING

Among the approaches to decrease undesirable surface roughness, a method for AM materials known as chemical finishing (CF) is commonly used. The CF method offers the advantage of acting on the entire surface, including the internal parts of complex surfaces, without the use of tools. There are numerous articles in the literature in which these techniques are applied for this purpose. However, there is also evidence that the nature of the material is altered, thus modifying other properties of the parts, such as their mechanical or thermal resistance [32].

As discussed above, many authors studied the influence of FDM process parameters on the surface roughness of manufactured parts [33, 34]. In general, layer thickness and part orientation were found to be the most significant factors that influence the surface finish of FDM parts [35].

However, chemical treatment achieves very high surface quality improvement without the limitation of mechanical approaches such as the need for specialized tooling or machinery. In particular, baths in dimethyl ketone (acetone) are the most widespread in this field, especially for treating acrylonitrile butadiene styrene (ABS) [36]. However, others, such as the manual painting of parts, are very commonly used by users, although less researched. Finally, techniques such as chemical machining or electro polishing are more commonly used in the field of metals.

Although other materials have been studied, ABS and poly lactic acid (PLA) have been the most widely used and analyzed, but the impact of chemical treatments on materials such as nylon has also been studied [37].

6.6.1 Manual painting

Traditionally, hand-painted coatings are widely used by 3D printer users (Figure 6.8). However, it is a technique that has been little studied scientifically, although it can have similar applications to, for example, electroplating, if painted with conductive materials. In fact, of the few authors who study it, some do not say with what material they paint it, how, nor what thickness of paint they use, so it is still not a methodological process [37].

To paint the part properly, it is usually sanded (to leave a more uniform surface finish) and given a coat of primer. There are various ways of applying paint by hand, such as using an airbrush, a paintbrush, or canned spray paint. In these cases, and not just for appearance purposes, it is important to consider the thickness of the topcoat, which can affect the properties of the material. Other applications of painting printed parts can be to obtain specific properties, such as antimicrobial properties for biomedical applications [38].

6.6.2 Electroplating/galvanizing

Electroplating is used industrially on polymeric materials. It is an important process that combines the beneficial properties of polymeric and metallic

Figure 6.8 Hand-painted 3D printed tortoise shell [14].

materials. Polymeric materials are usually lightweight and are characterized by being easily formable and less expensive [39]. However, they suffer from low temperature and chemical stability, poor abrasion resistance and unattractive appearance. In contrast, the surface of metallic materials can be corrosive, but is usually shiny and visually appealing. The disadvantages of metals are their high weight and cost. The application of electroplating polymers has specific decorative and technical purposes [40, 41].

Electroplating on polymeric materials includes a sequence of steps involving various chemical reagents. Each step creates the necessary modification in the surface topology of the polymer to enable a regular, well-bonded coating (Figure 6.9). After cleaning and degreasing, the first steps of the electroplating process provide the electrical conductivity of the polymer surface. In order to successfully electroplate their surface a complex preparation is necessary. It is also necessary that the surface of the polymer samples is impenetrable. This is achieved by the manufacturing process or by pre-treatment processes such as smoothing. So in order to carry out this treatment, the surface must first be smoothed and sealed. Subsequently, the surface of the material is etched to obtain a homogeneous surface roughness and to allow the anchoring of the Pd cores in the next step. In turn, these Pd cores certify the chemical precipitation of Ni. This process is known as Ni electrodeposition [42, 43]. The achieved electrical conductivity of this thin Ni layer is sufficient for subsequent electroplating. For industrial purposes, the first

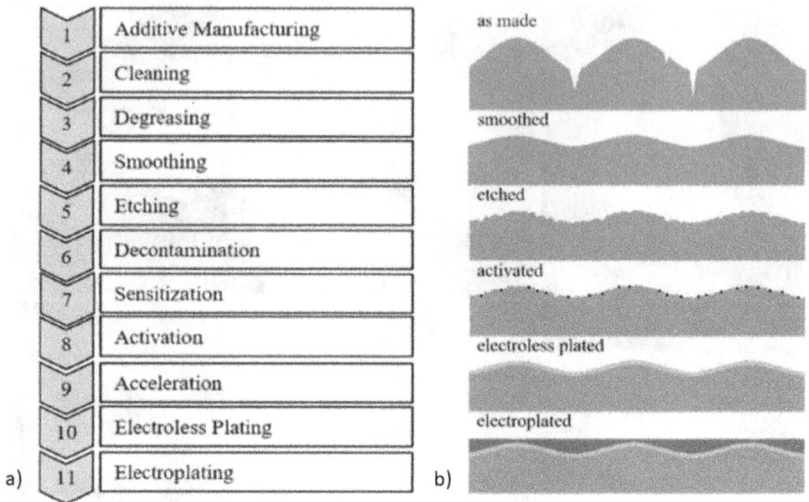

Figure 6.9 Process of electroplating of polymeric additive manufacturing: (a) general overview of single steps and (b) model of surface modification after significant steps [40].

electroplated metal is Cu, followed by Ni and Cr. Electroplated Cu is resistant to irregularities and homogeneous layers can be obtained in a short time. Ni can be seen as an intermediate layer. Cr provides good hardness and a shiny appearance.

This electroplating process can be carried out with commercially available reagent systems or with reagents produced in-house with compositions based on existing literature [45].

Electroforming, electroplating and electrotyping have been used for decades to replicate the complex shape of unique items and can be viable techniques to create complex metal shapes from FDM with polymeric materials. One of the main applications is to obtain porous metal composites from simpler and cheaper technology such as FDM. For example, the creation of metal scaffolds in medicine [46].

6.6.3 Solvents dipping/vapour smoothing

Baths in dimethyl ketone (acetone) are the most widespread in this field, especially for treating acrylonitrile butadiene styrene (ABS). Acetone can greatly reduce the surface roughness of ABS parts without significant changes in part dimensions. Immersion of an FDM-produced part in a solution of acetone and water, but the concentration and immersion time must be precisely optimized because this treatment is characterized by a very aggressive reaction very aggressive and fast reaction that could damage the sample [47]. There are several forms of treatment: immersion in liquid or with hot or cold acetone vapours.

In the hot vapour treatment, the vessel is heated to accelerate the kinetics of the sample. In the cold vapour treatment, no heat is applied, and the smoothing is more gradual. Hot steam smoothing is faster but more difficult to control, and an evaporation of steam can result in a surface treatment [48]. A post-treatment with cold acetone vapours was considered in a previous work, and the effects of treatment time as a function of different angles of manufacturing orientation were evidenced, Figure 6.10.

Alternatively, other authors have studied other similar organic solvents, such as ethyl acetate, and different ones such as chloroform, dichloromethane, tetrahydrofuran.

Acetone-like solvents are often used in ABS as well, with PLA showing less noticeable improvements. In contrast, chlorine-based solvents produce a marked improvement in the finish of PLA parts by radically reducing the staircase effect (Figure 6.11a) and obtaining improvements of up to 96% in terms of arithmetic mean roughness (Ra), or reaching values below one micrometre [50], Figure 6.11.

In addition, the interfacial and interlayer bonding of the parts can be radically improved with the use of these post-processing processes that slightly soften the polymer matrix. However, this occurs only in the outermost layers of the part, unless there are gaps through which the solvent can penetrate [25].

Nevertheless, the change in the structure, and therefore in the mechanical and thermal characteristics of the parts, has been evidenced in some

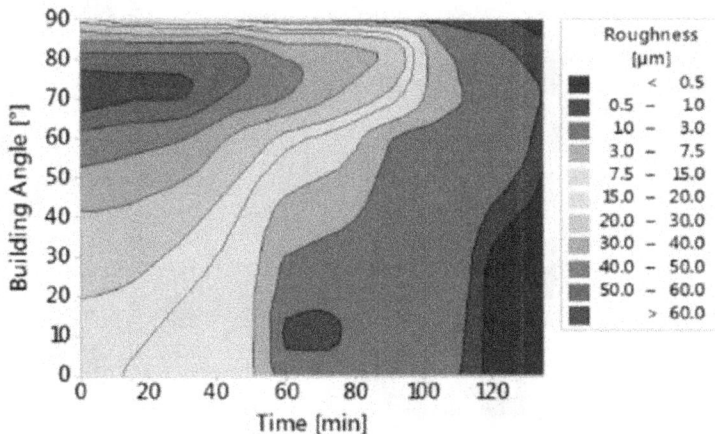

Figure 6.10 Effect of arithmetic mean roughness according to different manufacturing orientation angles and cold acetone vapour treatment time [49].

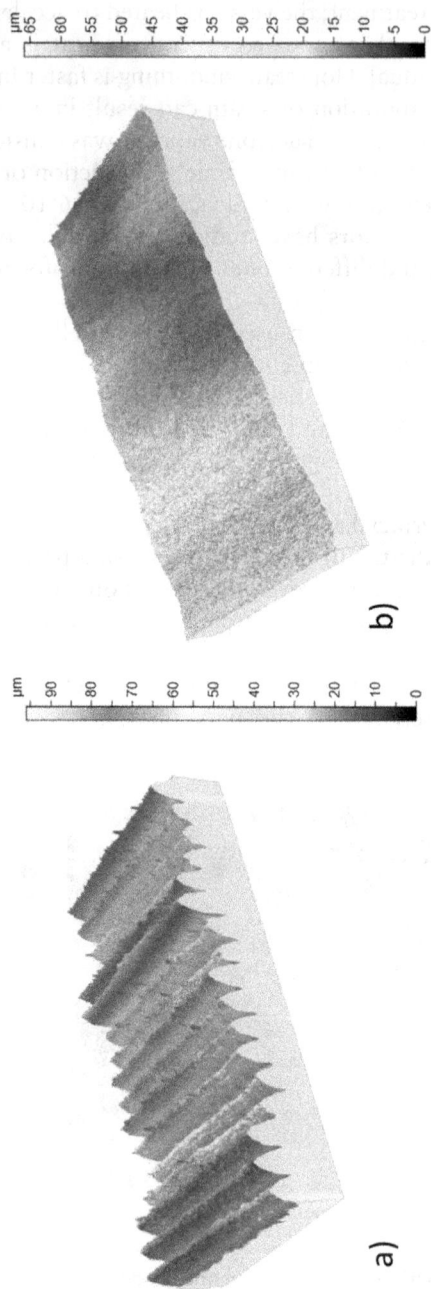

Figure 6.11 3D optical metrology of the surface finish of (a) untreated and (b) treated PLA part.

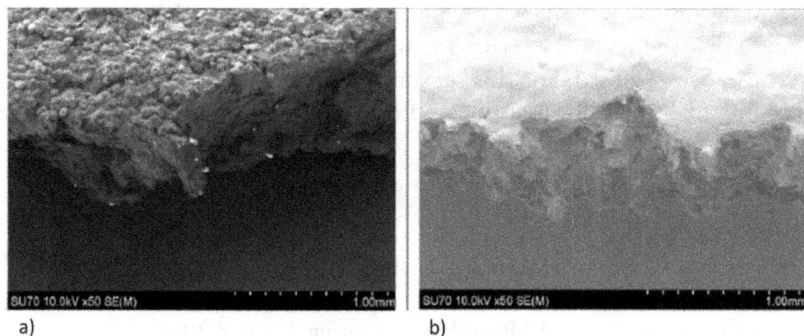

Figure 6.12 SEM images comparing the surface of flexure nylon tested specimens (a) untreated and (b) treated [37].

investigations. This means that in order to improve the surface finish with these techniques, the type of material to be used and, above all, the immersion times in the chemical baths must be taken into account [32].

Other materials, such as nylon, cannot be treated with the chemical treatment methods applied to PLA and ABS. The University of Sheffield has licensed a new chemical surface enhancement technique that can produce a smooth, shiny surface on other materials. This chemical treatment is called the PUSh™ process. This treatment reduces the surface roughness of sample parts by up to 70% Ra. The main advantage of this treatment is that it does not affect the dimensions of the parts, nor the flexural modulus, but it also has a significant impact on the mechanical properties of the tensile specimens if they are thin (Figure 6.12). Some authors studied the reduction of the ultimate tensile strength of nylon specimens, but the effects are negligible in larger specimens [51]. This treatment has been used so far only on sintered polymeric parts, so the impact on filament extrusion should be studied.

6.7 CONCLUSIONS

Fused Deposition Modelling is one of the most widely used processes due to its versatility, variety of materials, and the fact that it is an economical process. However, it has numerous manufacturing parameters that govern the process and affect the finish of the manufactured parts. In addition, one of the most characteristic defects, the ladder effect, is impossible to eliminate. There are numerous techniques for improving the surface finish of parts manufactured by Fused Deposition Modelling and drastically reducing this defect. Chemical treatments are one of the most widely used due to the speed of action on the surfaces. Many authors have studied different finishing improvement processes by chemical post-processing, all of them showing a very significant improvement of the surface quality. However, depending on the material and the process, changes in the dimensions of the part, as well as in the structure and properties of the treated material, have also been studied. For this reason, chemical treatments continue to be a broad and open field of study, and every year there are scientific studies on this subject.

REFERENCES

[1] J. Zhang, & Y.-G. Jung, *Additive manufacturing: Materials, processes, quantifications*, 1st ed., vol. Additive Manufacturing, Elsevier, 2018.

[2] Johannes, Karl Fink, 3d industrial printing with polymers, Wiley Online Library, 2019.

[3] R. B. Kristiawan, F. Imaduddin, D. Ariawan, Ubaidillah, Y, & Z. Arifin, «A review on the fused deposition modeling (FDM) 3D printing: Filament processing, materials, and printing parameters,» *Open Engineering*, vol. 11, n° 1, 2021.

[4] R. Jones, P. Haufe, E. Sells, P. Iravani, V. Olliver, C. Palmer, & A. Bowyer, «RepRap – the replicating rapid prototyper,» *Robotica*, vol. 29, n° 1, pp. 177–191, 2011.

[5] G. Wu, N. A. Langrana, R. Sadanji, & S. Danforth, «Solid freeform fabrication of metal components using fused deposition of metals,» *Materials & Design*, vol. 23, n° 1, pp. 97–105, 2002.

[6] M. Too, K. Leong, C. Chua, Z. Du, & S. Yan, «Investigation of 3D Non-random Porous Structures by Fused Deposition Modelling,» *International Journal of Advanced Manufacturing Technology*, vol. 19, pp. 217–223, 2002.

[7] C. Tong, K.-H. Ho, & T. Swee-Hin, «Scaffold, Design and in Vitro Study of Osteochondral Coculture in a Three-Dimensional Porous Polycaprolactone Scaffold Fabricated by Fused Deposition Modeling,» *Tissue Engineering*, vol. 9, pp. 103–120, 2003.

[8] K. Singh, «Experimental study to prevent the warping of 3D models in fused deposition modeling,» *International Journal of Plastics Technology*, vol. 22, p. 177–184, 2018.

[9] A. Dolenc, & I. Makela, «Slicing procedures for layered manufacturing,» *Computer-Aided Design*, vol. 26, n° 2, pp. 119–126, 1994.

[10] W. Rattanawong, S. Masood, & P. Iovenitti, «A volumetric approach to part-build orientations in rapid prototyping,» *Journal of Materials Processing Technology*, vol. 119, n° 1–3, pp. 348–353, 2001.

[11] H. Ko, S. Moon, &. J. Hwang, «Design for additive manufacturing in customized products,» *International Journal of Precision Engineering and Manufacturing-Green Technology*, vol. 16, n° 11, pp. 69–75, 2015.

[12] P. Nayyeri, K. Zareinia, & H. Bougherara, «Planar and nonplanar slicing algorithms for fused deposition modeling technology: a critical review,» *International Journal of Advanced Manufacturing Technology*, vol. 119, n° 5–6, pp. 2785–2810, 2022.

[13] D. Frunzaverde, V. Cojocaru, C.-R. Ciubotariu, C.-O. Miclosina, D. D. Ardeljan, E. Florin Ignat, & G. Marginean, «The Influence of the Printing Temperature and the Filament Color on the Dimensional Accuracy, Tensile Strength, and

Friction Performance of FFF-Printed PLA Specimens,» *Polymers*, vol. 14, n° 10, 2022.

[14] M. Batista, A. P. Valerga, J. Salguero, S. R. Fernandez-Vidal, & F. Girot, «State of the art of the fused deposition modeling using PLA: improving the performance,» de *Additive and Subtractive Manufacturing*, Aveiro, De Gruyter, 2019, pp. 59–112.

[15] J. Singh, & C. Singh, «Pre and post processing techniques to improve surface characteristics of FDM parts: a state of art review and future applications.,» *Rapid Prototyping Journal*, vol. 23, n° 3, 2017.

[16] Y. Ishida, D. Miura, & A. Shinya, «Application of fused deposition modeling technology for fabrication jigs of three-point bending test for dental composite resins,» *Journal of the Mechanical Behavior of Biomedical Materials*, vol. 130, 2022.

[17] B. T. Challa, S. K. Gummadi, K. Elhattab, J. Ahlstrom, & P. Sikder, «In-house processing of 3D printable polyetheretherketone (PEEK) filaments and the effect of fused deposition modeling parameters on 3D-printed PEEK structures,» *International Journal of Advanced Manufacturing Technology*, vol. (In press), 2022.

[18] F. Ning, W. Cong, Y. Hu, & H. Wang, «Additive Manufacturing of Carbon Fiber-Reinforced Plastic Composites Using Fused Deposition Modeling: Effects of Process Parameters on Tensile Properties,» *Composite Materials*, vol. 51, n° 4, pp. 451–462, 2017.

[19] M. R. Khosravani, Ž. Božić, A. Zolfagharian, & T. Reinicke, «Failure analysis of 3D-printed PLA components: Impact of manufacturing defects and thermal ageing,» *Engineering Failure Analysis*, vol. 136, 2022.

[20] K. L. Snapp, A. E. Gongora, & K. A. Brown, «Increasing Throughput in Fused Deposition Modeling by Modulating Bed Temperature,» *Journal of Manufacturing Science and Engineering-Transactions of the Asme*, vol. 143, n° 9, 2021.

[21] P. Sikder, B. T. Challa, & S. K. Gummadi, «A comprehensive analysis on the processing-structure-property relationships of FDM-based 3-D printed polyetheretherketone (PEEK) structures,» *Materialia*, vol. 22, 2022.

[22] P. Geng, J. Zhao, W. Wu, W. Ye, Y. Wang, S. Wang, & S. Zhang, «Effects of extrusion speed and printing speed on the 3D printing stability of extruded PEEK filament,» *Journal of Manufacturing Processes*, vol. 37, pp. 266–273, 2019.

[23] N. Aliheidari, J. Christ, R. Tripuraneni, S. Nadimpalli, & A. Ameli, «Interlayer Adhesion and Fracture Resistance of Polymers Printed Through Melt Extrusion Additive Manufacturing Process,» *Materials & Design*, vol. 156, pp. 351–361, 2018.

[24] A. Suhas, R. Rajpal, K. V. Gangadharan, & U. Pruthviraj, «An Experimental Study to Evaluate the Warpage and Cracking Issues in Fused Deposition Modeling,» de *Advances in Industrial and Production Engineering*, India, 2019.

[25] A. Valerga, S. Fernandez-Vidal, M. Batista, & F. Girot, «Fused deposition modelling interfacial and interlayer bonding in PLA post-processed parts,» *Rapid Prototyping Journal*, vol. 26, n° 3, pp. 585–592, 2020.

[26] P. Czyzewski, D. Marciniak, B. Nowinka, M. Borowiak, & M. Bielinski, «Influence of Extruder's Nozzle Diameter on the Improvement of Functional Properties of 3D-Printed PLA Products,» *Polymers*, vol. 14, n° 2, 2022.

[27] «Mechanical Properties and Characterization of Polylactic Acid/Carbon Fiber Composite Fabricated by Fused Deposition Modeling,» *Journal of Materials Engineering and Performance*, vol. 31, n° 6, 2022.

[28] S. Ambrus, R. A. Soporan,. N. Kazamer, D. T. Pascal, R. Muntean,. A. I. Dume, G. M. Marginean, & V. A. Serban, «Characterization and mechanical properties of fused deposited PLA material,» *Materials Today-Proceedings*, vol. 45, pp. 4356–4363, 2021.

[29] X. Tian, T. Liu, C. Yang, Q. Wang, & D. Li, «Interface and Performance of 3D Printed Continuous Carbon Fiber Reinforced PLA Composites,» *Composites: Part A*, pp. 198–205, 2016.

[30] S. Garzon-Hernandez, D. Garcia-Gonzalez, A. Jérusalem, & A. Arias, «Design of FDM 3D Printed Polymers: an Experimental-Modelling Methodology for the Prediction of Mechanical Properties,» *Materials & Design*, vol. 188, 2020.

[31] A. Szust, & G. Adamski, «Using thermal annealing and salt remelting to increase tensile properties of 3D FDM prints,» *Engineering Failure Analysis*, vol. 132, 2022.

[32] A. Valerga, S. Fernandez-Vidal, F. Girot, & A. Gamez, «On the Relationship between Mechanical Properties and Crystallisation of Chemically Post-Processed Additive Manufactured Polylactic Acid Pieces,» *Polymers*, vol. 12, n° 941, pp. 1–10, 2020.

[33] G. Krolczyk, P. Raos, & S. Legutko, «Experimental analysis of surface roughness and surface texture of machined and fused deposition modelled parts.,» *Tehnicki Vjesnik*, vol. 21, n° 1, p. 217–221, 2014.

[34] A. Fiorentino, G. Marenda, R. Marzi, E. Ceretti, D. Kemmoku, & J. Da Silva, «Rapid prototyping techniques for individualized medical prosthesis manufacturing.,» *de Innovative Developments in Virtual and Physical Prototyping. Proceedings of the 5th International Conference on Advanced Research in Virtual and Physical Prototyping*, Leiria, 2012.

[35] S. Rahmati, & E. Vahabli, «Evaluation of analytical modeling for improvement of surface roughness of FDM test part using measurement results,» *International Journal of Advanced Manufacturing Technology*, vol. 79, n° 5, p. 823–829, 2015.

[36] A. Colpani, A. Fiorentino, & E. Ceretti, «Characterization of chemical surface finishing with cold acetone vapours on ABS parts fabricated by FDM,» *Production Engineering*, vol. 13, pp. 437–447, 2019.

[37] N. Crane, Q. Ni, A. Ellis, & N. Hopkinson, «Impact of chemical finishing on laser-sintered nylon 12 materials,» *Additive Manufacturing*, vol. 13, pp. 149–155, 2017.

[38] Q. Guo, X. Cai, X. Wang, & J. Yang, «"Paintable" 3D printed structures via a post-ATRP process with antimicrobial function for biomedical applications,» *Journal of Material Chemistry B*, vol. 28, n° 1, pp. 6644–6649, 2013.

[39] N. Sathishkumar, N. Arunkumar, L. Balamurugan, L. Sabarish, & A. S. Shapiro Joseph, «Investigation of Mechanical Behaviour and Surface Roughness Properties on Copper Electroplated FDM High Impact Polystyrene Parts,» *de Advances in Additive Manufacturing and Joining. Lecture Notes on Multidisciplinary Industrial Engineering.*, Springer, Singapore, 2020.

[40] T. Maciąg, J. Wieczorek, & W. Kałsa, «Surface analysis of ABS 3D Prints subjected to cooper plating,» *Archives of Metallurgy and Materials*, vol. 64, n° 2, pp. 639–646, 2019.

[41] M. Mehdizadeh, M. Khorasanian, & S. M. Lari Baghal, "Direct electroplating of nickel on ABS plastic using polyaniline–silver surface composite synthesized using different acids", *Journal of Coatings Technology and Research*, vol. 15, pp. 1433–1442, 2018.

[42] A. Equbal, & A. Kumar Sood, "Investigations on metallization in FDM build ABS part using electroless deposition method", *Journal of Manufacturing Processes*, vol. 19, pp. 22–31, 2015.

[43] L. A. Cesar Teixeira, & M. Costa Santini, "Surface conditioning of ABS for metallization without the use of chromium baths," *Journal of Materials Processing Technology*, vol. 170, n° 1–2, pp. 37–41, 2005.

[44] C. Eßbach, D. Fischer, & D. Nickel, "Challenges in electroplating of additive manufactured ABS plastics", *Journal of Manufacturing Processes*, vol. 68, n° Part A, pp. 1378–1386, 2021.

[45] S. Olivera, H. Muralidhara, K. Venkatesh, K. Gopalakrishna, & C. S. Vivek, "Plating on acrylonitrile-butadiene-styrene (ABS) plastic: a review", *Journal of Materials Science*, vol. 51, pp. 3657–3674, 2016.

[46] M. Adam, F. Jordan, A. Madura, & D. Piovesan, "Use of Polymer Scaffolding and Electroplating to Create Porous Metal Structures", *Journal of Engineering Materials and Technology*, vol. 141, n° 3, 2019.

[47] A. Garg, A. Bhattacharya, & A. Batish, "Chemical vapor treatment of ABS parts built by FDM: analysis of surface finish and mechanical strength", *International Journal of Advanced Manufacturing Technology*, vol. 89, p. 2175–2191, 2017.

[48] A. Colpani, A. Fiorentino, & E. Ceretti, "3D printing for health & wealth: fabrication of custom-made medical devices through additive manufacturing.," *de AIP Conference Proceedings 1960, Article number 140006 21st International ESAFORM Conference on Material Forming, ESAFORM*, 2018.

[49] A. Colpani, A. Fiorentino, & E. Ceretti, "Characterization of chemical surface finishing with cold acetone vapours on ABS parts fabricated by FDM", *Production Engineering*, vol. 13, pp. 437–447, 2019.

[50] A. Valerga Puerta, M. Batista, J. Salguero, & F. Girot, "Post-processing of PLA parts after Additive Manufacturing by FDM technology", *Dyna*, vol. 93, n° 6, pp. 625–629, 2018.

[51] A. Ellis, R. Brown, & N. Hopkinson, "The effect of build orientation and surface modification on mechanical properties of high speed sintered parts", *Surface Topography: Metrology and Properties*, vol. 3, n° 3, 2015.

Chapter 7

Post welding cold forging and effect on mechanical properties of low-carbon mild steel wire arc additive manufacturing

Arvind Ganesh a/l Vasuthaven, Zarirah Karrim Wani, Ahmad Baharuddin Abdullah and Zuhailawati Hussain
USM Engineering Campus, Nibong Tebal, Penang, Malaysia

CONTENTS

7.1 INTRODUCTION

Additive manufacturing (AM) or additive layer manufacturing is also known as 3D printing. Metal arc AM is used widely in diverse applications such as the aerospace, marine, and automotive industries. The most common types of material categories used in AM are polymer, ceramic, and metal. AM technologies offer a new way of manufacturing near-net-shape metallic and polymer parts with complex geometries at a low cost [1]. Industries have many types of metal arc AM which includes selective laser sintering, selective

DOI: 10.1201/9781003288619-7

laser melting, electron beam melting, direct energy deposition (DED), and laser-enabled net shaping [2]. Wire arc additive manufacturing (WAAM) is one of the metal AM categorized under the DED technique. It also known as 3D welding because the main process involved is welding but to deposit a material in 3D profile. However, the profile is produced at a near-net shape, which requires additional processes such as machining to obtain the final part with good dimensional accuracy. Although very flexible, machining takes a longer time to finish. Another alternative is forging. Theoretically, forging offers a quick solution and may improve part strength.

At present, broken parts are typically dumped in a landfill. This practice is not sustainable because a new part requires additional time, machines, and materials to be consumed before it can be used. Metal AM via welding or WAAM is a fast, flexible method, which allows building a complex 3D profile at an intended location. However, properties such as hardness and impact toughness that are required for a certain part to function are affected by the heat produced during welding. There are various methods performed to enhance the properties of welded material. The next section will further elaborate the methods in detail.

7.2 ENHANCEMENT OF PROPERTIES IN WAAM

Porosity is one of the main problems in WAAM of aluminium alloys which can severely limit the mechanical properties of the part. Porosity is generated as a result of several factors including arc welding, process parameters, interpass temperature, wire quality, and alloy composition [3]. In a multiple-layer WAAM, the heat input of a new superposed layer can contribute to the growth of pores. The presence of porosity is accompanied by a decrease in mechanical properties such as the strength and ductility of the material. Generally, the presence of pores has a detrimental effect on the wear performance of materials [4]. Pores act as pre-existing incipient cracks in the subsurface layer, waiting to become unstable at an appropriate stress level [3]. Therefore, WAAM parts require post process treatment to improve material properties, reduce surface roughness and porosity, and remove residual stress and distortions. The majority of issues that relate to the properties including the wear of the material can be addressed by the appropriate application of post processing. Section 7.2.1 further discusses few post processing techniques to improve material properties.

7.2.1 Post process heat treatment

Post process heat treatment is widely used in WAAM to reduce residual stress, enhance material strength, and control hardness. The selection of a suitable heat treatment depends on the target materials, applications, AM method, working temperature, and heat treatment conditions [5]. If the heat treatment parameters are set incorrectly, the probability of cracking increases under mechanical loading as the combination of existing residual stress with load stress exceeds the material's design limitation. Balasubramanian et al. [6] studied the microstructure and mechanical properties of precipitation-hardened arc-welded joints of aluminium alloy 2219. These studies showed that the tensile strength of these joints can be improved to various levels by different heat treatment regimes [7].

The accumulation of residual stresses is reduced by preheating, decrease in heat input and increase in welding speed [8]. Interpass temperature also has a major effect when producing objects from materials other than steel. Derekar et al. [9] argued that a higher interpass temperature is even desired when producing aluminium alloy components, as it leads to reduced porosity.

7.2.2 Interpass cold rolling

Rolling of the weld bead between each deposited layer reduces residual stresses and distortion [10]. Interpass cold rolling not only lowers residual stress but also brings more homogeneous material properties. Moreover, interpass cold rolling can affect SW reduction [11], compressive stress induction, and the effect on static and fatigue properties [12] as well as improve microstructural properties and reduce porosity [13].

Two methods for enhancing surface quality and geometrical accuracy are typically used: (i) vertical rolling over the wall; and (ii) side rolling on the wall's side surface. Although vertical rolling can increase SW, it is smoothened by side rolling [14]. Colegrove et al. [15] reported the application of this interlayer rolling technique during building WAAM steel structures. Similarly, Colegrove et al. [15] used this technique during the deposition of Ti–6Al–4V alloy. Cunningham et al. [13] used interpass cold rolling in the refinement of grains, improvement of mechanical properties, and reduction of distortion and residual stresses. However, this interlayer rolling has not been applied to WAAM aluminium alloy. Williams et al. [16] and Hönnige et al. [17] stated that interpass rolling results in increased wall width, uniform layer height, and improved mechanical properties with reduced grain size despite increased manufacturing time.

7.2.3 Peening and ultrasonic impact treatment

Peening and ultrasonic impact treatments (UIT) have been used in welding applications to reduce local residual stress and improve weld mechanical properties. Both techniques are cold mechanical treatments that affect the weld surface using high-energy media to release tensile stress by imposing compressive stress on the treatment surface [6] (Wu et al., 2018). Li et al. [18] showed that after using UIT, the surface residual stress of WAAM-fabricated Ti6Al4V part can be reduced to 58%, and micro hardness can be increased by 28% compared with the as-fabricated sample. Moreover, the surface-modified layers undergo plastic deformation with remarkable grain refinement and dense dislocations.

7.2.4 Interpass cooling

Using such rapid cooling, in-situ layer temperature and heat cycle can be controlled within a range to obtain the desired microstructure and mechanical properties. Spencer et al. [19] were among the first to use a pyrometer to measure and control interpass temperature when producing simple objects from mild steel in gas metal arc welding (GMAW). They showed that a higher interpass temperature leads to increased SW, whereas a lower interpass temperature leads to reduced productivity due to longer dwell times between successive layers. Xiong et al. [20] performed a thermal analysis for thin walls made of low-carbon steel using infrared thermography. They found that longer interlayer dwell times lead to a decreased mean temperature of the part and a smaller high temperature area while the temperature gradient and the cooling rate of the deposited part increase, resulting in an improved surface quality and a reduction in total height difference. Using numerical simulations and single-layer mild steel experiments, Zhao et al. [21] showed that the magnitude of residual stresses could be reduced by increasing the dwell times, thus lowering the interpass temperature. Moreover, Wu et al. [6] investigated the effects of interpass temperature and forced interpass cooling using carbon dioxide (CO_2) on Ti-6Al-4V alloy components. They reported better surface finish and increased tensile strength at lower interpass temperatures (100° C or less). Henckell et al. [22] also investigated the effects of interpass cooling during a continuous build-up of rotational objects from mild steel. They experimented with different positions of the cooling gas nozzles and several cooling gases. Their forced-cooling experiments resulted in a higher productivity and an improved object geometry, a refined microstructure, and a homogeneous hardness across all layers, with the best results achieved with a gas mixture of nitrogen with 5% hydrogen.

The main objective of this paper is to evaluate the effect of mechanical deformation via cold forging of additive manufactured part through various tests including impact toughness test, hardness test, and microstructure observation. This work aims to identify the potential of WAAM in part repair.

7.3 METHODOLOGY

The methodology is divided into two parts. The first is to prepare the specimen by MIG welding and then proceed with surface finishing and testing. The second is to prepare five mild steel samples by cutting and then proceed with surface finishing and testing.

7.3.1 Parameters for MIG welding

Three different levels with three different parameters namely voltage, filler speed, and feed rate were used for this study as listed in Table 7.1. A total of 27 combinations of the levels and parameters can be obtained. The function of Taguchi is to reduce the variance in a process through robust design experiments. To simplify the experiment, 27 initial combinations can be reduced to nine equivalent sets by using Taguchi, as summarized in Table 7.2. These nine experiments were initially set during MIG welding. Voltage (A) and filler speed (B) were set in the MIG welding machine whereas feed rate (C) was set by part program to control the MIG movement.

7.3.2 Specimen preparation

Before the material deposition starts, several steps are followed according to specified parameters. Figure 7.1 shows the torch of the MIG welding machine that deposits the molten mild steel on the aluminium plate. Figure 7.2 shows the six deposited mild steel beads. The same process was performed for the eight other sets and each set required six specimens to be produced. Two methods of specimen preparation namely cold forging

Table 7.1 Selected parameters and their levels

	Parameters	L1	L2	L3
A	Voltage (V)	2	4	6
B	Filler speed (mm/sec)	1	3.5	7
C	Feed rate (mm/min)	50	100	150

Table 7.2 Initial sets combination for the analysis

Set	A, B, C
1	2, 1, 50
2	2, 3.5, 100
3	2, 7, 50
4	4, 1, 50
5	4, 3.5, 150
6	4, 7, 50
7	4, 3.5, 150
8	6, 7, 150
9	6, 7, 50

Figure 7.1 Deposition of material using MIG based 3D welding machine.

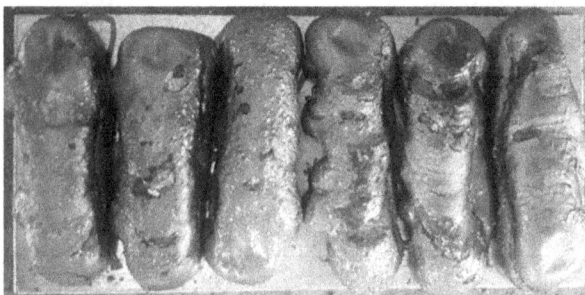

Figure 7.2 Deposited mild steel on aluminium substrate.

and machining (milling) were carried out in this study before sending for characterization:

a. Cold forging

After all the specimens were fabricated, three samples from each set were cold forged using a 100-ton forging machine. The specimens were forged several times to obtain an exact thickness of 10 mm. The forged specimens are then machined to obtain a 10 mm width. Figure 7.3 shows the forged specimens.

b. Machining

The three other specimens from each set that had not been forged were used for machining where the thickness and the width had to be machined until 10 mm. The material was removed layer by layer, and each removed layer had a thickness of less than 1 mm, to ensure that the surface of the specimen are in a good condition. Figure 7.4 shows the milled specimens.

Figure 7.3 Specimen after cold forging process.

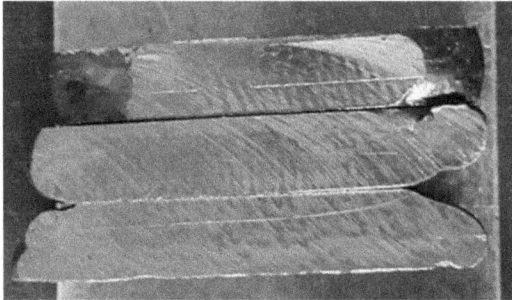

Figure 7.4 Specimen prepared by milling process.

7.3.3 Characterization

The manufactured bead was characterized based on three output parameters: tensile strength, hardness, and SEM observation. The next section will further elaborate the steps.

a. Tensile test

   Figure 7.5 shows the specimen for the tensile test (a) before; and (b) after the test. The tests were repeated three times for each setting.

b. Hardness test

   After completing the tensile test for all the nine sets from the MIG welded specimens and the parent mild steel samples, the hardness test is done. All the specimens are cut into smaller pieces of approximately 10 × 10 × 20 mm because if the top and bottom surfaces are irregularly shaped, taking the hardness reading on an irregular surface is impossible. All the specimens are cut using a cutting tool. Once the specimens were cut into a proper flat surface (transverse surface), they

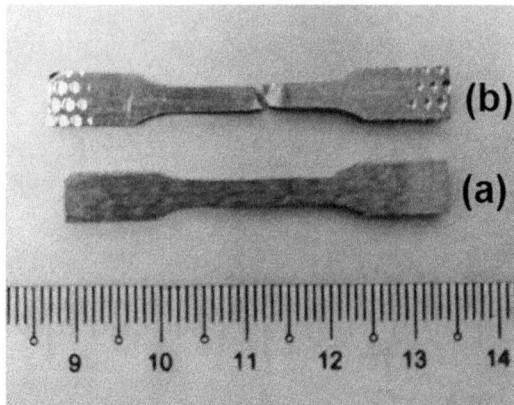

Figure 7.5 Tensile test specimen (a) before and (b) after the test.

were ready for the hardness test. After the specimen preparation, the scale in the Rockwell Hardness tester machine is set. The scale used in this study is HRC, where the force applied on the specimen was 150 kg with diamond cone to make an indentation on the specimen. The hardness test was conducted on two surfaces for each specimen: transverse and longitudinal. In this study, the top surface was taken as the longitudinal view while the cross-sectional area was considered as the transverse view. An average of five hardness results were taken for each surface.

c. Microstructure observation

The microstructure of the best specimen from the weld-forged, weld-machined, and parent mild steel were observed under an S-3400N model SEM. The method used to polish the specimen before SEM is the same as the method stated in the next section. Prior to SEM, the specimens is coated to reduce the charging of the surface oxide. Once the polishing and coating work is done, the specimen is ready for SEM. The specimens used for SEM analysis are from the highest impact test value of weld-forged, weld-machined, and parent mild steel. For each specimen, the microstructure analysis using SEM is done on the transverse and longitudinal surfaces.

d. Impact toughness

After completing the forging and milling for all the nine sets with six specimens in each set, having a dimension of 10 × 10 × 65 mm, the impact test is performed. Before conducting the impact test, a notch approximately 2 mm with 45° is produced. In this study, Izod impact test is used. Figure 7.6 shows the notch on the specimen. After the specimens are notched, the scale in the Izod impact test is set at 150 J.

Figure 7.6 Mild steel specimen after notching for impact toughness test.

7.4 RESULTS AND DISCUSSION

The results focus on three main aspects namely tensile properties, hardness, and impact toughness. The microstructure is elaborated to support to the findings made in the previously stated observations.

7.4.1 Tensile test

In general, the unforged specimen had a higher yield stress than the cold forged specimen; the former was approximately 460 Mpa, whereas the latter was 345 Mpa or 33.33% reduction. Furthermore, the unforged specimen was more ductile compared with the forged one. However, the UTS depicted a tremendous change for forged specimen as it achieved 948 Mpa compared with unforged, which achieved up to 588 Mpa only or 61.2% increment as shown in Figure 7.7. This observation was due to the strain hardening as a result cold forging. An engaging finding made on the yield strength of the material was it reduced even with the increase of UTS. The result was possibly due to the crack propagation after cold forging in some places, causing necking to start easily. This result may be supported with the microstructure observation, discussed later.

7.4.2 Hardness test

Table 7.3 summarizes the hardness test results for milling and forging for the nine sets based on longitudinal (HRC) and transverse (HRC) views. The average was calculated based on five readings for each surface of the

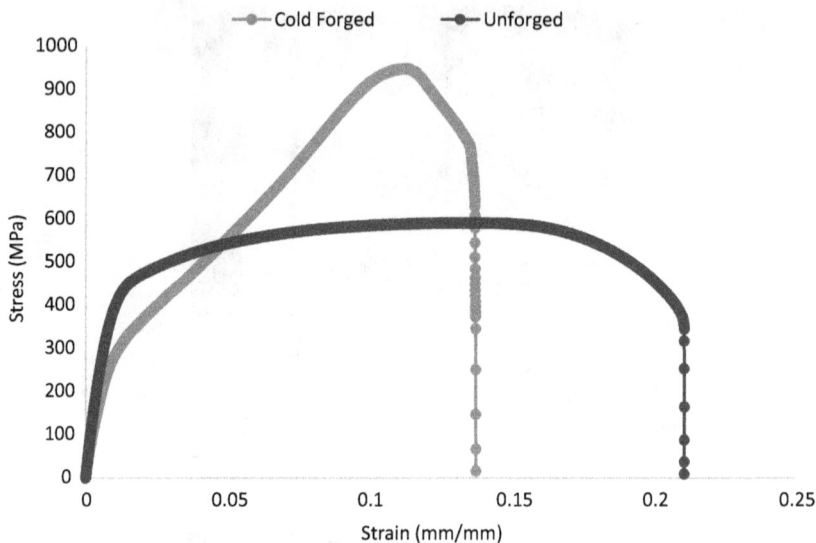

Figure 7.7 Stress strain curve of both the cold forged and unforged specimens.

Table 7.3 Hardness test results of the forged and unforged specimen from longitudinal and transverse cross-sectional views

	Unforged		Cold Forged	
Set	*Longitudinal (HRC)*	*Transverse (HRC)*	*Longitudinal (HRC)*	*Transverse (HRC)*
1	56.93	55.40	59.50	59.57
2	58.27	59.93	58.47	61.67
3	59.87	56.27	67.33	70.20
4	56.20	54.70	64.50	64.03
5	57.33	56.24	68.00	69.83
6	57.70	56.37	59.93	60.73
7	56.83	58.07	65.13	67.03
8	55.20	57.33	65.10	68.33
9	57.03	57.90	60.60	61.77

specimen. For each set, the average of three specimens in the same set was calculated. The average hardness value of the parent mild steel on the longitudinal and transverse surfaces were 58.08 and 58.02 HRC, respectively. The hardness test results in Table 7.3 show that the weld-forged specimen is harder than the weld-machined specimen. The reading for the weld-machined is almost same for the transverse and longitudinal surfaces for all the sets, which is between 54.70 HRC (Set 4, transverse) and 59.93 HRC

(Set 2, transverse). The highest hardness value for the weld-forged specimen is 70.20 HRC (Set 3, transverse). This result shows that the weld-forged specimen is slightly better in terms of hardness. Clearly, the forged specimen has a better hardness test value.

However, the hardness of all the weld-forged specimens show positive results. The average hardness value is more than that of the parent mild steel specimen, and the average hardness value for the weld-machined one. Hardness value and tensile strength can be linked, as according to Peltonen [23], where both variables are indications of metal resistance to plastic deformation. Based on the hardness data, this approximation may be used to define tensile strength. Usually, when the specimen has a low hardness value, it is known as a ductile specimen. The weld-machined specimens are more ductile than the weld-forged and absorbed more impact energy. The ductility and brittleness of the specimen are explained in more detail in the microstructure analysis. However, the parent metal is not the hardest. The hardest specimens based on the hardness test results is the weld-forged specimen. The hardness value of the parent mild steel metal is in the range of 58 HRC, which is between the weld-forged material (high) and weld-machined (low) material. This result shows that the parent mild steel material has a lower hardness value than the weld-forged material.

7.4.3 Microstructure observation

Figure 7.8 shows the image from the optical Olympus microscope with 500× magnification, and Figure 7.9 shows the image from the SEM with the same magnification. The images obtained from the Olympus microscope are not as clear as that from the SEM. The image from the SEM shows the grain and the grain boundary in detail. Some minute gaps are also observed in the parent metal, and the small black spots are the gaps between the grains. This gap is one of the factors that influence the impact test result. The black spot formed because electron conduction did not occur at that location, possibly

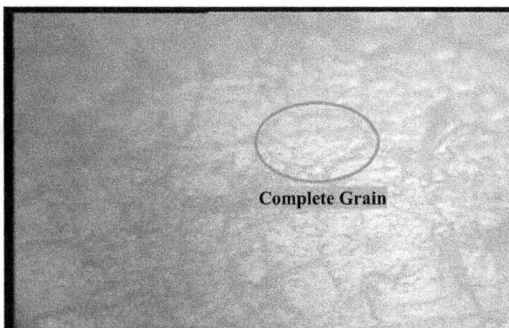

Figure 7.8 Longitudinal parent mild steel specimen.

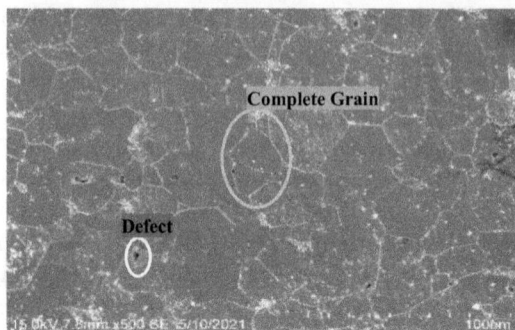

Figure 7.9 Transverse parent mild steel specimen.

due to the improper coating of the conductor on the specimen, and the white spot indicates the electron flow was high at that location. This white and black spot is not visible in Figure 7.8 because the image was captured using a light source.

Figures 7.10 and 7.11 show the longitudinal and transverse microstructure images, respectively of the forged specimen from Set 1 and the best impact value reading in the forged specimen. The voltage, filler speed, and feed rate used are 2 V, 1 mm/s, and 50 mm/min, respectively. Based on these images, the forged specimen has some gaps in between the sample. This reason might explain why the specimen failed during the impact test, and this can also be a reason for the specimen to be brittle and have a low impact value. Comparing the grain size with the parent mild steel, the grain size is smaller than the parent mild steel. The grain size is one of the most important factors that influence the hardness test. As the grain size increases, the hardness value decreases. However, a different result was reported in Martina et al. [24] as the grain size of the specimen increases, the hardness value of the specimen should increase. This behaviour is influenced by the porosity level. According to Colegrove et al. [25], the porosity of the specimen decreases, and the hardness increases. Some possibilities for reducing

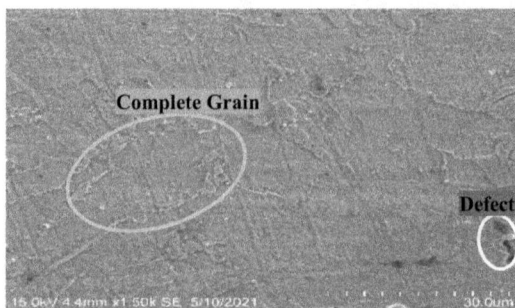

Figure 7.10 Longitudinal surface of the forged specimen.

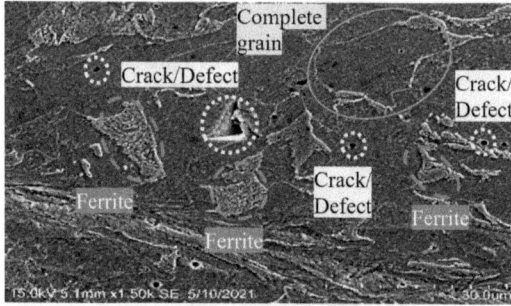

Figure 7.11 Transverse surface of the forged specimen.

porosity by forging exist where the specimen becomes compacted by high force. Moreover, this weld-forged specimen has the smallest grain size compared with the weld-machined and parent mild steel, as shown in Figure 7.10, Figures 7.12–7.14. The analysis clearly shows that a smaller grain size provides a harder, brittle mechanical property. The dashed line circles in Figure 7.11 indicates the pearlite, the thin-line circle shows the grain size, and the dotted-line circles indicates the internal gaps/cracks in the specimen. Pearlite is essentially a composite microstructure consisting of cementite layers (which are hard and brittle) sandwiched between ferrite layers (which are soft and ductile). This pearlite made the forged specimen harder and stronger. However, the impact strength of the forged specimen was reduced mainly due to the yellow circle, which indicated the internal gap in the specimen.

Figures 7.12 and 7.13 show the longitudinal and transverse microstructure images, respectively, of the machined specimen from Set 8. The voltage, filler speed, and feed rate were 6 V, 3.5 mm/s, and 100 mm/min, respectively. These images show the grain size is slightly larger than the parent mild steel and forged specimens. The empty spaces/gaps are much less than those in the forged specimen. This factor might explain the machined specimen achieving a higher impact test value than the forged specimen. This factor

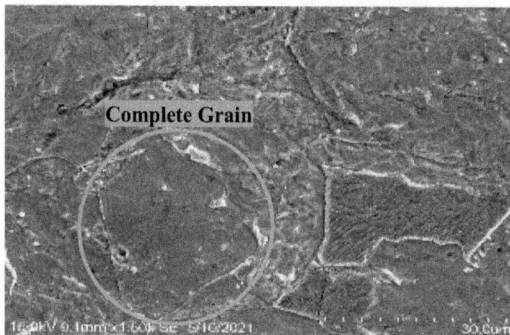

Figure 7.12 Longitudinal surface of the machine specimen.

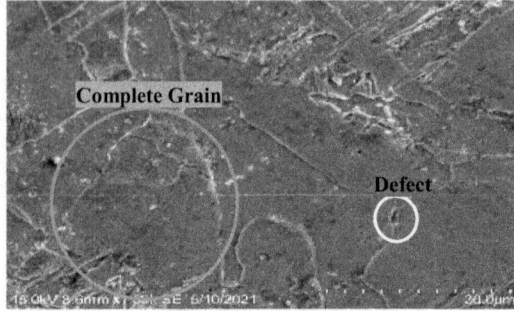

Figure 7.13 Transverse surface of the machine specimen.

shows that the machined specimen has a much better quality than the forged specimen. The grain size in the welded machined specimen shows the largest grain size among weld-forged and parent mild steel. When the grain size increases, the hardness of the material decreases, and for a smaller grain size, the hardness result is higher as the result is obtained in the weld-forged and weld-machined materials. Usually, when the grain size is large, the material ductility increases.

Figures 7.14 and 7.15 show the microstructure of the parent mild steel by using SEM. Figure 7.14 shows the longitudinal microstructure, whereas Figure 7.15 shows the transverse microstructure. This grain was used to compare the characteristic of the material with the forged and machined specimens. The grain structure appears more organized compared with the welded forged and welded machined specimens. This grain structure analysis clearly shows that the welded specimen has an inorganized grain structure mainly due to the recrystallization of the molten mild steel wire in the forged and machined specimens.

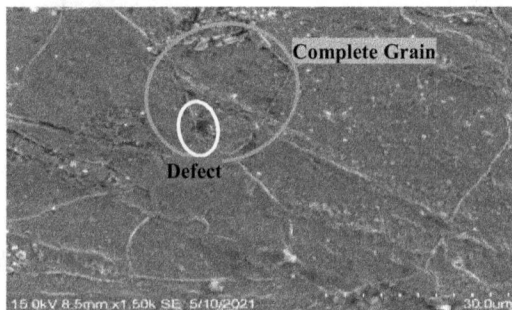

Figure 7.14 Longitudinal surface of the parent specimen.

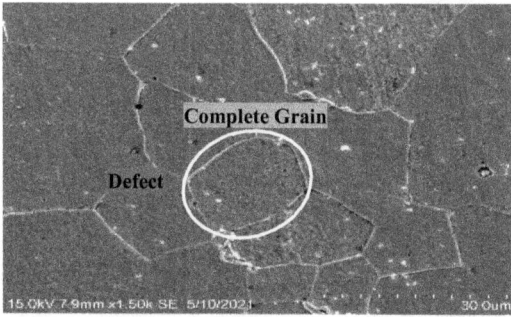

Figure 7.15 Transverse surface of the parent specimen.

7.4.4 Impact toughness

Table 7.4 summarizes the impact test results for all the nine sets. The impact value of the weld-forged specimen is much lower than that of the weld-machined specimen. The highest weld-machined value is from the average of Set 8 that is 113 J, and the lowest impact value for weld-machined value is from the average of Set 5 that is 48.33 J. The highest weld-forged value is from the average of Set 1 that is 56.33 J, and the lowest impact value for weld-forged value is from the average of Set 9 that is 10.67 J. However, the average impact test value for parent mild steel is 100.4 J.

Based on these results, microstructure analysis was performed. The selected sets for the microstructure were from Sets 1 and 8, which achieved the highest impact values for weld-machined and weld-forged specimens, respectively.

a. **Forged**

Theoretically, when the specimen is forged, the hardness and impact value should be higher. However, in this study, the impact value is

Table 7.4 Result of impact test for forged and unforged specimens

Set	Unforged (J)	Cold Forged (J)
1	80.67	56.33
2	71.00	34.00
3	79.67	27.33
4	86.33	37.57
5	48.33	43.33
6	92.00	44.33
7	103.00	19.67
8	113.00	13.00
9	100.67	10.67

much lower for the weld-forged specimen than for the machined specimen, possibly because of some defect in the forged specimen. The most common defect in the welded forged specimen is internal cracking, as clearly shown in the microstructure analysis in Figure 7.15. It is brittle because the impact value is low and the hardness value is high for the weld-forged specimen. This result shows that weld-forged specimens are brittle, hence absorbing a small amount of impact energy.

b. **Machined**

The results in Table 7.4 summarize that the impact results for the machined specimen for Sets 7, 8, and 9 are much higher than those of the parent mild steel specimen. It is brittle because the impact value is low, and the hardness value is high for the weld-forged specimen. Usually, when the specimen has a low hardness value, it is known as a ductile specimen. This paper shows that weld-forged specimens are brittle, hence absorbing a small amount of impact energy.

c. **Parent Metal**

The parent metal is not the strongest. The strongest specimen based on the impact test result is the welded machined specimen. The impact test value of the parent metal is 100.4 J, which is between the weld-machined (high) and weld-forged (low) specimens. This result shows that the parent mild steel material has a lower impact value than the weld-machined material. The weld-forged, weld-machined, and parent mild steel were compared based on three different aspects namely hardness, impact, and microstructure analysis. Among the highest from the impact test, the percentage errors of weld-forged and weld-machined specimens based on the parent mild steel are –46.93% and 6.60%, respectively. This result shows the weld-forged material provides a negative response towards the impact test, whereas the weld-machined material yields a positive response. Therefore, the weld-forged material is not strong in toughness, and it is not applicable for part repair usage. Among the highest results from the hardness test on the transverse surface, the percentage errors of weld-forged and weld-machined materials based on the parent mild steel are 14.33% and –2.39%, respectively. The weld-forged specimen is harder than the weld-machined sample. Theoretically, the impact and hardness values for the weld-forged sample should be higher than those of the weld-machined sample. These impact value and hardness value are supported when the specimens' microstructure was observed under the SEM microscope. However, due to some defects in the specimen, which is shown in Figure 7.11, an internal crack forms in the sample. Based on the analysis, these internal cracks/gaps in the weld-forged specimen are the primary cause of the porosity in the specimen. Hence, the samples fail to withstand the impact force. The positive response in the hardness comparison among the weld-forged and parent mild steel specimen is because of the presence of ferrite and the size of the grains in the microstructure analysis, as shown in Figure 7.11. When the amount of ferrite is high and the grain size is small, the mechanical property of the specimen increases. Hence, the mechanical property of the weld-forged should be increased, and the specimen failed to withstand the impact force due to the internal cracks.

7.5 CONCLUSIONS

The strength of the welded specimen does not depends on only the welding parameters. In comparison, milling and forging have more influence on the mechanical strength of the specimens. In this study, the impact test shows that weld-forged specimens absorb less energy from the impact test than the weld-machined and parent mild steel specimens due to the internal cracking in the weld-forged specimen. Usually, all-welded specimens have some minute internal cracks. During forging, the internal cracks' propagation grows and easily cracks during the impact test on the specimen. Based on the microstructure analysis, the welded specimen has an irregular grain structure compared with the parent mild steel specimens because the grain formation at irregular temperature during the layer-by-layer welding. During the layer-by layer deposition, the specimen is affected by HAZ. WAAM is suitable for use in part repair but not for forging the repaired part because crack propagation and internal stress increase when cold forging is done on the MIG welded specimen and facilitate fracture. In the future, the specimen should undergo heat treatment first for better grain formation, because grain structure in one of the most important factors that influence the hardness and the strength of the specimen. Next, the best way to avoid or minimize the internal crack in the welded specimen must be identified. Performing heat treatment for the specimen before testing is recommended for future work. The welding needs to be conducted in a single-layer analysis.

ACKNOWLEDGEMENT

This work was supported by the Universiti Sains Malaysia under STG (Matching Grant). (304/PMEKANIK/6315605).

REFERENCES

[1] Bruno P. G., (2013), Fatigue Behaviour on MAG and Hybrid Welding Geometrical and Hardness Analysis, Master Thesis, Luleå University of Technology.
[2] Ortega, A. G., Galvan, L. C., Salem, M., Moussaoui, K., Segonds, S., Rouquette, S. & Deschaux-Beaume, F., (2019). Characterisation of 4043 aluminium alloy deposits obtained by wire and arc additive manufacturing using a Cold Metal

Transfer process, *Science and Technology of Welding and Joining*, vol. 24, no. 6, pp. 538–547.

[3] Sinha, A., & Farhat, Z. (2015). Effect of surface porosity on tribological properties of sintered pure Al and Al 6061. *Materials Sciences and Applications*, 06(06), 549–566.

[4] Raghukiran, N., & Kumar, R. (2013). Processing and dry sliding wear performance of spray deposited hyper-eutectic aluminum–silicon alloys. *Journal of Materials Processing Technology*, 213(3), 401–410.

[5] Wu, B., Pan, Z., Ding, D., Cuiuri, D., Li, H., Xu, J., & Norrish, J. (2018). A review of the wire arc additive manufacturing of metals: Properties, defects and quality improvement. *Journal of Manufacturing Processes*, 35, 127–139.

[6] Balasubramanian, V., Ravisankar, V., & Reddy, G. M. (2007). Effect of pulsed current and post weld aging treatment on tensile properties of argon arc welded high strength aluminium alloy. *Materials Science and Engineering: A*, 459(1–2), 19–34.

[7] Malarvizhi, S., Raghukandan, K., & Viswanathan, N. (2008). Effect of post weld aging treatment on tensile properties of electron beam welded AA2219 aluminum alloy. *The International Journal of Advanced Manufacturing Technology*, 37(3–4), 294–301.

[8] Wu, Q., Mukherjee, T., De, A., & DebRoy, T. (2020). Residual stresses in wire-arc additive manufacturing – Hierarchy of influential variables. *Additive Manufacturing*, 35, 101355.

[9] Derekar, K., Lawrence, J., Melton, G., Addison, A., Zhang, X., & Xu, L. (2019). Influence of interpass temperature on wire arc additive manufacturing (WAAM) of aluminium alloy components. *MATEC Web of Conferences*, 269, 05001.

[10] Colegrove, P. A., Donoghue, J., Martina, F., Gu, J., Prangnell, P., & Hönnige, J. (2017). Application of bulk deformation methods for microstructural and material property improvement and residual stress and distortion control in additively manufactured components. *Scripta Materialia*, 135, 111–118.

[11] Dirisu, P., Supriyo, G., Martina, F., Xu, X., & Williams, S. (2020). Wire plus arc additive manufactured functional steel surfaces enhanced by rolling. *International Journal of Fatigue*, 130, 105237.

[12] Busachi, A., Erkoyuncu, J., Colegrove, P., Martina, F., & Ding, J. (2015). Designing a WAAM based manufacturing system for defence applications. *Procedia CIRP*, 37, 48–53.

[13] Cunningham, C. R., Wikshåland, S., Xu, F., Kemakolam, N., Shokrani, A., Dhokia, V., & Newman, S. T. (2017). Cost modelling and sensitivity analysis of wire and arc additive manufacturing. *Procedia Manufacturing*, 11, 650–657.

[14] Hönnige, J. R., Colegrove, P. A., Ahmad, B., Fitzpatrick, M. E., Ganguly, S., Lee, T. L., & Williams, S. W. (2018). Residual stress and texture control in Ti-6Al-4V wire + arc additively manufactured intersections by stress relief and rolling. *Materials & Design*, 150, 193–205.

[15] Colegrove, P. A., Coules, H. E., Fairman, J., Martina, F., Kashoob, T., Mamash, H., & Cozzolino, L. D. (2013). Microstructure and residual stress improvement in wire and arc additively manufactured parts through high-pressure rolling. *Journal of Materials Processing Technology*, 213(10), 1782–1791.

[16] Williams, S. W., Martina, F., Addison, A. C., Ding, J., Pardal, G., & Colegrove, P. (2016). Wire + arc additive manufacturing. *Materials Science and Technology*, 32(7), 641–647.

[17] Hönnige, J. R., Colegrove, P. A., Ahmad, B., Fitzpatrick, M. E., Ganguly, S., Lee, T. L., & Williams, S. W. (2018). Residual stress and texture control in Ti-6Al-4V wire + arc additively manufactured intersections by stress relief and rolling. *Materials & Design, 150,* 193–205.

[18] Li, G., Qu, S., Xie, M., & Li, X. (2017). Effect of ultrasonic surface rolling at low temperatures on surface layer microstructure and properties of HIP Ti-6Al-4V alloy. *Surface and Coatings Technology, 316,* 75–84.

[19] Spencer, J. D., Dickens, P. M., & Wykes, C. M. (1998). Rapid prototyping of metal parts by three-dimensional welding. *Proceedings of the Institution of Mechanical Engineers, Part B: Journal of Engineering Manufacture, 212*(3), 175–182.

[20] Xiong, J., Lei, Y., Chen, H., & Zhang, G. (2017). Fabrication of inclined thin-walled parts in multi-layer single-pass GMAW-based additive manufacturing with flat position deposition. *Journal of Materials Processing Technology, 240,* 397–403.

[21] Zhao, H., Zhang, G., Yin, Z., & Wu, L. (2013). Effects of interpass idle time on thermal stresses in multipass multilayer weld-based rapid prototyping. *Journal of Manufacturing Science and Engineering, 135*(1), 011016.

[22] Henckell, P., Günther, K., Ali, Y., Bergmann, J. P., Scholz, J., & Forêt, P. (2017). The influence of gas cooling in context of wire arc additive manufacturing—a novel strategy of affecting grain structure and size. In T. M. Tms Metals &. Materials So (Ed.), *TMS 2017 146th Annual Meeting & Exhibition Supplemental Proceedings* (pp. 147–156). Springer International Publishing.

[23] Peltonen, M. (2014). Weldability of high-strength steels using conventional welding methods, Master Thesis, Aalto University.

[24] Martina, F., Ding, J., Williams, S., Caballero, A., Pardal, G. and Quintino, L. (2019). Tandem metal inert gas process for high productivity wire arc additive manufacturing in stainless steel, *Additive Manufacturing, 25*(July), 545–550.

[25] Colegrove, P. A., Coules, H. E., Fairman, J., Martina, F., Kashoob, T., Mamash, H. and Cozzolino, L. D. (2013). Microstructure and residual stress improvement in wire and arc additively manufactured parts through high-pressure rolling, *Journal of Materials Processing Technology, 213,* 1782–1791.

Chapter 8

Severe plastic deformation for fatigue strength of additively manufactured components

Abeer Mithal

Nanyang Technological University, Singapore

Advanced Remanufacturing and Technology Centre, Agency for Science Technology and Research, Singapore

Niroj Maharjan

Advanced Remanufacturing and Technology Centre, Agency for Science Technology and Research, Singapore

Sridhar Idapalapati

Nanyang Technological University, Singapore

CONTENTS

8.1 IMPORTANCE OF FATIGUE

Most engineering components fail due to fatigue loading, which is probabilistic in nature depending on a range of material, processing, geometrical, and loading parameters. Fatigue is defined as the failure of engineering components at operating stresses well below the material nominal strength due to time varying loads. The maximum load amplitude at which a material

DOI: 10.1201/9781003288619-8

can sustain 10^6 cycles is usually termed as the fatigue or endurance strength. It is commonly used to screen materials for typical engineering applications.

There are two broad phases in fatigue failure: crack initiation and crack propagation. In the crack initiation phase, microcracks nucleate from stress singularities at or near the surface of the specimen [1]. These cracks propagate until eventually they become macroscopic in size (spanning multiple grains). This signals the start of the crack propagation phase when the crack growth is stable, and its growth direction is normal to the loading direction [2]. This crack growth is governed by the Paris Law and continues until final catastrophic failure.

Fatigue strength considerations add a significant level of complexity to the design of components. It is not sufficient to merely design the parts such that the stress level remains below the yield stress; rather, the response of the material to various stress levels under repeated loading needs to be understood. This enables a service life to be set for the component after which it should be replaced or refurbished.

Fatigue is normally measured and assessed in terms of the stress and number of cycles (S–N) curve which plots stress amplitude against number of loading and unloading cycles to failure as shown in Figure 8.1, for a variety of materials processed via various manufacturing methods. SN curves can also be generated as a function of mean stress and other operating environmental conditions for screening purpose.

8.2 WHY IS FATIGUE STRENGTH AN ISSUE IN ADDITIVE MANUFACTURING?

The importance of fatigue strength has been well established over the last century in part due to some unfortunate accidents and their subsequent investigations [7]. The question then becomes what is particular about additively manufactured (AM) specimens that warrants its own discussion on fatigue life? Why are the considerations for fatigue life of AM parts any different from those made conventionally by casting, forging, or machining? The answers lie in the unique nature of the AM processes which lead to peculiar material structures, often with high anisotropy and micro-scale cellular sub-grain structures resulting from extremely high solidification rates (especially in laser-based AM).

There are some additional quality concerns, inherent in AM processes, that can have a major impact on fatigue strength:

1. **Surface Finish:** As with conventionally manufactured parts, high surface roughness can give rise to an abundance of crack initiation sites which can reduce the fatigue life of AM parts. Typically, parts built via AM processes do not have good surface roughness due to the following three aspects [8]:
   a. Staircase effect
   b. Balling or adherence of unmelted particles to surfaces
   c. Incomplete melting

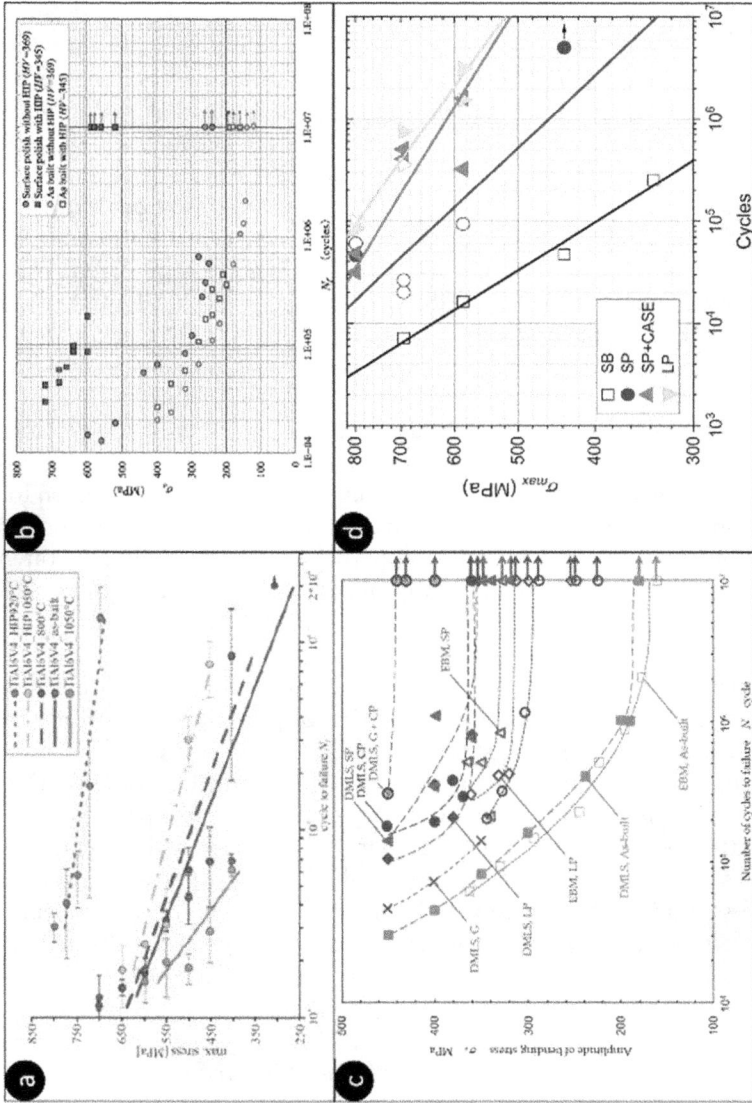

Figure 8.1 Typical examples of S-N curves for additively manufactured specimens: (a) from [3]; (b) from [4]; (c) from [5]; and (d) from [6]. For colour versions of these graphs, readers are referred to the original sources.

2. **Tensile Residual Stresses (TRS):** In all thermally driven AM processes, including Directed Energy Deposition (DED), Selective Laser Melting (SLM), Laser Powder Bed Fusion (LPBF), Direct Metal Laser Sintering (DMLS) and Laser Metal Deposition (LMD), differential thermal expansion and contraction between layers leads to the rise of tensile residual stresses in the deposited layers. As the deposited layer cools, contraction occurs. However, the substrate and previously solidified layers restrict it from contracting. This leads to the development of TRS. The temperature gradient and cooling rates are the primary determinants of the residual stress profile developed in AM processes.

   As with conventional parts, these residual stresses can have quite a significant effect on the fatigue life. Due to the superposition of stresses, TRS, especially near surface regions, are detrimental to fatigue life through crack initiation and propagation. On the other hand, compressive residual stresses (CRS) can aid in fatigue life improvement by suppressing the cracking process. As discussed subsequently, this is what 'peening' techniques aim to achieve via localized severe plastic deformation.

3. **Internal porosity or voids:** While the above two factors are relevant to fatigue life of conventionally manufactured parts as well, the presence of porosities/voids is not something that is typically a concern for wrought, forged, or cast parts. However, in AM, due to factors such as entrapment of gases or unmelted particles, porosity, and voids often develop. These defects can compromise the fatigue strength of AM parts by acting as crack initiation sites [9].

Unfortunately for AM, with the current state of technology, it is not possible to mitigate these factors solely by altering/optimizing the AM process parameters. Although several studies have been carried out to reduce the above fatigue failure-inducing defects via AM parameters optimization, post-processing is still very much necessary to obtain usable parts [10–13]. Most of the post-processes which aim to improve fatigue, will target one or more of the abovementioned factors (residual stresses, porosities, surface roughness). Additionally, there may be some microstructural effects which could affect the fatigue performance as well. The following sections summarize the various post-processes used to improve fatigue strength of AM parts.

8.3 FATIGUE LIFE ENHANCEMENT PROCESSES

In general, improved surface finish, elimination of stress concentrators, surface hardening techniques (for example carburizing or nitriding in steels) and induction of surface compressive residual stresses (CRS) all enhance fatigue life. In the following sections, severe plastic deformation-based (induction of work hardening and CRS) techniques are elaborated.

8.3.1 Shot peening

Shot peening (SP) is a severe plastic deformation (SPD) process in which small, hard, particles (often spherical balls) are fired with high velocity at the component to be peened. On impact with the component, the particles impart some of their high kinetic energy to the component in the form of plastic deformation. This plastic deformation (stretching of the surface) causes work hardening and leads to the development of compressive residual stresses in the surface. Of all the peening processes, shot peening is the most well-established, and is commonly employed by industries such as the aerospace and automotive industries, to improve the service life of components.

SP can be used to improve the fatigue life of parts for AM processes as well. The most commonly stated benefit of SP for AM parts is the development of near-surface CRS, which can lower the mean stress level and retard crack propagation [14]. Figure 8.2 shows the mechanism of residual stress development after shot peening. Note that the same principle can be applied to any peening process with minor differences.

Figure 8.3 shows the residual stresses developed in components as reported in previous studies. It can be seen quite clearly that SP (as well as other peening processes) can transform the tensile surface residual stresses

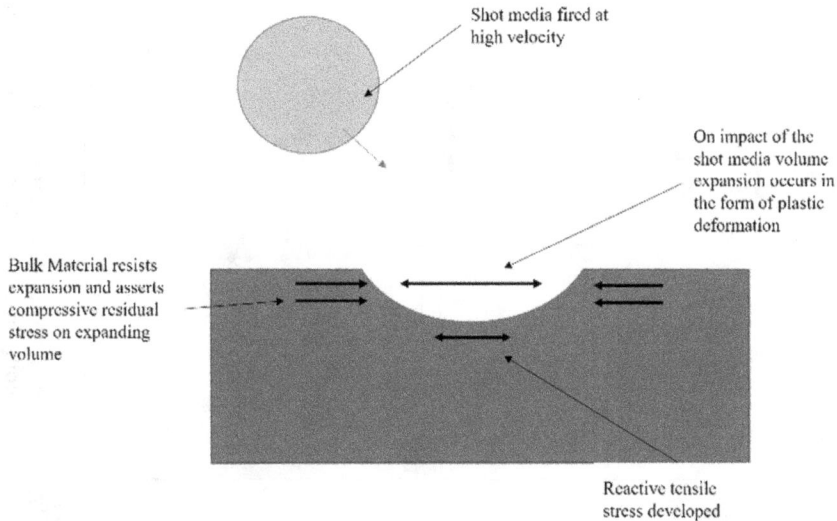

Shot media fired at high velocity

On impact of the shot media volume expansion occurs in the form of plastic deformation

Bulk Material resists expansion and asserts compressive residual stress on expanding volume

Reactive tensile stress developed

Figure 8.2 Residual stress evolution in shot peening.

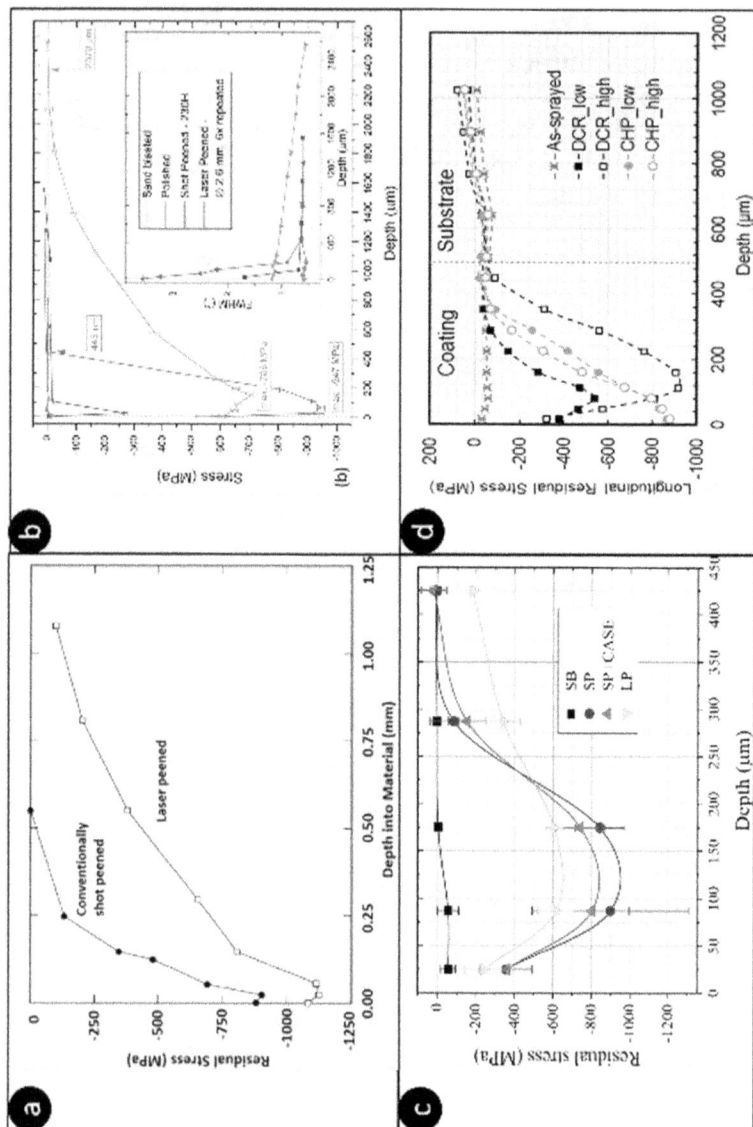

Figure 8.3 Typical residual stress depth profiles for various peening processes: (a) from [15]; (b) from [16]; (c) from [6]; and (d) from [17]. For colour images readers are referred to the original sources of the images.

typical of AM processes into compressive ones at and near the surface. This conversion from tensile to compressive residual stress is largely due to the localized lateral expansion of material near the surface, when compressed in the normal direction by the peening process. The lateral expansion will be resisted by the bulk of the material, leading to compressive residual stresses being formed [8], as shown schematically in Figure 8.2.

One factor of concern for SP of conventional parts is that it tends to degrade the surface finish of components. A machined surface which has been subjected to shot peening may have a greater roughness after the peening process due to the many indentations created by SP, which might compromise fatigue performance. For AM this is less of a concern as most metallic AM processes generate surfaces with relatively poor surface roughness. It has been shown that surface roughness parameters for AM components remain unchanged or even improve after SP [18–20].

Almen intensity, which is a function of shot media type, mass, size, velocity, and hardness, is often considered to be the key descriptor of the SP process under various conditions. Generally, a higher Almen intensity would mean greater plastic deformation and hence greater work hardening and CRS in the surface. However, higher Almen intensity does not always lead to better fatigue life due to the occurrence of potential microcracking [21, 22]. Even at the same Almen intensity, the hardness and the size of the shots can influence the fatigue life as shown by Maleki et al. [23]. Their study used steel and ceramic media of 0.43 mm and 0.1–0.15 mm diameter, respectively, and found that despite higher Almen intensity with the steel media, the samples subject to ceramic media SP showed better fatigue strength. At the same time, some studies propose that, where fatigue strength is dominated by residual stresses as compared to surface condition, larger shot media is better as it induces greater CRS [24].

In the selection of SP media for AM components the main consideration is the hardness of the substrate to be peened. Use of the same media as the conventional counterpart of the AM sample under consideration may not be sufficient due to the typically higher hardness of AM components. The SP media needs to be carefully selected to ensure the hardness of the media is of sufficient hardness to induce the desired effects.

A variation of the SP process which has recently been applied to AM parts is Ultrasonic Shot Peening (USP) [25, 26]. The main difference between the two is how kinetic energy is imparted to the particles. Unlike conventional SP which generally uses air flow or centrifugal force, USP relies on a vibrating surface or a sonotrode to impart energy to the impinging media. However, the fundamental working principle and the effect on material substrates is similar.

Given the conversion of tensile residual stresses without significant increase in surface roughness (and no effect on internal voids), it would be expected that fatigue life of shot peened samples would be better than as-built AM samples. This has been shown on a variety of engineering alloys as summarized in Table 8.1.

Table 8.1 Summary of fatigue life enhancement due to SP

AM process	Material and reference	Key findings relating to fatigue improvement
SLM	SS316 L [20]	• Improvement of Ra from 21.1 μm in as-built condition to 4.1 μm after SP. • SP improved fatigue strength by ~45% at 10^7 cycles. • Combination of Ra improvement + CRS equivalent to much better Ra (from machining) in terms of fatigue life.
SLM	AlSi10Mg [23]	• Removal of surface defects after SP to more regular homogenous surface (even though Sa increased from 6.4 μm to 11.0 μm). • Peak CRS of ~150 MPa achieved by SP. • Surface hardening and CRS more effective in improving fatigue life than microstructure homogenization through heat treatment. SP improved fatigue strength of notched specimens at 3×10^6 cycles by >2 times compared to heat-treated condition and >15 times compared to as-built condition.
SLM	AlSi10Mg [21]	• Ra and Rq both reduced after SP. • Peak CRS of ~150 MPa achieved by SP. • Surface hardening and CRS more effective in improving fatigue life than microstructure homogenization through heat treatment. SP improved fatigue strength at 3×10^6 cycles by >3 times compared to as-built condition.
EBM	Ti6Al4V [27]	• SP improved Ra from 19 μm to 5 μm. • Surface RS induced by SP was 348 MPa. • At an applied load of 400 MPa SP improved the number of cycles to failure by ~4 times as compared to as-built sample.
DMLS	IN 718 [28]	• SP caused improvement in number of cycles to failure from ~176,000 to ~268,000. • Post SP polishing further improved number of cycles to failure to ~348,000.
PTA-SFFF	Ti–TiB [29]	• SP caused localized hardening up to a depth of 1 mm from the surface. Peak hardness value was 400.4HV which occurred on the surface. • SP increased the endurance limit 318.3 MPa as compared to 247.8 MPa in the unpeened condition.
SLM	Ti6Al4V [16]	• SP induced a peak CRS of 947 MPa. • Ra improved from 16.19 μm in the as-built condition to 4.62 μm after SP. • Fatigue life enhancement attributed to CRS. Surface roughness found to have lower influence on fatigue performance.

Process	Material	Notes
LPBF	IN 718 [30]	• SP applied post solution heat treatment cycle. • Peak CRS of ~1000 MPa observed at a depth of ~200 μm from the peened surface. • Fatigue life improved of SP samples was higher than unpeened sample however SP applied along with HIP worsened fatigue life. This was driven by interaction between balancing TRS due to SP and sub-surface defects.
SLM	AlSi10Mg [31]	• SP found to reduce overall porosity by 0.1–0.3% however this did not influence the fatigue performance. • SP did not affect the magnitude of maximum residual stresses but increased the depth of CRS. • SP improved the fatigue strength at 10^7 cycles from 60 to 80 MPa.
SLM	Ti6Al4V [6]	• SP induced a peak CRS of ~900MPa. CRS were found to a depth of 300 μm. • High cycle fatigue strength improved 75% due to SP. This was attributed to CRS and surface roughness reduction. • Applying CASE (a chemical assisted vibratory finishing technique) after SP increased fatigue life more due to further reduction of surface roughness.
LPBF	A357 [24]	• Sa reduced from 25.9 μm for the as-built condition to between 7.97 and 12.8 μm for the SP conditions (depending on the SP parameters used). • SP improved fatigue strength at 2×10^6 cycles by up to 80%. • SP with larger shots causes greater CRS and hence greater fatigue strength enhancement.
EBM	Ti6Al4V [25]	• EBM + HIP + USP improved fatigue life but not as much as EBM + HIP + machining. This was explained by the presence of sub-surface defects which the USP failed to 'heal'. Maximum failure stress at 10^5 cycles was 220 MPa HIP condition, 500 MPa for HIP + USP MPa condition and 700 for HIP + machining condition. • USP induced peak CRS pf ~400 MPa and improved Ra from 46.2 to 15.7 μm. • Fatigue strength attributed mainly to surface roughness improvement along with CRS, grain refinement, and work hardening.
EBM and DMLS	Ti6Al4V [5]	• SP reduced Ra from 12 to 4 μm for DMLS sample and from 19 to 5 μm for EBM sample. • SP induced CRS on the surface of both DMLS and EBM samples. Magnitude of the CRS was ~200 and ~150 MPa respectively. • Fatigue life were 4.7 and 1.95 times better respectively for the SP'ed DMLS and EBM samples as compared to the corresponding as-built conditions. • This enhancement was explained by reduction in roughness, surface hardening, and residual stresses.
SLM	Ti6Al4V [32]	• SP improved Ra from 11.7 μm in as-built condition to 4.6 μm in post SP condition. Subsequent chemical assisted vibratory polishing method (CASE) further improved Ra to 2.7 μm. • SP improved fatigue life by ~10 times. SP + Case provided even greater fatigue life enhancement. • Indicated that surface roughness and CRS more important for fatigue life than porosities.

8.3.2 Hammer peening

Hammer peening (HP), as shown in Figure 8.4, is a relatively new surface enhancement technology that relies on a high hardness tool to generate localized plastic deformation and the associated material enhancements via direct impact on the surface [33]. Like SP, the fundamental mechanism behind material property enhancement is localized plastic deformation. However, instead of loose media, a rigid tool is used to create multiple surface impacts. This means that rather than random, stochastic impacts as in SP, HP can create well controlled and uniform impacts on the surface to be peened. The key aspect of HP is transfer of kinetic energy from a 'hammering tool' to the workpiece which is generated by vibrating the tool at high frequencies [33].

When compared to other shot peening processes, HP has a few benefits. Unlike SP, it generally produces a much smoother surface due to the controlled, evenly spaced nature of the impacts. It also leaves little-to-no debris on the workpiece being peened. Although laser shock peening (LSP) has the same advantages, the process requires the presence of a water confinement layer and in some cases an ablative coating as well. The HP process does not require any additional interface. All these advantages make HP an interesting choice for AM post-processing.

Due to its relative novelty, there has been little direct work on how HP can impact fatigue life of AM parts. However, there have been some investigations on HP of AM parts which have yielded results of the same nature as SP (CRS, grain refinement, surface hardening). Following similar reasoning to the effects of SP, HP should enhance fatigue life of AM parts as summarized in Table 8.2. It is expected in the coming years, further investigations will be carried out to study the direct relationship between fatigue strength and HP.

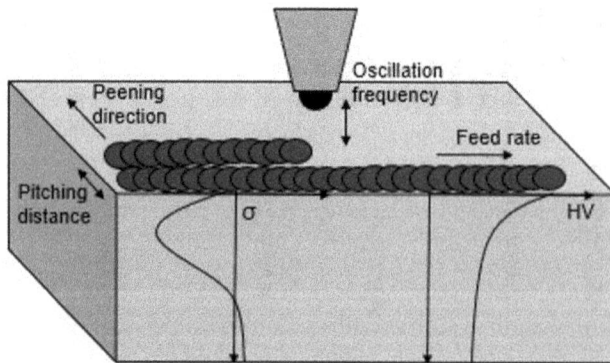

Figure 8.4 Schematic of the hammer peening process [33], where σ is residual stress and HV is the hardness distribution.

Table 8.2 Hammer peening for AM

AM type	Material and reference	Peak CRS	CRS depth	Roughness as built	Roughness after HP	Microstructural effects	Fatigue life reported
Cold Sprayed	Ti6Al4V [17]	900 MPa	400 μm	0.5 μm Ra	1.1 μm Ra	No significant hardening	N
Laser Deposition	SS 431 [34]	2100 MPa	>210 μm	0.568 μm	0.397 μm	Increased hardness	N
LPBF	Hastelloy X [35]	NR (not reported)	NR	13.2 μm Sa	7 μm Sa	NR (focused on microstructural effects from heat treatment and electro spark deposition)	N
SLM	AlSi10Mg [36]	150 MPa	2 mm	NR	NR	NR	N
DMLS	Ti6Al4V [37]	190 MPa	NR	10 μm Ra	3 μm Ra	No significant microstructural changes observed	Y
SLM	AlSi10Mg [38]	170 MPa	2 mm	NR	NR	Hardness increases up to a depth of 1.5 mm	N

Rather than applying HP as a post-process to achieve material property enhancement, a more common use of hammer peening has been to use it as a 'hybrid process' with AM [39–41]. In these studies, HP is applied intermittently through the AM process (for example, after every layer deposited) to enhance microstructure and properties. A detailed description of this is beyond the scope of this chapter.

8.3.3 Deep cold rolling/burnishing

Deep Cold Rolling (DCR), also referred to as low plasticity burnishing (LPB), is another tool-based surface enhancement technique which utilizes rollers or balls as the tool to induce surface and sub-surface plastic deformation. Unlike HP, where the tool intermittently impacts the specimen, in DCR, the employed tool is continuously in contact with the specimen over the length of a track, to achieve a pressing/rolling type action as shown in Figure 8.5.

The key process parameters in DCR are tool diameter, rolling speed, and normal pressure, all of which have an influence on the CRS profile, the work hardening of the surface as well as the roughness achieved [42].

Compared to other peening processes, DCR has the advantage of deeper CRS as well as a much higher quality surface finish. The better surface

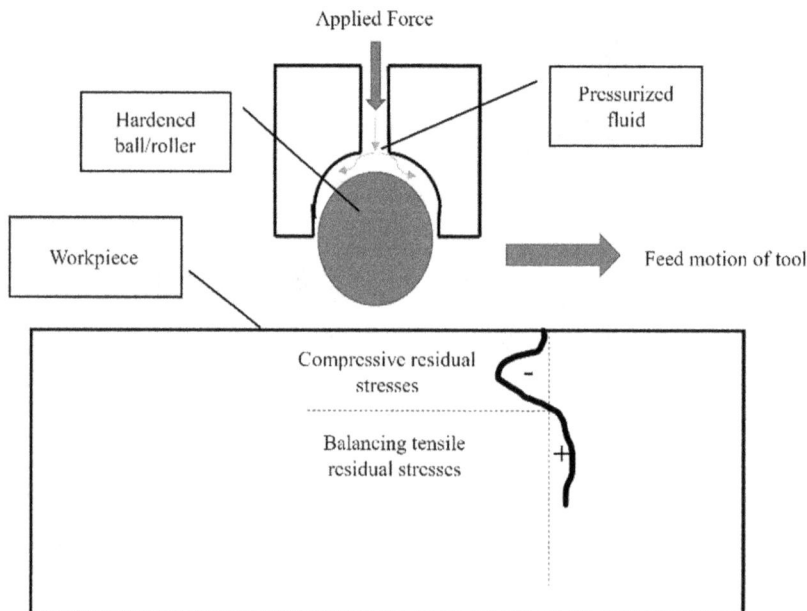

Figure 8.5 Schematic of deep cold rolling (DCR) or low plasticity burnishing (LPB) process.

quality from DCR is a result of the continuous nature of the process. Other peening processes all rely on intermittent processing. A typical witness of this lack of continuity is visible indentation marks in SP and HP and ablative marks in LSP.

Like HP, due to the novelty of DCR there are at present few direct studies on fatigue enhancement of AM parts by DCR. However, several promising results have been reported in terms of improvement of surface characteristics (CRS, work hardening, surface quality), as discussed in Table 8.3. It is expected in future many more studies will be done showing the enhancement of fatigue life through DCR post-processing.

Although direct links between fatigue life and DCR have not yet been clearly established, initial studies of DCR on AM components indicate promising results in terms of CRS and reduction of surface roughness. The surface roughness improvement in particular is a major benefit of DCR when compared to other peening processes. It is quite common for DCR to be able to achieve sub-micron roughness. In the extreme case, DCR has also been shown capable of reducing or removing the roughness due to the staircase effect [49]. Overall, DCR could be useful in applications where fatigue life is primarily governed by surface crack initiation due to roughness inherent in the AM process.

8.3.4 Laser shock peening

Laser shock peening (LSP) is a peening process which induces a high intensity shockwave into the target material via a high energy pulsed laser. When a high-energy pulsed laser is incident on the surface, it generates a high-energy plasma. This plasma, when confined by a transparent tamping layer (such as a film of water), is unable to escape, and instead generates a mechanical pressure wave into the material. These pressure waves cause localized plastic deformation and induce compressive stresses into the sub-surface regions of the material [50] as shown in Figure 8.6. In some cases, an additional thermal protective coating, also known as an ablative layer, is applied to prevent surface melting effects, although this is not mandatory and many LSP applications today do not use this coating. This is also referred to as laser shock peening without coating (LSPwC).

A key process variable in LSP is the peak power density. This is a function of the incident laser radiation (total pulse energy and temporal pulse width) as well as the beam spot diameter. Typical power densities used are in the range of 1–10 GW/cm^2. In general, higher power density will lead to a larger pressure wave in the material and correspondingly greater and deeper depth of influence. However, above a certain threshold power density (for example ~2 GW/cm^2 for A356-T6 alloy), while the magnitude of residual stresses along the depth will continue to increase, the surface residual stresses will start to decrease [51].

Table 8.3 Deep cold rolling for AM

AM process	Material and reference	Peak CRS	CRS depth	Roughness as built	Roughness after DCR	Microstructural effects	Fatigue life reported
Cold Sprayed	Ti6Al4V [17]	900 MPa	500 μm	0.5 μm Ra	0.2 μm Ra	No significant hardening	N
EBM	IN 718 [43]	900 MPa	NR	21.2 μm Ra	0.8 μm Ra	Increased microhardness	N
LPBF	GPI SS [44]	600 MPa	>400 μm	11 μm Ra	0.2 μm Ra	Deformed grains and increased hardness	Y
SLS	SS 316 L [45]	NR	NR	7.39 μm Ra	0.55 μm Ra	Hardness increases by ~70%	N
SLM	AlSi2 [46]	NR	NR	23 μm Ra	2.5 μm Ra	Microcracks seen in the rolled specimens	Y
Laser Cladding	Cr–Ni SS [47]	250 MPa	NR	0.35–0.4 μm Ra Note: This is after turning (machining)	0.1–0.15 μm Ra	Surface hardening up to a depth of 80 μm reported	N
Cold Spray	17-4 PH Steel [48]	200 MPa	> 180 μm	NR	NR	Microhardness increase reported in affected depth	N
Laser Cladding	17-4 PH Steel [48]	900 MPa	> 300 μm	NR	NR	Microhardness increase reported in affected depth	N

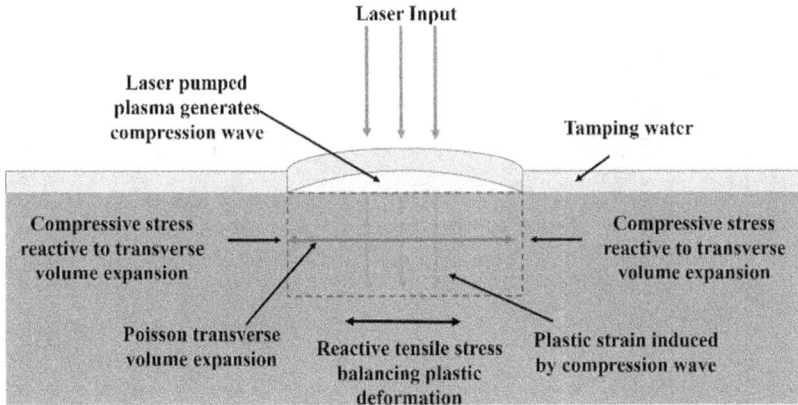

Figure 8.6 Schematic of LSP process [50]. For colour image readers are referred to the original source of the image.

Although LSP is significantly more expensive than the established conventional SP, it has a couple of key benefits. First, the magnitude and depth of CRS is much greater from LSP. LSP has been shown to induce a greater compressive residual stress at both the surface and in-depth (near-surface regions). Second, it also causes a much lower level of surface roughening and strain hardening [15, 50].

In much the same way as SP, the deep induced CRS, along with no worsening of surface roughness, would be expected to have a positive impact on fatigue life of AM samples. This has shown to be the case in multiple studies which are summarized in Table 8.4.

8.3.5 Other SPD techniques

There are also a few other niche SPD processes that could be applied to enhance the fatigue life of AM components such as waterjet peening, vibropeening, ultrasonic burnishing, or cavitation peening. However, these have seen limited application to AM components, and hence further discussion is not considered.

8.3.6 Hot isostatic pressing

Hot Isostatic Pressing (HIP) is a thermomechanical technique for enhancement of fatigue strength in AM parts. Although generally classified more as a thermomechanical post-process than an SPD post-process, it has been included here due to its ability to target internal voids or porosities. The techniques described till now achieve fatigue life enhancement via a combination of surface roughness improvement, inducement of CRS and

Table 8.4 Fatigue life enhancement for AM through LSP

AM process	Material and reference	Key findings on fatigue
SLM	SS 316 L [50]	• LSP improved the fatigue strength by 60% for notched condition and 80% for unnotched condition. • Laser Peening + notched sample had same fatigue life as built + unnotched sample. LSP was able to recover the detrimental fatigue effect of a notch. • Fatigue enhancement was attributed to deep CRS induced.
SLM	Ti6Al4V [16]	• LSP was able to induce CRS up to a depth of more than 2 mm. Peak CRS was ~700 MPa. • Fatigue life enhancement was more significant for LSP than for SP. This was attributed to CRS distribution. Surface roughness appeared to have minimum effect.
SLM	Ti6Al4V [6]	• LSP samples showed better fatigue life than SP samples despite having lower peak CRS and rougher surface due to larger depth of CRS. LSP samples had peak CRS of ~600 MPa (as compared to ~900 MPa for SP). However, depth of CRS was >400 µm (as compared to ~300 µm for SP). • Most crack initiation for LSP samples was in sub-surface.
EBM and DMLS	Ti6Al4V [5]	• LSP slightly reduced the Ra from 12 to 10.5 µm for DMLS sample and while for the EBM sample the Ra remain unchanged at 19 µm. • LSP induced TRS of 25 MPa on the surface of the DLMS sample and negligible magnitude CRS on the surface of the EBM sample. • Fatigue lives were 3.5 and 1.87 times better for the LSP'ed DMLS and EBM samples as compared to the corresponding as-built conditions. • Fatigue Strength enhancement was explained by reduction in roughness, surface hardening, and residual stresses. • Fatigue improvement from SP and LSP was not very different even though SP produced a much lower roughness surface.
SLM	Ti6Al4V [32]	• LSP marginally improved Ra from 11.7 µm in as-built condition to 10.2 µm in post LSP condition. • LSP induced greater fatigue performance enhancement as compared to SP and HIP processes. • The fatigue life enhancement due to LSP was attributed to the greater depth of the CRS field which inhibited surface crack growth.

EBM	Ti6Al4V [52]	• LSP increased the surface hardness from ~305 HV to ~355 HV. • LSP increased the fatigue strength by 17% from 600 MPa to 700 MPa at 2,000,000 cycles. • CRS from LSP suppressed the initiation and propagation of cracks, which enhanced fatigue strength. Additionally pre-existing cracks were closed. • Grain refinement of Alpha phase led to increased difficulty for crack propagation again contributing to fatigue strength enhancement.
LAM	Ti6Al4V [53]	• LSP improved the fatigue strength by 23.6% from 365 to 451 MPa. • This was attributed to two effects: presence of CRS and surface hardening. • CRS delayed crack initiation and reduced crack propagation rate by lowering the effective mean stress. Surface hardening created a barrier to the movement of dislocation thereby increasing the number of required cycles to failure.
DMLS	IN718 [28]	• LSP improved number of cycles to failure from ~176,000 to ~390,000. • Post LSP polishing further improved number of cycles to failure 1,352,000. • After exposure to high temperature (600° C for 50hrs) LSP + polished samples showed large drop in fatigue life. Number of cycles to failure reduced to ~420,000. • LSP followed by annealing iteratively (3 times) showed good stability of fatigue life even after exposure to high temperature (600° C for 50hrs). Number of cycles to failure was ~1,450,000.

microstructural changes. HIP is currently the only established process that can target the last bit of the puzzle i.e. sub-surface voids and porosities.

Defect closure in HIP is achieved by the simultaneous application of thermal and pressure loads to the component. The thermal loads seek to soften the material and promote solid-state diffusion while the pressure loads tend to cause pore collapse and densification [8, 54]. Sintering due to thermal load also aids in densification.

HIP generally has no effect on surface roughness but rather targets internal voids in the AM material. It is unable to close surface connected voids but rather only those completely inside the volume of the part [8]. It cannot improve the poor surface finish of as-built AM components. This is a shortcoming of HIP as far as fatigue life is concerned when compared to SPD processes. Poor surface finish means that cracks can easily initiate from the surface. In such a case, even if HIP reduces internal defects, it may not have much effect on the fatigue strength.

An important consideration in HIP treatments for fatigue enhancement is the temperature applied and the time of the process. The exposure to elevated temperatures can have the same effects as a heat treatment cycle, leading to alleviation of residual stresses and microstructural changes such as grain coarsening.

There is a school of thought that HIP is more effective at fatigue enhancement when combined with a surface finishing technique. Since HIP alone does not affect the surface roughness, surface cracks will still develop relatively early, leading to a low fatigue life. However, when combined with a surface finishing technique, the synergistic nature of low surface roughness and low near-surface porosities can have a large impact on fatigue strength.

At present, there is no consensus on the effects of HIP on fatigue life. While it is agreed that internal defects *can* act as crack initiation sites and that HIP can close these internal defects, the impact of this on fatigue life is still debated. Different studies have shown positive, negative and zero impact on fatigue life of AM components arising from HIP post-processing as summarized in Tables 8.5 and 8.6.

Considering the multiple aspects of HIP (temperature, time) and the different microstructural considerations for different materials, it is difficult to make a generalized statement about the impact of HIP on the fatigue life of AM components. The role of porosities in fatigue life is still debated and is likely to vary by material as well as AM process type. Nonetheless, it is clear that there are situations where HIP, whether alone or in combination with other processes, will contribute to an enhancement of fatigue life due to pore closure, and stress relief effects. On the other hand, care needs to be taken to avoid some of the beneficial mechanical and material properties that arise out of AM processes (cellular substructures, high dislocation densities, higher yield and tensile strength) being erased by grain growth in the HIP process.

Table 8.5 Studies concluding zero or negative impact on fatigue life from HIP

AM type	Surface finishing	Material and reference	Key findings
LPBF	Yes: EDM, Machining and surface finishing	IN 718 [30]	• No Difference in fatigue life between heat treatment and HIP despite reduction in porosity from HIP. • This was attributed to a) inability of HIP to remove brittle phases and inclusions and b) removal of Υ'' precipitates by solutionization during HIP.
SLM	Yes: Sandblasting	Ti6Al4V [32]	• No difference in fatigue life between annealed and HIPed specimens. • Attributed to poor surface finish which meant cracks initiated easily.
SLM	Yes: Machined out	SS 316 L [3]	• 316 L has a high monotonic strength in as-built condition due to presence of dislocation cells. • Due to high strength and ductility, stress risers such as pores can be compensated; therefore, not so relevant for fatigue life. • On the other hand, as-built 316 L SS has higher yield and tensile strength due to dislocation substructures. HIP treatment reduces this and hence worsens fatigue performance.
EBM and DMLS	No	Ti6Al4V [55]	• Minimal impact of HIP on fatigue life for both EBM and DMLS samples. • Surface roughness dominates fatigue life. • Note: With milling found HIP improved fatigue life. See next table.
SLM	Yes: machining and polishing	AlSi10Mg [56]	• Stress relief + HIPing reduced the fatigue life due to a reduction in mechanical properties and grain coarsening. • Exposure to elevated temperatures caused the fine cellular structure to disintegrate and instead separated Si particles formed leading to loss of strength and hardness. • Interface of Si-Al could also act as a site for crack initiation.

(Continued)

Table 8.5 (Continued)

AM type	Surface finishing	Material and reference	Key findings
LPBF	No	Ti6Al4V [57]	• Without surface finishing surface roughness and microcracks still governed fatigue life. • Note: With surface machining and polishing found HIP improved fatigue life. See next table.
EBM and LS	No	Ti6Al4V [58]	• No improvement in fatigue life when HIP applied without surface finishing for both EBM and LS specimens. • Note: With surface machining and polishing found HIP improved fatigue life. See next table.
SLM	Yes: Machining	AlSi7Mg0.6 [59]	• HIP showed no improvement in fatigue life compared to T6 heat treatment as the crack initiated at sites present similarly in both.

Table 8.6 Studies concluding positive effective of HIP on fatigue life

AM type	Surface finishing	Material and reference	Key findings
SLM	Yes: Specimens machined out	Ti6Al4V [60]	• HIP samples showed greater fatigue life than as-built and heat-treated samples. Fatigue strength of 620 MPa was achieved due to HIP. • Fatigue life enhancement was attributed to HIP's ability to close pores (stress concentrators) and alleviate residual stresses. • HIP samples delayed crack initiation but had no effect on crack propagation.
DMLS and EBM	No	Ti6Al4V [4]	• HIP improved fatigue life of both DMLS and EBM samples. The fatigue strength was 190 MPa with HIP compared to 140 MPa for as-built EBM sample and 195 MPa with HIP compared to 155 MPa for as-built DMLS sample. • Surface Polishing without HIP improved fatigue life more than HIP without surface polishing. • Defects most important aspect for fatigue of Ti64 (surface roughness and porosity).
EBM and DMLS	Yes: Milling	Ti6Al4V [55]	• HIP improved fatigue life for both EBM and DMLS samples when surfaces where milled. • High cycle fatigue (HCF) dominated by inherent surface roughness from EBM and DMLS process. • When surface finish is improved porosities play a role in fatigue strength.
LPBF	Yes: Polishing	Hastelloy [61]	• HIP achieved pore and microcrack closure which contributed to fatigue life enhancement in high cycle regime as compared to as-built condition. • At higher applied stress level, the improvement in fatigue life due to HIP was much lower. • Yield Strength and UTS were lower due to grain coarsening by recrystallization.

(Continued)

Table 8.6 (Continued)

AM type	Surface finishing	Material and reference	Key findings
SLM	No	Ti6Al4V [62]	• HIP reduced microhardness and yield strength by 20% and 30% respectively due to microstructural changes. • HIP improved the fatigue endurance ratio at 10^6 cycles from 0.3 to 0.55. • Fatigue life enhancement attributed to phase change from brittle α' martensite to tough α+β mixture.
LPBF	Yes: machining and polishing	Ti6Al4V [57]	• Fatigue life improvement of HIPed specimens attributed to shrinkage of internal defects (when surface roughness was removed to acceptable level), relaxation of residual stress and increased ductility.
EBM and LS	Yes: machining and polishing	Ti6Al4V [58]	• Significant level of fatigue life enhancement when surface finishing and HIP applied. EBM specimens fatigue life comparable to wrought specimens. • Fatigue life enhancement attributed to closure of internal defects.
EBM and SLM	Yes: machining for some specimens	IN 718 [63]	• HIP applied along with heat treatment (HT) cycle. As compared to HT only HIPed specimens showed improvement in fatigue life. • The fatigue life improvement was present for both machined and as-built specimens and for both SLM and EBM specimens. • This was attributed to reduction in defects due to HIP. • Fracture in HT specimens appeared to be ductile while those in HIPed specimens appeared to be cleavage fracture.
LPBF	Yes: Machined and polished	IN 718 [64]	• HIPing allowed material to regain its fatigue strength against the detrimental effects of induced pores. • The pore closure of HIP was credited as the reason for fatigue life recovery.

8.3.7 Summary of processes

In Sections 8.3.1–8.3.4, four different SPD techniques have been discussed as methods of fatigue life enhancement. The fundamental principle is the same for all the peening techniques i.e. to create local plastic deformation in the surface and sub-surface regions. This leads to near-surface CRS, strain hardening and grain refinement which can benefit the fatigue life of the parts by delaying crack initiation and retarding crack propagation. In addition, where AM is specifically concerned, SPD processes can also improve the generally rough surface finish of as-built parts and provide further fatigue enhancement.

SPD post-processing can also be combined with other post-processing techniques such as machining, polishing, or grinding, which can further improve the surface quality of the parts. However, in such situations care must be taken as each post-process will have an impact on the results of the previous post-processing treatment. If SPD is done followed by surface finishing, then there is a chance that the beneficial CRS induced by the SPD process will be removed. If, on the other hand, surface finishing is done first followed by SPD process, then the high-quality surface achieved, may be damaged by the SPD process (for example, SP) causing an increase in surface roughness. These factors mean the combination of post-processing techniques needs to be carefully considered to avoid surprising and undesirable reductions in fatigue life such as those found by [65].

Another post-process that can be combined with SPD techniques is heat treatment. Heat treatment is commonly used to stress relieve AM parts. Again, the process design must be carefully considered as heat treatments can lead to a relief of the beneficial CRS induced by the SPD processes. Additionally, they can lead to thermal softening due to grain coarsening and reduce the yield and tensile strength of the material.

Further, Section 8.3.5 also discussed HIP as a post-processing treatment for fatigue life enhancement of AM components, which can target internal voids and defects, common in many AM components.

Table 8.7 compares the five processes discussed in Sections 8.3.1–8.3.4 and 8.3.6, in terms of various characteristics.

From a practical point of view, the varied nature of the SPD post-processes means that the specific application must be carefully considered while selecting a post-process. This is likely to depend on both the specific material as well as the loading and component geometry. For applications where fatigue life is dominated by crack initiation at surface roughness, HP and DCR could be excellent options as they provide a high-quality surface finish. On the other hand, for high temperature applications, such as in gas turbine engine components, LSP would be a better option due to its better high temperature stability.

Table 8.7 Comparison of various surface modification processes

	SP	LSP	HP	DCR	HIP
Surface quality	Modifies the surface significantly. For AM this can be reduction in surface roughness	Not much effect on surface roughness	Modifies the surface significantly. For AM this can be reduction in surface roughness	Best surface roughness due to continuous nature of process	No effect
Strain hardening	High	Low	High	Low	None
Noise	Noisy	Noisy	Noisy	Quiet	Quiet
High temp stability of effects	Unstable	More stable than SP, but still unstable at prolonged high temperature exposure	Unknown	Unknown	Stable
Compressive residual stresses depth	Low Typical 100–300 µm	High Typical 500–2000 µm	High Typical 500–2000 µm	Moderate–high Typical 400–1500 µm	None: Stress relief effect only
Cost	Cheap	Expensive	Moderate	Moderate	Moderate
Consumables required	Shot peening media, compressed air	Ablative layer, tamping medium	Tool	Tool	None
Commercial availability	Widespread	Moderate	Uncommon	Uncommon	Widespread
Chance of thermal effects	None	Moderate if no ablative layer used	None	None	Certain
Ability to close pores	Initial studies have demonstrated flattening [66] and shrinkage [31] of pores but still being researched	Initial studies have demonstrated pore closure possibilities in near-surface region [67] but still being researched	Not demonstrated	Not demonstrated	Yes

8.4 CONCLUSIONS

Currently, post-processing is still required to ensure acceptable fatigue lives of additively manufactured parts. Many researchers have demonstrated SPD based post-processing to improve fatigue life of AM parts. The general mechanisms of this improvement have also been revealed to a large degree and relate to the introduction of CRS, microstructural changes, improvement of surface finish and closure of porosities. However, the following aspects are yet to be explored:

i. Evaluation of DCR and HP on fatigue of AM parts and process parameter optimization.
ii. Clarification on the context in which HIP can improve/worsen fatigue of AM parts.

The current phase of development seems to be investigating the impact of the entire manufacturing route which may involve multiple post-processing steps. Figure 8.7 shows the predicted trajectory for SPD post-processing applied to AM.

The requirement to post-process AM parts means that the potential for AM to enable complex, intricate geometries cannot be fully unlocked as the design will be limited by the post-process itself. One possible solution to this, as suggested by Ye et al. [8], could be to implement targeted post-processing wherein only areas of the AM part, susceptible to fatigue crack growth would be post processed. Another solution, which is recently gaining more interest, is to integrate the post-process into the AM setup itself to achieve a hybrid or 'in-situ post-process'. This would not only remove one whole operational step, which is beneficial for production, but could also enable tailored property distribution throughout the part. However,

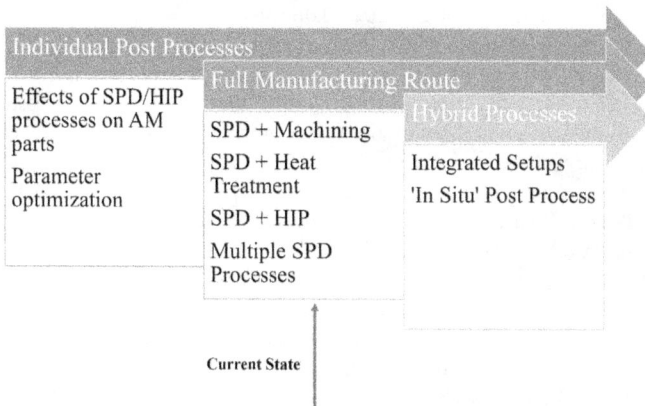

Figure 8.7 Past and expected development of SPD post-processing for AM.

significant development work is still required to make this a reality. It is expected that in the coming years, significant research effort will be focused in this direction.

Overall, it is clear that SPD post-processing is often necessary to obtain usable metallic AM parts with acceptable service lives. However, these post-processes need to be designed very carefully to ensure that they do not detract from the benefits of AM.

REFERENCES

[1] C. S. Pande, "Fundamentals of fatigue crack initiation and propagation: Some thoughts", in *Fatigue of Materials II: Advances and Emergences in Understanding*, T. S. Srivatsan, M. A. Imam, and R. Srinivasan Eds. Cham: Springer International Publishing, 2016, pp. 3–15.

[2] "Fatigue as a phenomenon in the material", in *Fatigue of Structures and Materials*, J. Schijve Ed. Dordrecht: Springer Netherlands, 2009, pp. 13–58.

[3] S. Leuders, T. Lieneke, S. Lammers, T. Tröster, and T. Niendorf, "On the fatigue properties of metals manufactured by selective laser melting – The role of ductility", *Journal of Materials Research*, vol. 29, no. 17, pp. 1911–1919, 2014, doi: 10.1557/jmr.2014.157

[4] H. Masuo et al., "Influence of defects, surface roughness and HIP on the fatigue strength of Ti-6Al-4 V manufactured by additive manufacturing", *International Journal of Fatigue*, vol. 117, pp. 163–179, 2018, doi: 10.1016/j.ijfatigue.2018.07.020

[5] H. Soyama and F. Takeo, "Effect of various peening methods on the fatigue properties of titanium alloy Ti6Al4V manufactured by direct metal laser sintering and electron beam melting", *Materials (Basel)*, vol. 13, no. 10, May 12 2020, doi: 10.3390/ma13102216

[6] S. Aguado-Montero, C. Navarro, J. Vázquez, F. Lasagni, S. Slawik, and J. Domínguez, "Fatigue behaviour of PBF additive manufactured TI6AL4V alloy after shot and laser peening", *International Journal of Fatigue*, vol. 154, 2022, doi: 10.1016/j.ijfatigue.2021.106536

[7] "Introduction to fatigue of structures and materials", in *Fatigue of Structures and Materials*, J. Schijve Ed. Dordrecht: Springer Netherlands, 2009, pp. 1–9.

[8] C. Ye, C. Zhang, J. Zhao, and Y. Dong, "Effects of post-processing on the surface finish, porosity, residual stresses, and fatigue performance of additive manufactured metals: A review", *The Journal of Materials Engineering and Performance*, pp. 1–19, Jul 26 2021, doi: 10.1007/s11665-021-06021-7

[9] T. H. Becker, P. Kumar, and U. Ramamurty, "Fracture and fatigue in additively manufactured metals", *Acta Materialia*, vol. 219, p. 117240, 2021/10/15/ 2021, doi: https://doi.org/10.1016/j.actamat.2021.117240

[10] D. Ramos, F. Belblidia, and J. Sienz, "New scanning strategy to reduce warpage in additive manufacturing", *Additive Manufacturing*, vol. 28, pp. 554–564, 2019, doi: 10.1016/j.addma.2019.05.016

[11] N. T. Aboulkhair, N. M. Everitt, I. Ashcroft, and C. Tuck, "Reducing porosity in AlSi10Mg parts processed by selective laser melting", *Additive Manufacturing*, vol. 1–4, pp. 77–86, 2014, doi: 10.1016/j.addma.2014.08.001

[12] J. Gockel, L. Sheridan, B. Koerper, and B. Whip, "The influence of additive manufacturing processing parameters on surface roughness and fatigue life", *International Journal of Fatigue*, vol. 124, pp. 380–388, 2019, doi: 10.1016/j.ijfatigue.2019.03.025.

[13] J.-P. Kruth, J. Deckers, E. Yasa, and R. Wauthlé, "Assessing and comparing influencing factors of residual stresses in selective laser melting using a novel analysis method", *Proceedings of the Institution of Mechanical Engineers, Part B: Journal of Engineering Manufacture*, vol. 226, no. 6, pp. 980–991, 2012, doi: 10.1177/0954405412437085

[14] "Surface treatments", in *Fatigue of Structures and Materials*, J. Schijve Ed. Dordrecht: Springer Netherlands, 2009, pp. 421–436.

[15] A. K. Gujba and M. Medraj, "Laser peening process and its impact on materials properties in comparison with shot peening and ultrasonic impact peening", *Materials*, vol. 7, no. 12, pp. 7925–7974, 2014. [Online]. Available: https://www.mdpi.com/1996-1944/7/12/7925

[16] S. Slawik et al., "Microstructural analysis of selective laser melted Ti6Al4V modified by laser peening and shot peening for enhanced fatigue characteristics", *Materials Characterization*, vol. 173, 2021, doi: 10.1016/j.matchar.2021.110935.

[17] N. Maharjan, A. Bhowmik, C. Kum, J. Hu, Y. Yang, and W. Zhou, "Post-processing of cold sprayed Ti-6Al-4 V coatings by mechanical peening", *Metals*, vol. 11, no. 7, 2021, doi: 10.3390/met11071038

[18] M. Walczak and M. Szala, "Effect of shot peening on the surface properties, corrosion and wear performance of 17-4PH steel produced by DMLS additive manufacturing", *Archives of Civil and Mechanical Engineering*, vol. 21, no. 4, 2021, doi: 10.1007/s43452-021-00306-3

[19] M. Sugavaneswaran, A. V. Jebaraj, M. D. B. Kumar, K. Lokesh, and A. J. Rajan, "Enhancement of surface characteristics of direct metal laser sintered stainless steel 316 L by shot peening", *Surfaces and Interfaces*, vol. 12, pp. 31–40, 2018, doi: 10.1016/j.surfin.2018.04.010

[20] P. Wood, T. Libura, Z. L. Kowalewski, G. Williams, and A. Serjouei, "Influences of horizontal and vertical build orientations and post-fabrication processes on the fatigue behavior of stainless steel 316 L produced by selective laser melting", *Materials (Basel)*, vol. 12, no. 24, Dec 14 2019, doi: 10.3390/ma12244203

[21] S. Bagherifard, N. Beretta, S. Monti, M. Riccio, M. Bandini, and M. Guagliano, "On the fatigue strength enhancement of additive manufactured AlSi10Mg parts by mechanical and thermal post-processing", *Materials & Design*, vol. 145, pp. 28–41, 2018, doi: 10.1016/j.matdes.2018.02.055

[22] A. Maamoun, M. Elbestawi, and S. Veldhuis, "Influence of shot peening on AlSi10Mg parts fabricated by additive manufacturing", *Journal of Manufacturing and Materials Processing*, vol. 2, no. 3, 2018, doi: 10.3390/jmmp2030040

[23] E. Maleki et al., "Fatigue behaviour of notched laser powder bed fusion AlSi10Mg after thermal and mechanical surface post-processing", *Materials Science and Engineering: A*, vol. 829, 2022, doi: 10.1016/j.msea.2021.142145.

[24] A. Gatto, A. Sola, and E. Tognoli, "Effect of shot peening conditions on the fatigue life of additively manufactured A357.0 parts", *SAE International*

Journal of Materials and Manufacturing, vol. 13, no. 2, 2020, doi: 10.4271/05-13-02-0009

[25] T. Persenot, A. Burr, E. Plancher, J.-Y. Buffière, R. Dendievel, and G. Martin, "Effect of ultrasonic shot peening on the surface defects of thin struts built by electron beam melting: Consequences on fatigue resistance", *Additive Manufacturing*, vol. 28, pp. 821–830, 2019, doi: 10.1016/j.addma.2019.06.014

[26] N. Alharbi, "Shot peening of selective laser-melted SS316L with ultrasonic frequency", *The International Journal of Advanced Manufacturing Technology*, 2021, doi: 10.1007/s00170-021-08398-0

[27] Y. Okura, H. Sasaki, and H. Soyama, "Effect of mechanical properties on fatigue life enhancement of additive manufactured titanium alloy treated by various peening methods", in *International Conference on Advanced Surface Enhancement*, 2019: Springer, pp. 88–96.

[28] L. Hackel, J. Fuhr, M. Sharma, J. Rankin, V. Sherman, and K. Davami, "Test results for wrought and AM In718 treated by shot peening and laser peening plus thermal microstructure engineering", *Procedia Structural Integrity*, vol. 19, pp. 452–462, 2019/01/01/ 2019, doi: https://doi.org/10.1016/j.prostr.2019.12.049

[29] L.-A. DiCecco, M. Mehdi, and A. Edrisy, "Fatigue improvement of additive manufactured Ti–TiB material through shot peening", *Metals*, vol. 11, no. 9, 2021, doi: 10.3390/met11091423

[30] D. T. Ardi, L. Guowei, N. Maharjan, B. Mutiargo, S. H. Leng, and R. Srinivasan, "Effects of post-processing route on fatigue performance of laser powder bed fusion Inconel 718", *Additive Manufacturing*, vol. 36, 2020, doi: 10.1016/j.addma.2020.101442.

[31] J. Damon, S. Dietrich, F. Vollert, J. Gibmeier, and V. Schulze, "Process dependent porosity and the influence of shot peening on porosity morphology regarding selective laser melted AlSi10Mg parts", *Additive Manufacturing*, vol. 20, pp. 77–89, 2018, doi: 10.1016/j.addma.2018.01.001

[32] P. Carlos Navarro et al., "Effect of surface treatment on the fatigue strength of additive manufactured Ti6Al4V alloy", *Frattura ed Integrità Strutturale*, vol. 14, no. 53, pp. 337–344, 2020, doi: 10.3221/igf-esis.53.26

[33] W. L. Chan and H. K. F. Cheng, "Hammer peening technology—the past, present, and future", *The International Journal of Advanced Manufacturing Technology*, 2021, doi: 10.1007/s00170-021-07993-5

[34] H. Liu, C. K. I. Tan, X. Dong, T. L. Meng, J. Cao, and Y. Wei, "Laser-cladding and robotic hammer peening of stainless steel 431 on low alloy steel 4140 for surface enhancement and corrosion protections", *Journal of Adhesion Science and Technology*, pp. 1–15, 2021, doi: 10.1080/01694243.2021.2011657

[35] P. D. Enrique et al., "Enhancing fatigue life of additive manufactured parts with electrospark deposition post-processing", *Additive Manufacturing*, vol. 36, 2020, doi: 10.1016/j.addma.2020.101526

[36] X. Xing, X. Duan, X. Sun, H. Gong, L. Wang, and F. Jiang, "Modification of residual stresses in laser additive manufactured AlSi10Mg specimens using an ultrasonic peening technique", *Materials (Basel)*, vol. 12, no. 3, Feb 1 2019, doi: 10.3390/ma12030455

[37] Walker, Malz, Trudel, Nosir, ElSayed, and Kok, "Effects of ultrasonic impact treatment on the stress-controlled fatigue performance of additively

manufactured DMLS Ti-6Al-4 V alloy", *Applied Sciences*, vol. 9, no. 22, 2019, doi: 10.3390/app9224787

[38] X. Xing, X. Duan, T. Jiang, J. Wang, and F. Jiang, "Ultrasonic peening treatment used to improve stress corrosion resistance of AlSi10Mg components fabricated using selective laser melting", *Metals*, vol. 9, no. 1, 2019, doi: 10.3390/met9010103

[39] J. R. Hönnige, P. Colegrove, and S. Williams, "Improvement of microstructure and mechanical properties in Wire + Arc additively manufactured Ti-6Al-4 V with machine hammer peening", *Procedia Engineering*, vol. 216, pp. 8–17, 2017/01/01/ 2017, doi: https://doi.org/10.1016/j.proeng.2018.02.083

[40] L. Neto, S. Williams, J. Ding, J. Hönnige, and F. Martina, "Mechanical properties enhancement of additive manufactured Ti-6Al-4 V by machine hammer peening", in *Advanced Surface Enhancement*, Singapore, S. Itoh and S. Shukla, Eds., Singapore: Springer, 2020, pp. 121–132.

[41] Y. Wang and J. Shi, "Recrystallization behavior and tensile properties of laser metal deposited Inconel 718 upon in-situ ultrasonic impact peening and heat treatment", *Materials Science and Engineering: A*, vol. 786, 2020, doi: 10.1016/j.msea.2020.139434

[42] C. Y. Seemikeri, P. K. Brahmankar, and S. B. Mahagaonkar, "Low plasticity burnishing: An innovative manufacturing method for biomedical applications", *Journal of Manufacturing Science and Engineering*, vol. 130, no. 2, 2008, doi: 10.1115/1.2896121

[43] R. Karthick Raaj et al., "Exploring grinding and burnishing as surface post-treatment options for electron beam additive manufactured Alloy 718", *Surface and Coatings Technology*, vol. 397, 2020, doi: 10.1016/j.surfcoat.2020.126063

[44] G. Rotella, L. Filice, and F. Micari, "Improving surface integrity of additively manufactured GP1 stainless steel by roller burnishing", *CIRP Annals*, vol. 69, no. 1, pp. 513–516, 2020, doi: 10.1016/j.cirp.2020.04.015

[45] M. Salmi, J. Huuki, and I. F. Ituarte, "The ultrasonic burnishing of cobalt-chrome and stainless steel surface made by additive manufacturing", *Progress in Additive Manufacturing*, vol. 2, no. 1–2, pp. 31–41, 2017, doi: 10.1007/s40964-017-0017-z

[46] S. Greuling, W. Weise, D. Fetzer, K. Müller-Lohmeier, and M.-M. Speckle, *Impact of Various Surface Treatments on the S-N Curve of Additive Manufactured AlSi12*, Cham: Springer International Publishing, in Mechanical Fatigue of Metals, 2019, pp. 83–89.

[47] P. Zhang and Z. Liu, "Effect of sequential turning and burnishing on the surface integrity of Cr–Ni-based stainless steel formed by laser cladding process", *Surface and Coatings Technology*, vol. 276, pp. 327–335, 2015, doi: 10.1016/j.surfcoat.2015.07.026

[48] C. Courbon et al., "Near surface transformations of stainless steel cold spray and laser cladding deposits after turning and ball-burnishing", *Surface and Coatings Technology*, vol. 371, pp. 235–244, 2019, doi: 10.1016/j.surfcoat.2019.01.092.

[49] L. Hiegemann, C. Agarwal, C. Weddeling, and A. E. Tekkaya, "Reducing the stair step effect of layer manufactured surfaces by ball burnishing", in *AIP Conference Proceedings*, 2016, October: AIP Publishing LLC, Vol. 1769, No. 1, p. 190002.

[50] L. Hackel, J. R. Rankin, A. Rubenchik, W. E. King, and M. Matthews, "Laser peening: A tool for additive manufacturing post-processing", *Additive Manufacturing*, vol. 24, pp. 67–75, 2018, doi: 10.1016/j.addma.2018.09.013

[51] C. S. Montross, T. Wei, L. Ye, G. Clark, and Y.-W. Mai, "Laser shock processing and its effects on microstructure and properties of metal alloys: A review", *International Journal of Fatigue*, vol. 24, no. 10, pp. 1021–1036, 2002, doi: https://doi.org/10.1016/S0142-1123(02)00022-1

[52] X. Jin, L. Lan, S. Gao, B. He, and Y. Rong, "Effects of laser shock peening on microstructure and fatigue behavior of Ti–6Al–4 V alloy fabricated via electron beam melting", *Materials Science and Engineering: A*, vol. 780, 2020, doi: 10.1016/j.msea.2020.139199

[53] S. Luo, W. He, K. Chen, X. Nie, L. Zhou, and Y. Li, "Regain the fatigue strength of laser additive manufactured Ti alloy via laser shock peening", *Journal of Alloys and Compounds*, vol. 750, pp. 626–635, 2018, doi: 10.1016/j.jallcom.2018.04.029.

[54] H. V. Atkinson and S. Davies, "Fundamental aspects of hot isostatic pressing: An overview", *Metallurgical and Materials Transactions A*, vol. 31, no. 12, pp. 2981–3000, 2000, doi: 10.1007/s11661-000-0078-2

[55] D. Greitemeier, F. Palm, F. Syassen, and T. Melz, "Fatigue performance of additive manufactured TiAl6V4 using electron and laser beam melting", *International Journal of Fatigue*, vol. 94, pp. 211–217, 2017, doi: 10.1016/j.ijfatigue.2016.05.001.

[56] N. E. Uzan, R. Shneck, O. Yeheskel, and N. Frage, "Fatigue of AlSi10Mg specimens fabricated by additive manufacturing selective laser melting (AM-SLM)," *Materials Science and Engineering: A*, vol. 704, pp. 229–237, 2017, doi: 10.1016/j.msea.2017.08.027

[57] R. Molaei, A. Fatemi, and N. Phan, "Significance of hot isostatic pressing (HIP) on multiaxial deformation and fatigue behaviors of additive manufactured Ti-6Al-4 V including build orientation and surface roughness effects", *International Journal of Fatigue*, vol. 117, pp. 352–370, 2018, doi: 10.1016/j.ijfatigue.2018.07.035

[58] M. Kahlin, H. Ansell, and J. J. Moverare, "Fatigue behaviour of notched additive manufactured Ti6Al4V with as-built surfaces", *International Journal of Fatigue*, vol. 101, pp. 51–60, 2017, doi: 10.1016/j.ijfatigue.2017.04.009

[59] M. Bonneric, C. Brugger, and N. Saintier, "Effect of hot isostatic pressing on the critical defect size distribution in AlSi7Mg0.6 alloy obtained by selective laser melting", *International Journal of Fatigue*, vol. 140, 2020, doi: 10.1016/j.ijfatigue.2020.105797.

[60] S. Leuders et al., "On the mechanical behaviour of titanium alloy TiAl6V4 manufactured by selective laser melting: Fatigue resistance and crack growth performance", *International Journal of Fatigue*, vol. 48, pp. 300–307, 2013, doi: 10.1016/j.ijfatigue.2012.11.011

[61] Q. Han et al., "Laser powder bed fusion of Hastelloy X: Effects of hot isostatic pressing and the hot cracking mechanism", *Materials Science and Engineering: A*, vol. 732, pp. 228–239, 2018, doi: 10.1016/j.msea.2018.07.008

[62] M.-W. Wu, J.-K. Chen, B.-H. Lin, and P.-H. Chiang, "Improved fatigue endurance ratio of additive manufactured Ti-6Al-4 V lattice by hot isostatic pressing", *Materials & Design*, vol. 134, pp. 163–170, 2017, doi: 10.1016/j.matdes.2017.08.048.

[63] A. R. Balachandramurthi, J. Moverare, N. Dixit, and R. Pederson, "Influence of defects and as-built surface roughness on fatigue properties of additively manufactured Alloy 718", *Materials Science and Engineering: A*, vol. 735, pp. 463–474, 2018, doi: 10.1016/j.msea.2018.08.072

[64] E. Strandh, J. Gårdstam, S. D.-L. Goff, and P. Mellin, "Artificial porosity introduced during L-PBF of IN718, and its effect on fatigue performance before and after HIP", *Powder Metallurgy*, vol. 64, no. 5, pp. 434–443, 2021, doi: 10.1080/00325899.2021.1971870

[65] E. Wycisk, C. Emmelmann, S. Siddique, and F. Walther, "High cycle fatigue (HCF) performance of Ti-6Al-4 V alloy processed by selective laser melting", in *Advanced Materials Research*, 2013, vol. 816: Trans Tech Publ, pp. 134–139.

[66] D. A. Lesyk, V. V. Dzhemelinskyi, S. Martinez, B. N. Mordyuk, and A. Lamikiz, "Surface shot peening post-processing of inconel 718 alloy parts printed by laser powder bed fusion additive manufacturing", *Journal of Materials Engineering and Performance*, vol. 30, no. 9, pp. 6982–6995, 2021, doi: 10.1007/s11665-021-06103-6

[67] A. du Plessis et al., "Pore closure effect of laser shock peening of additively manufactured AlSi10Mg", *3D Printing and Additive Manufacturing*, vol. 6, no. 5, pp. 245–252, 2019, doi: 10.1089/3dp.2019.0064

Comparison of effects of ultrasonic and shot peening treatments on surface properties of L-PBF-manufactured superalloy subjected to HIP combined with heat treatments

D.A. Lesyk

National Technical University of Ukraine, Igor Sikorsky Kyiv Polytechnic Institute Kyiv, Ukraine

G.V. Kurdyumov Institute for Metal Physics of the NAS of Ukraine, Kyiv, Ukraine

West Pomeranian University of Technology, Szczecin, Poland

S. Martinez

CFAA – University of the Basque Country, Zamudio, Spain

B.N. Mordyuk

G.V. Kurdyumov Institute for Metal Physics of the NAS of Ukraine, Kyiv, Ukraine

O.O. Pedash

MOTOR SICH JSC, Zaporizhzhia, Ukraine

V.V. Dzhemelinskyi

National Technical University of Ukraine Igor Sikorsky Kyiv Polytechnic Institute, Kyiv, Ukraine

A. Lamikiz

University of the Basque Country, Bilbao, Spain

CFAA – University of the Basque Country, Zamudio, Spain

DOI: 10.1201/9781003288619-9

CONTENTS

9.1 INTRODUCTION

The high-strength, corrosion-resistant, and oxidation-resistant nickel (Ni) chromium alloys are widely used for the production of critical parts (stationary gas turbines and turbojet engines), which work in extreme environments (from –217° to 700° C). It is renowned that superalloys are difficult metals to shape or machine using conventional production methods (forging, casting, and powder metallurgy) owing to high-speed strain hardening [1, 2]. Currently, additive manufacturing (AM) methods contribute to the expansion of the manufacturing facility, including age-hardenable superalloys. The current results show that the complexly shaped Inconel (IN) components manufactured by the powder bed fusion (PBF) techniques contain good mechanical strength and high-temperature performance, making superalloys an attractive choice for high-temperature applications [3–7]. Consequently, the PBF-fabricated Ni-based alloys are favourable for the aviation [8, 9], aerospace [10], and nuclear [11] applications.

Currently, the powder bed fusion [5, 9, 12–15], direct energy deposition [4, 16–18], binder jetting, or material jetting [19–21] techniques are used and compared to manufacture the Ni-based alloy parts that are impossible or difficult to obtain using conventional manufacturing processes. Generally, it can be concluded that PBF, including the laser powder bed fusion (L-PBF) [23, 24] and electron beam melting (EBM) [25] techniques, is one of the most promising three-dimensional (3D) printing methods to fabricate the high-density metallic components of complex shape. The L-PBF process can sinter much more precisely compared to the EBM process. The density of the L-PBF-fabricated superalloy is about 99.8% [13]. Moreover, the L-PBF technology provides an avenue to produce crack and defect-free parts by controlling the thermo-mechanical conditions [26]. Nevertheless, it is also important to point out that the L-PBF-printed super-alloys still suffer from an inhomogeneous grain microstructure [27–32], rough and different surface [33–36], surface and buck defects [37–40], and the presentence of tensile residual macrostresses in the subsurface layers [41, 42]. The abovementioned manufacturing defects can be eliminated or decreased by the L-PBF parameters optimization combined with the application of the post-processing.

Nowadays, the post-treatment issue is one of the main difficulties to use the L-PBF-built Ni-based alloys with good fatigue, tensile, creep, and rupture strength in a wide range of applications. Consequently, thermal, mechanical, and chemical post-processing techniques are mostly needed to improve the surface finish and microstructure, eliminating the residual porosity and tensile stresses in the Ni-based alloy components manufactured by L-PBF. The use of hot isostatic pressing (HIP) at the heating temperature of 1,120–1,240° C combined with heat treatments (solution annealing at the heating temperature of 930–1,040° C or homogenization at the heating temperature of 1,040–1,200° C [43–46], and double ageing hardening at the heating temperature of ~720° C/~620° C [45, 47]) provides the required stress state, structure, grain size, phase composition, and material density in the L-PBF-built superalloys. The effect of the heat treatment on the microstructural characteristics of the Ni-based alloys is widely investigated in the following works [48–53]. The ageing heat treatment at the final stage provides the precipitation of secondary phases (γ' and γ'') into the metal matrix via heat treatment in the temperature range of 600 to 700° C [54–56]. Thus, the abovementioned heat treatments affect the fatigue performance of the L-PBF-built superalloys, increasing the microstructure stability and reducing the porosity at the HIP treatment and tensile residual stress at the stress relieving and annealing treatment [57].

The surface post-processing is the subsequent/final stage in the additive manufacturing routes. The chemical (polishing and etching) or electrochemical (polishing and plating), mechanical (blasting, peening, tumbling, machining, and polishing) surface treatments, or laser surface modification are selected based on the geometry complexity and the size of product design, operational properties, and surface quality requirements. The surface post-processing of the L-PBF-built metallic parts often requires expensive or enhanced solutions. In particular, the vibratory finishing [58], barrel finishing [59, 60], ultrasonic shot peening [35, 61], sandblasting or shot peening [35, 62–64], water jet shot peening [65], electrochemical polishing [66], and magnetic polishing [67] can be applied to surface post-processing of the complexly shaped and small-sized nickel-based alloy components. To process the large-scale superalloy parts, the shot or cavitation peening, laser shock peening [65, 68–70], ultrasonic impact peening/finishing [35, 71], and ball burnishing [72] can be used in additive manufacturing processes. Recently, various mechanical surface treatments, such as machining [53, 58, 73–77], laser shock peening [68], shot peening (SP) [44, 61, 63, 64], and mono-pin or multi-pin ultrasonic impact treatment (UIT) [44, 61, 71], were combined with heat treatments to enhance the surface properties of the PBF-fabricated Ni-based alloys. The results indicated that mechanical treatments can effectively modify the surface texture, enhance surface hardness and induce compressive residual stresses.

This work aimed to compare the effects of the HIP, SP, UIT, and combined post-treatments on the microstructure, crystallite size, lattice microstrains, phase composition, macrohardness, near-surface microhardness distribution, residual macrostresses, subsurface porosity, surface topography, roughness, and waviness in the IN 718 alloy parts fabricated by the L-PBF technology. Particular attention was paid to the study of the near-surface microstructural features induced by the ultrasonic and shot surface mechanical treatments after heat treatment.

9.2 MATERIAL AND METHODS

9.2.1 Laser powder bed fusion details

The material studied is a commercial-grade Ni-based IN 718 superalloy. The gas-atomized IN 718 powder used with a particle size distribution in a range of 10–55 µm was characterized by predominantly spherical-shaped particles (Table 9.1) [78]. The powder size and morphology were estimated due to scanning electron microscopy (SEM) and image analysis [79]. The morphology of the IN 718 powder is presented in Figure 9.1a. It can be seen that satelliting is also present throughout the powders. The chemical composition of the used IN 718 powder is given in Table 9.1. The powder corresponds to the specific chemical composition of the IN 718 alloy in accordance with AMS 5663/ASTM B637 standard.

The plane parts with dimensions 40 mm × 3.5 mm × 30 mm (X×Y×Z; Z is the build direction (BD), XY is the scan direction (SD)) were manufactured by the laser powder bed fusion (L-PBF) process using an industrial Renishaw AM400 machine (Figure 9.1b). The L-PBF additive manufacturing system

Table 9.1 Nominal chemical composition and particle size of the IN 718 powder

Nominal chemical composition, % by weight

Ni	Cr	Nb	Mo	Ti	Co	Al	Mn	Si	C	Fe
50...55	17...21	4.75...5.5	2.8...3.3	0.65...1.2	≤1.0	0.2...0.8	≤0.35	≤0.35	≤0.08	Balance

Particle size, µm

$D_1 = 10; D_{10} = 17.1; D_{50} = 32.6; D_{90} = 54.8$

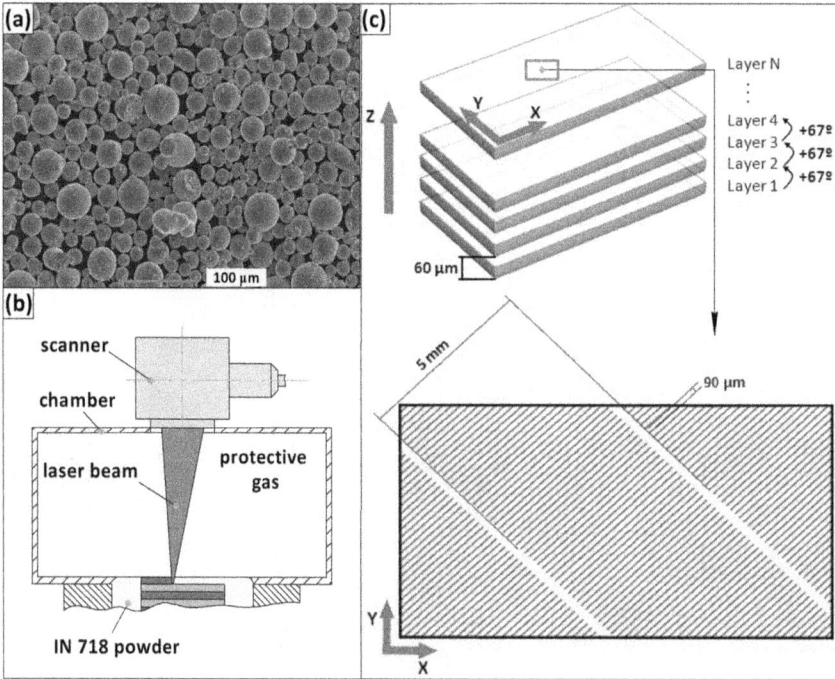

Figure 9.1 SEM micrograph of the pre-alloyed gas-atomized IN 718 powder (a); scheme of the L-PBF process (b); and scheme of the scanning strategy used (c).

consisted of a scanning optics with a maximum scanning speed of 7000 mm/s and ytterbium fibre laser with a maximum power of 400 W. To reduce thermal stresses, the stripe pattern scanning strategy with a rotation angle of 67°, line spacings of 90 µm, and stripe spacings of 5 mm were used (Figure 9.1c). The L-PBF parameters are listed in Table 9.2.

The test parts were manufactured simultaneously in a group of 28 parts with the same dimensions. After L-PBF fabrication, the build was removed from the base plate by an electrical discharge machining (EDM).

9.2.2 Heat post-processing

9.2.2.1 Hot isostatic pressing

After the L-PBF fabrication, the IN 718 alloy parts were heat-treated under different heat treatments shown in Table 9.2. A modern hot isostatic press QIH 0.9 × 1.5-2070-1400 MURC was used for the hot isostatic pressing (HIP) of the laser powder bed fused IN 718 alloy parts taking into account ASTM F3055 standard [61]. The L-PBF-fabricated parts were hot isostatic pressed for 3 hours at an argon pressure of 160 MPa and a heating tempera-ture of 1160° C (Table 9.2). It should also be noted that the forced-convec-tion high-speed cooling of the highly pressurized argon gas was activated

Table 9.2 L-PBF additive manufacturing and post-processing parameters

L-PBF parameters

Laser power, W	Laser spot size, µm	Scanning speed, mm/s	Stripes spacing, mm	Hatch angle, °	Layer thickness, µm
200	70	700	5	67	60

Heat post-processing parameters

Heat treatment conditions: temperature (°C) per time (h)

	HIP	Homogenization	Ageing #1	Ageing #2
HIP	1160/3	-	-	-
HT1	1160/3	-	720/8	620/9
HT2	1160/3	1180/1	720/8	620/9

Mechanical post-processing parameters

UIT parameters

Vibration frequency of ultrasonic horn (kHz)	Vibration amplitude of ultrasonic horn (µm)	Load of ultrasonic tool (N)	Rotation speed of impact head (rpm)	Treatment duration (s)
21.6	18	50	76	120

SP parameters

Stand-off distance between the treated parts and nozzle (mm)	Peening pressure (MPa)	Feed rate of nozzle (mm/s)	Media size (mm)	Treatment duration (s)
180	0.5	10	0.5	60

inside the pressure vessel during the cooling of the L-PBF-built parts to increase the cooling rate in the HIP furnace.

9.2.2.2 Heat post-processing

Subsequently, the HIP-treated IN 718 alloy parts were treated by HIP followed by a two-step ageing (HT1 treatment) and HIP followed by a homogenization and two-step ageing (HT2 treatment) [61]. An SNV 80 vacuum furnace was used to perform the double ageing heat treatment using the recommended conditions. The ageing was conducted at an operating temperature of 720° C for a dwell time of 8 hours and then cooled down to a temperature of 620° C and holding for 9 hours (Table 9.2). The cooling was carried out in the furnace to 80° C in argon. An IPSEN T2T vacuum furnace was used to perform the homogenization heat treatment of the HIP-treated parts for a holding time of 1 hour at the heating temperature of 1180° C (Table 9.2). The IN 718 alloy parts were cooled in the furnace.

9.2.3 Mechanical surface post-processing

9.2.3.1 Ultrasonic impact peening

The surface of the L-PBF-fabricated and heat-treated parts was mechanically modified by ultrasonic impact treatment (UIT) [71]. The used UIT device contained an ultrasonic generator with a frequency of 21.6 kHz and a power output of 0.8 kW (Table 9.2) and the piezoceramic vibration system equipped with a multi-pin impact head. This impact head was positioned on the step-like horn tip and served as a guide for seven cylindrical pins of 5 mm in diameter. The pins produced the high-frequency sliding impacts (1 ± 0.5 kHz) following the impact head forcedly rotated during the UIT treatment (a rotation speed of 76 rpm), providing the lateral component of the load. The acquired kinetic energy of the pins is transferred from the ultrasonic horn tip and then spent on the produced impacts by the treated surface (it is the normal (vertical) component of the load). In this study, the UIT duration was chosen to be 120 s, the amplitude of the ultrasonic horn was 18 μm, and the static load of the ultrasonic tool was 50 N (Table 9.2).

9.2.3.2 Shot peening

After heat treatments, both L-PBF-fabricated and heat-treated parts were also subjected to conventional shot peening (SP). A commercial shot peening system was used to perform the surface severe plastic deformation [60, 80]. The AISI 52100 bearing steel shots of 0.5 mm in diameter was driven perpendicular direction to the part surface by the compressed air pressure of 0.5 MPa through the nozzle moving transversally to this surface. The stand-off distance between the treated part and the nozzle was set to ~180 mm. The SP treatment lasted for 60 s, providing full surface coverage (Table 9.2).

9.2.4 Material examination methods

9.2.4.1 Microstructural characterization

Two cross-sections in two different orientations were mechanically cut from the L-PBF-built and post-processed parts, i.e. a vertical section perpendicular to the fused layers and a horizontal section parallel to the fused layers. Then, standard metallographic samples were prepared to perform etching and microstructural analysis using a Leica MEF4A optical microscope equipped with a digital camera. The Image J152 software was used for statistical analysis of residual porosity in the unetched specimens at the depth from the surface of ~100 μm.

Further, a Lucas' reagent consisting of lactic acid of 50 ml, hydrochloric acid of 150 ml, and oxalic acid of 3 g was used for electrolytic etching of the samples (the voltage and current were 2.5 A and 2.5 V, respectively, and duration of 5–10 s). A LEICA Reichert Polyvar 2 light optical microscope (LOM) was used for observations of the bulk and surface microstructures. An XL 30 ESEM scanning electron microscope (SEM) was used for the microstructure analysis at higher magnification and energy-dispersive X-ray spectroscopy (EDS) detector was employed for revealing the elemental chemical composition.

A Rigaku Ultima IV diffractometer with a *CuKα*-radiation (the voltage of 30 kV, current of 30 mA, and scanning speed of 2°/min) was used for the X-ray diffraction (XRD) analysis in the surface layer of about 3.5–8.5 μm thick. To complete the XRD peak analysis by applying Gaussian fitting and peak deconvolution, an OriginPro 9.0 software for Windows was used.

A Scherrer's equation expresses the physical broadening of the peak (β) through the crystallite/grain size (D), the lattice microstrains (η), the wavelength (λ) of X-ray radiation, and the angle of diffraction (θ), were applied [81]:

$$\beta = \left(K\lambda / D\cos\theta \right) + \eta \tan\theta \tag{9.1}$$

where $K \approx 0.9$ is the constant, $\beta = (B^2 - b^2)^{1/2}$, B is the FWHM obtained from the modified layer, and b is the FWHM of instrumental broadening. The full widths at half maximum (FWHM) of the (220) reflections were analyzed.

The residual macrostresses σ_{x+y} in the γ solid solution in the subsurface layers of the surface-modified specimens were estimated using the well-known formula $\sigma_{x+y} = -E\Delta d/\mu d_0$, accounting for the γ lattice spacing d_0 in the L-PBF or HT1/HT2 conditions and its change Δd after SP/UIT, as well as the mechanical properties of IN718 alloy at room temperature, i.e. Young's modulus E = 200 GPa and Poisson's ratio μ = 0.29 for IN 718 alloy at room temperature. The change in the lattice spacing Δd was assessed based on a shift in the diffraction sub-maximums (220) of the γ phase from their angular positions observed for the undeformed samples.

9.2.4.2 Hardness measurements

A hardness tester DIA TESTOR 2Rc operated at a load of 10 kgf (HV_{10}) was used to register the surface macrohardness of each specimen (the average of 5–7 measurements were reported). The data scatter was less than 2–5%. A tester Leica VMHT with a Vickers indenter loaded by 0.025 kg ($HV_{0.025}$) for 15 s was used to observe the microhardness depth profiles in the building direction cross-section of the specimens starting from the depth of 15–20 µm from the top surface. The standard deviation for these measurements was ±1 HV.

9.2.4.3 Topography observation

A 3D Taylor Hobson Form Talysurf 120 tester was used to assess the surface texture, roughness, and waviness parameters. The roughness (Ra) and waviness (Wa) profile parameters were selected to estimate the arithmetical mean height for surface roughness and waviness profiles along the sampling length of 0.8 mm, respectively. The surface roughness and waviness profile parameters were evaluated and averaged in both X and Y axes in accordance with ISO 25178 standard. The area roughness parameters, such as Sa (the arithmetical mean height of a line to the 3D surface area), Sp (the height of the highest peak within the defined area), Sv (the absolute value of the height of the largest pit/valley within the defined area), and Sz (the maximum profile height of a line to the 3D surface area), were analyzed in accordance with ISO 4287 standard.

9.3 RESULTS AND DISCUSSION

Post-processing of the additively manufactured materials is naturally directed on the mechanical properties' enhancement. On the one hand, considering the IN 718 alloy, a twofold goal of the post-processing procedure is usually pursued, namely, the thermal post-processing allows reaching the precipitation strengthening, and the mechanical post-processing (the surface finishing) permits eliminating the surface defects, subsurface porosity, surface roughness, and texture. On the other hand, post-processing results can be considered in two following aspects: (i) microstructural changes comprising grain structure and phase composition; and (ii) the surface-related characteristics and residual porosity affecting the quantity of the surface/bulk stress risers. Both aspects are responsible for the mechanical properties and operation life of the material. These aspects are considered in detail in the following sub-sections.

9.3.1 Microstructure examination

The effects of thermo-mechanical (HIP) and thermal (HT1/HT2) post-processes on the microstructural changes in the L-PBF-fabricated IN 718

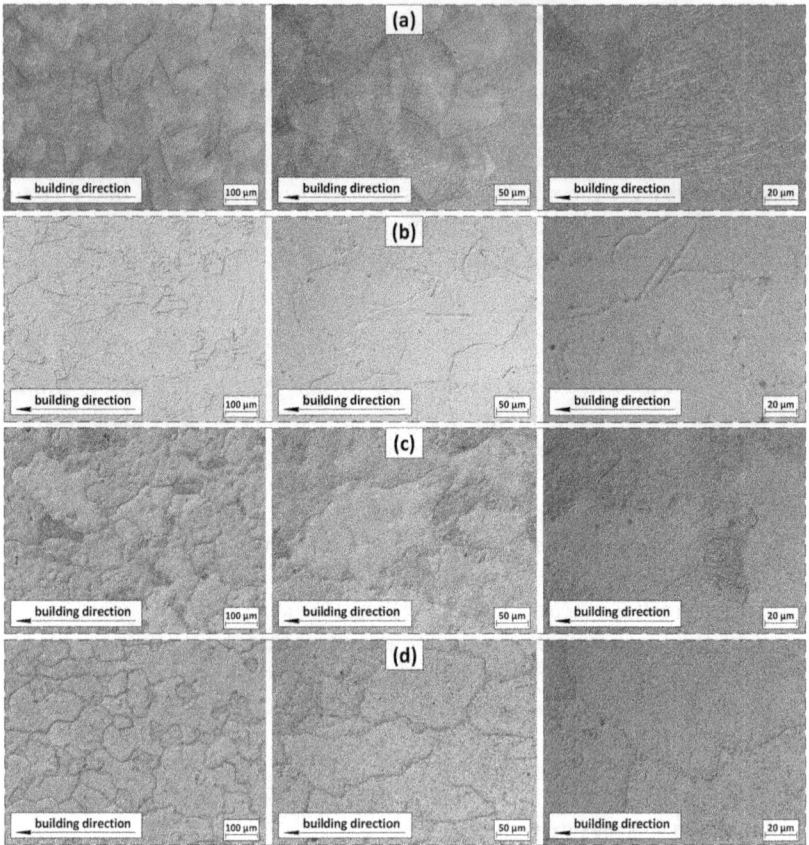

Figure 9.2 LOM micrographs of microstructure in the L-PBF-fabricated (a) and HIP (b), HT1 (c) and HT2-treated (d) IN 718 alloy parts.

alloy samples were analyzed by LOM (Figure 9.2) and XRD analysis (Figure 9.3). As usual, the microstructure of the L-PBF-manufactured alloy contains the overlapped semi-circular melt pool patterns. These patterns are formed by the elongated colonies of cells epitaxially grown across multiple layers from the substrate in the heat flow direction (Figure 9.2a).

Application of the HIP (Figure 9.2a), HT1 (Figure 9.2c), and HT2 (Figure 9.2d) thermal post-treatments results in the formation of secondary phases fixing the grain boundaries and stimulating the equiaxed/homogeneous morphology of the grain microstructure that also contains some portion of the annealing twins. The microstructures formed after HT1 and HT2 post-processes are almost similar except for the grain size of γ-matrix, which appears slightly larger after HT2 as compared to that in the HT1 condition. A high enough temperature (1,160–1,180° C) used at the HIP and HT1 treatments facilitate the dissolution of Nb-rich Laves and δ phases while

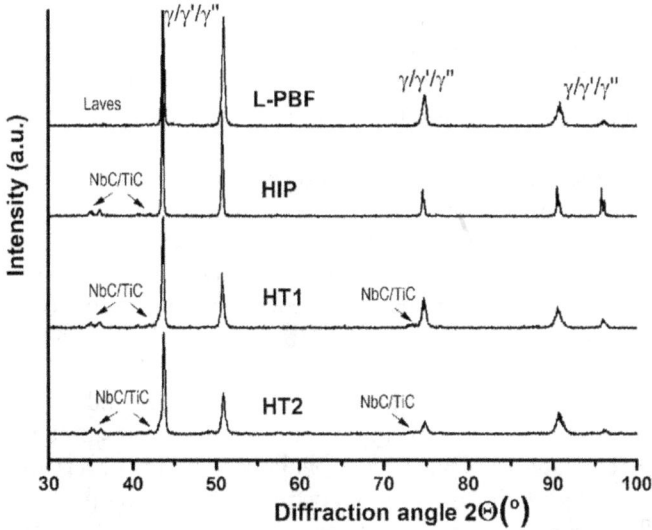

Figure 9.3 XRD patterns of the L-PBF-fabricated IN 718 alloy parts in the as-built, HIP-processed and heat-treated (HT1 or HT2) conditions.

some portion of niobium/titanium carbides still retains to be concentrated on the grain boundaries. This microstructural state correlates well with the literature reports [16, 46, 55, 56, 60, 82] and our XRD data (Figure 9.3).

Figure 9.3 shows the X-ray diffraction (XRD) patterns of the L-PBF-built IN 718 alloy specimens in the as-fabricated, HIP-processed, and heat-treated (HT1 or HT2) conditions. These patterns demonstrate that the main peaks belonging to fcc solid solution (γ) and accompanying γ' (Ni_3(Ti, Al) and γ'' (Ni_3Nb)) phases only slightly changed their intensity and widths. The main difference relates to the HIP and HT1/HT2-induced dissolution of the Laves and δ phases formed at the L-PBF manufacturing. Appeared additional XRD peaks related to NbC/TiC carbides, which dissolution temperatures are higher than those applied at the HIP (1,160°C) [83]. Evidently, the XRD peaks of the γ (γ', γ'') phase are noticeably thinner for the HIP-processed sample. It is well explainable. Indeed, the L-PBF-fabricated parts that are well-known to usually contain cellular structure (dendritic/interdendritic areas) with residual defects (pores, dislocations, and twins) and residual stresses due to relatively rapid solidification and cyclic reheating of the progressively added layers [16, 83, 84], i.e. the XRD peaks are significantly broadened. Conversely, the following HIP allows for eliminating most of the L-PBF-formed defects and stresses and thus resulting in thinner XRD peaks. Additionally, the L-PBF-formed dendritic cellular microstructure is naturally transformed into homogeneous/equiaxed grain microstructure with precipitations of secondary phases appearing along the grain boundaries (Figure 9.2). This observation is in line with the literature data regarding the

HIP-induced changes in the phase composition of IN 718 alloy observed by neutron diffraction [85]. Subsequent heat treatments containing two-step annealing (HT1) and homogenization followed by two-step annealing (HT2) led to the precipitation of fine γ' and γ'' strengthening particles into the γ-phase matrix.

The XRD analysis was also used to reveal the changes in the crystalline size, lattice microstrains, phase composition, and residual macrostresses in more detail. For this purpose, the (220) $\gamma/\gamma'/\gamma''$ XRD peak analysis and deconvolution were performed for all the studied specimens [55, 60]. Figures 9.4–9.6 show the XRD pattern fragments with the deconvoluted (220) $\gamma/\gamma'/\gamma''$ XRD peaks registered for the HIP-treated, HT1/HT2-treated, and surface-modified specimens. Considering the increase in the integral intensity of the γ'/γ'' sub-peaks after HT1/HT2 it can be concluded that the volume fractions of fine γ'/γ'' precipitates are increased as compared to the HIP-processed state (see Figures 9.4a, 9.5a, 9.6a). This observation is in a good agreement with the literature reporting the phase composition changes in the HIP-processed IN 718 alloy observed by neutron diffraction [84] and a "time–temperature–transformation" diagram of IN 718 alloy [71, 84–87]. Both HT1/HT2 thermal post-processes resulted in a shift of the (220) γ sub-peak to larger diffraction angles in comparison with the position after HIP. This shift can be explained well by the ageing-promoted precipitation of the phases (γ' ($Ni_3(Al,Ti)$), γ'' (Ni_3Nb), and $(Nb,Ti)C$ carbides) with a high Nb content that is naturally accompanied by the Nb depletion of the γ solid solution (the decrease in the γ lattice parameter). Additionally, the higher content of more uniformly distributed Nb atoms would facilitate the formation of the higher volume fractions of the smaller γ'' and γ' precipitates [44, 71, 83, 84, 86]. Experimental data presented in Figures 9.4–9.6 also confirm that the broader sub-peaks of the γ'' and γ' phases were observed after HT2 than those after HT1, i.e. γ'/γ'' precipitates might be smaller in size in the HT2 condition. After annealing, the increased volume fractions of the NbC/TiC carbides are also seen. Moreover, this increase is higher for the samples after the HT2 process (Figure 9.6) in comparison with that observed in the HT1 condition (Figure 9.5). Additionally, a good correlation is seen to the earlier reported phase composition and NbC/TiC carbides observations of the forged Inconel 718 alloy [88].

The above-discussed microstructure changes can be observed in Figure 9.7 demonstrating the SEM images of the microstructure of the L-PBF-built, HIP, and SP/UIT post-processed IN 718 alloy.

On the contrary to the HT1/HT2 heat post-processing, the mechanical surface post-processing by SP or UIT has a much lower influence on the precipitate's quantity. The sub-peaks obtained for the precipitating phases are also broadened enough (Figures 9.5 and 9.6). However, in the case of the γ solid solution, this broadening is naturally related to the applied plastic deformation.

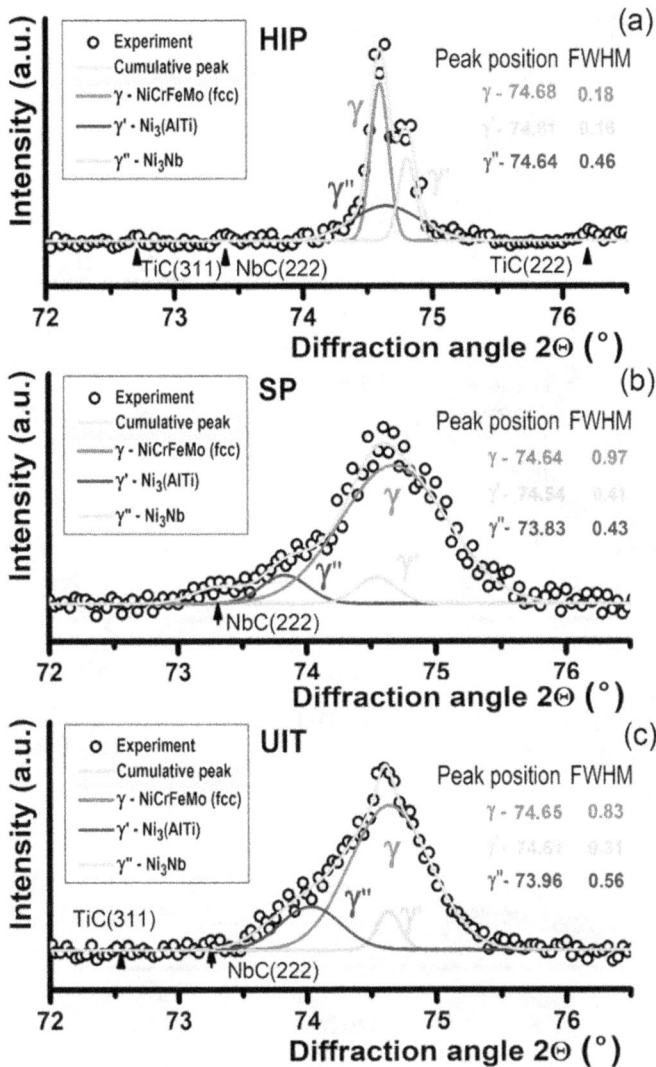

Figure 9.4 Deconvolution of (220) XRD peak of the LPBF-fabricated and HIP (a), SP (b), and UIT-treated (c) IN 718 alloy parts.

The effects of the deformation post-processing by SP and UIT on the L-PBF-built (Figure 9.4), HT1-processed (Figure 9.5), and HT2-processed (Figure 9.6) samples are evident. Severe plastic deformation induced by the SP or UIT processes resulted in a significant broadening of the XRD peaks of the γ solid solution and their shift towards the lower diffraction angles regardless of the initial state of the surface-finished samples. Normally, the physical widths of the XRD peaks relate to the crystalline size/lattice

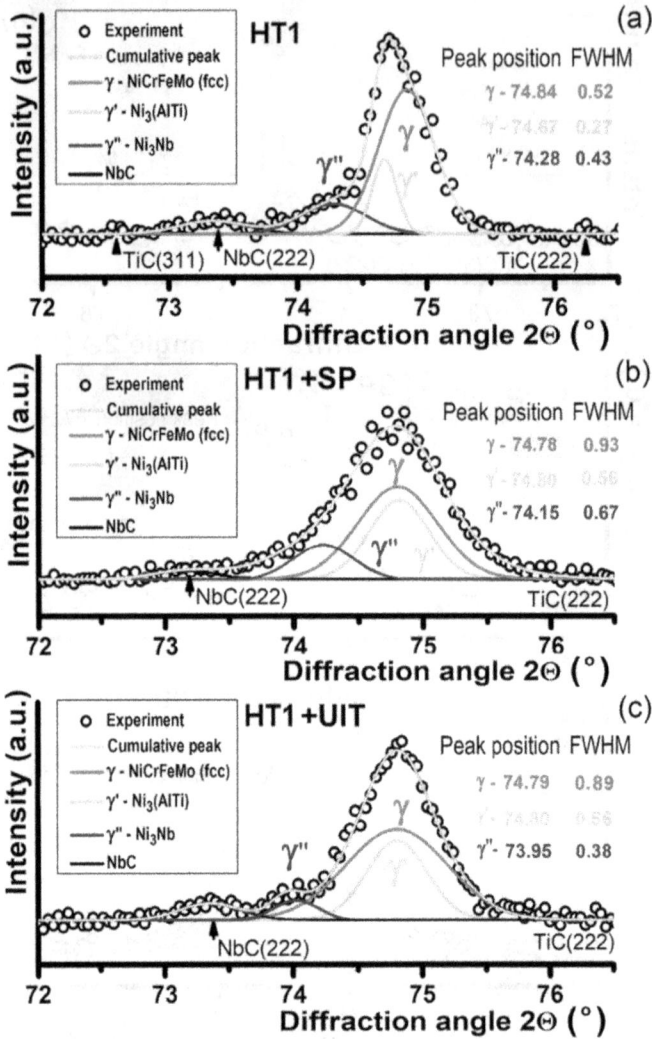

Figure 9.5 Deconvolution of (220) XRD peak of the L-PBF-fabricated and HT1 (a), HT1+SP (b), HT1+UIT-treated (c) IN 718 alloy parts.

microstrains and their angular positions allow making conclusions regarding the residual stresses formed. Analyzing the γ sub-peaks one can characterize the γ matrix solid solution.

Such analysis was performed for the cases of the SP-finished (Figure 9.8a) and UIT-finished (Figure 9.8b) surfaces. As shown in Figure 9.8, the cell/crystallite size of the L-PBF-fabricated samples becomes different after the heating to a high temperature during HIP and subsequent heat treatments. The cell/crystallite size demonstrates some decrease during single HIP

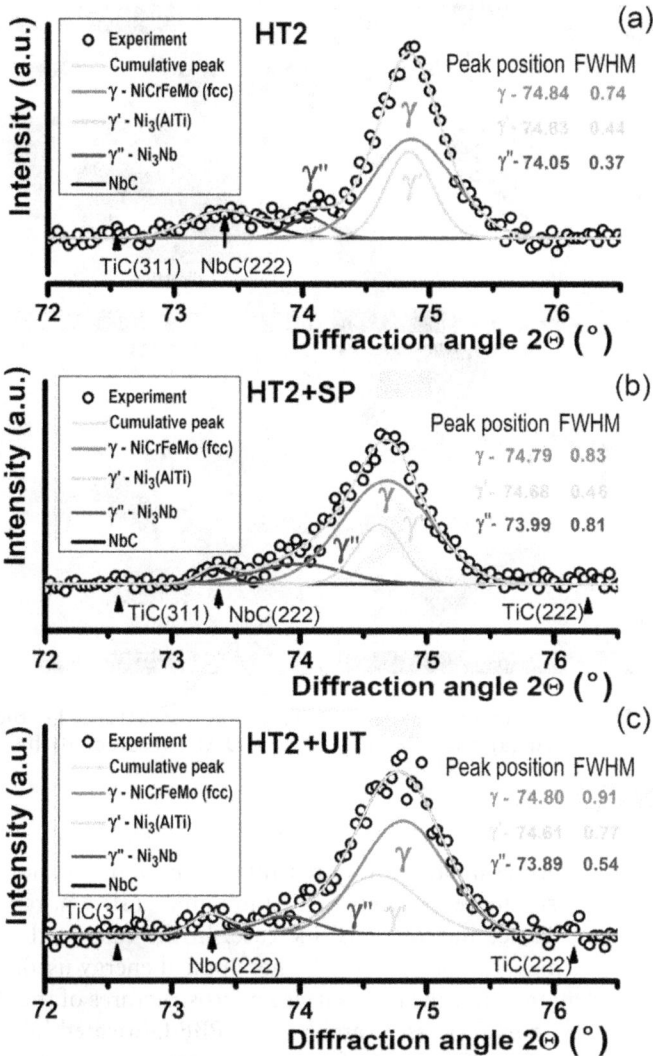

Figure 9.6 Deconvolution of (220) XRD peak of the L-PBF-fabricated and HT2 (a), HT2+SP (b), HT2+UIT-treated (c) IN 718 alloy parts.

because of the applied compression, while it grows during following HT1/ HT2 post-processes although at different rates. The HT2 post-process containing homogenization and annealing led to the formation of larger crystallites. Conversely, the lattice microstrains behave in the opposite way: the microstrains relieving occurs at the HIP that leads to dislocation annihilation, while the following HT1/HT2 post-processes inducing the precipitates-matrix misfit cause a rather high increase in the lattice strain. Additionally, severe plastic deformation at following mechanical post-processing by SP or

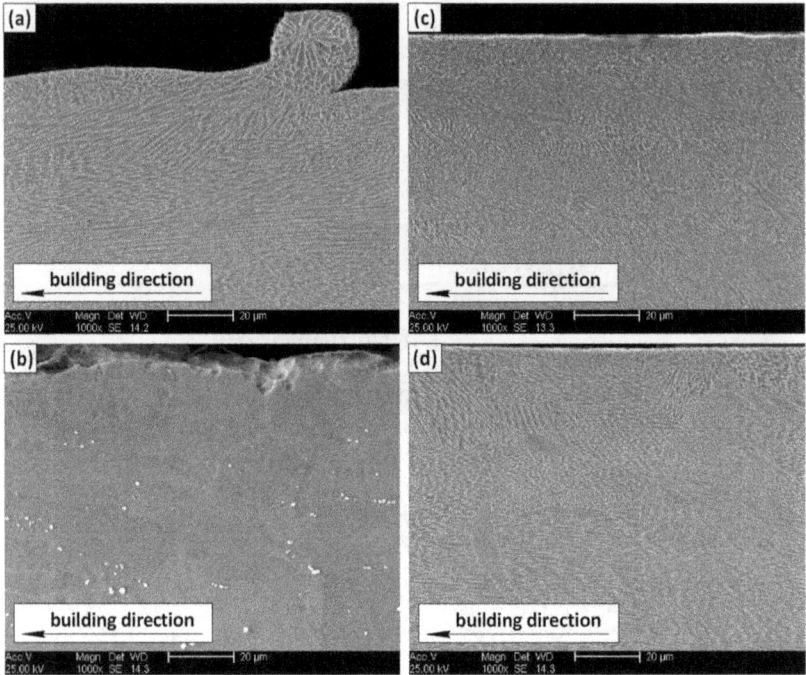

Figure 9.7 SEM micrographs of near-surface microstructure in the L-PBF-fabricated (a) and HIP (b), SP (c) and UIT-treated (d) IN 718 alloy parts.

UIT results in a further multiplication/rearrangement of dislocations leading to significant crystallite refinement (to ~80 nm and ~110 nm, respectively) and a moderate lattice microstrains' increase. In the case of the SP used, these effects are higher because of the higher applied energy used.

Figure 9.9 demonstrates the near-surface microstructures of the thermally and mechanically post-processed samples of L-PBF-fabricated IN 718 alloy. The SEM images of typical near-surface microstructures are in line with the XRD analysis results. They also confirm the HIP/HT1/HT2 induced formation of the homogeneous grain structure instead of the dendritic grain structure observed in the L-PBF-built sample, and the grain boundaries become fixed by precipitations and carbides (Figure 9.9a). Additionally, there are visible effects of the SP (Figure 9.9e) and UIT (Figure 9.9d) post-processing on the grains' shape in the subsurface layers: their thickness decreases smaller as the grains acquire the plate-like shape elongated along the shot-peened or ultrasonically treated surfaces.

Besides, the sub-peaks of the γ solid solutions in the samples mechanically post-processed by SP or UIT change their angular positions (Figures 9.5 and 9.6). The shift of the XRD peak to the lower angles in the diffraction pattern

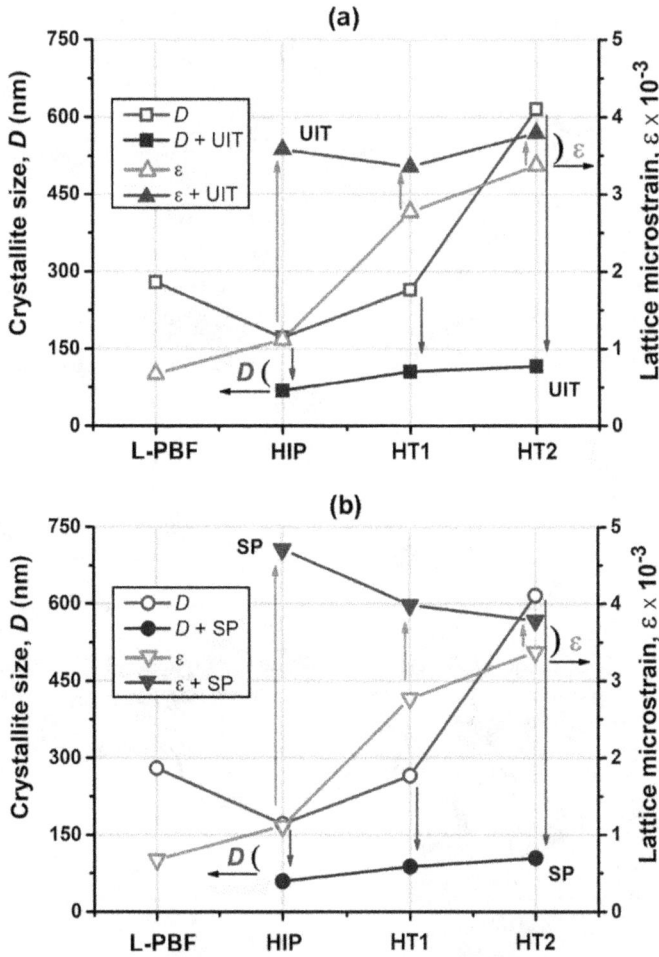

Figure 9.8 Surface crystallite size of the L-PBF-fabricated IN 718 alloy parts in the as-built, HIP-processed and heat-treated (HT1 or HT2) conditions.

is known, and can be induced by compressive residual stresses formed. This shall be addressed in the next section.

9.3.2 Residual stress and hardness

Notice that the assessment of the residual stresses in the case of such a complex system as the γ solid solution with variated Nb content using the analysis of the shift of the XRD peaks should be used with some precautions. Indeed, the heat treatments (HT1/HT2) used can also result in the shift of the XRD peaks due to the alters in the phase composition of the studied alloy. The dissolution of the Nb-rich phases (Laves or δ phases)

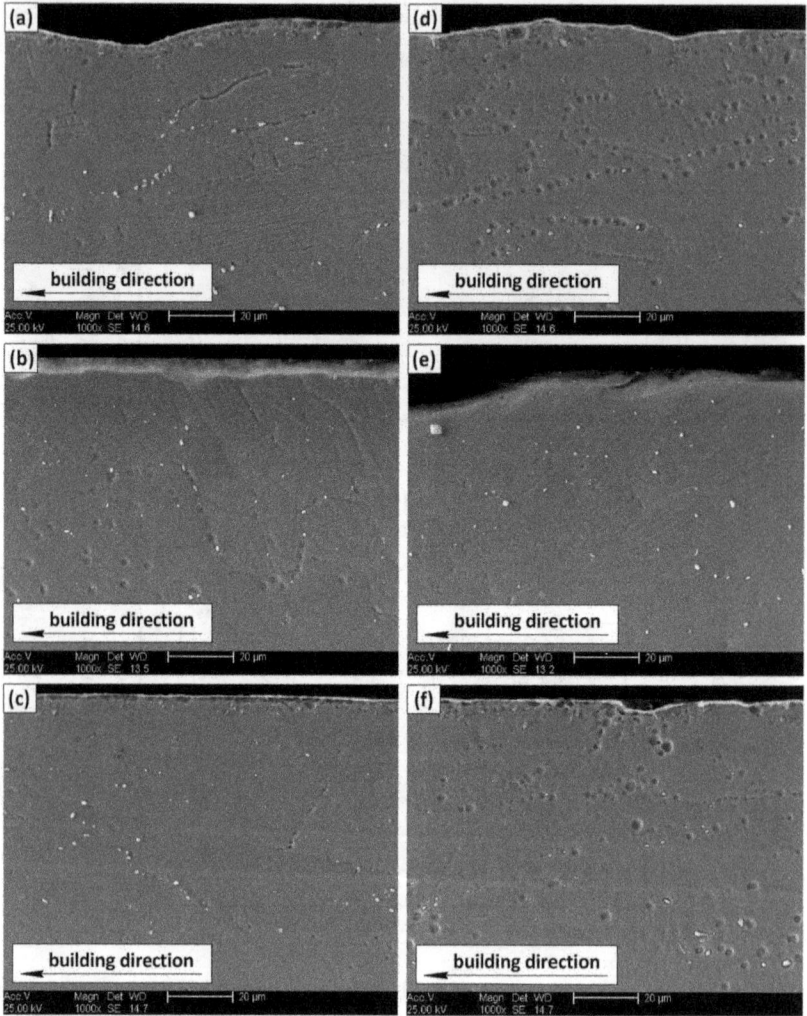

Figure 9.9 SEM micrographs of near-surface microstructure in the post-processed L-PBF-fabricated IN 718 alloy parts by HTI (a), HTI+SP (b), HTI+UIT (c), HT2 (d), HT2+SP (e), HT2+UIT (f).

accompanied by the Nb-enrichment of the γ solid solution would shift the XRD peaks towards lower angles and, vice-versa, the Nb depletion of the γ solid solution owing to precipitation of the Nb-rich γ'/γ'' phases would cause the reverse to shift the XRD peaks.

However, in our case, the probability of the deformation-induced redistribution of Nb atoms is rather low owing to the relatively high strength/low plasticity of the HT1/HT2 post-processed samples. Therefore, the observed peak shift and appropriate compression stresses are caused by the

UIT/SP-induced surface severe plastic deformation. Magnitudes of these strain-induced compressive residual macrostresses σ_{x+y} formed in the L-PBF-built and HT1/HT2 samples after plastic deformation were evaluated using the determined shift of the XRD peaks registered for the mechanically-finished surfaces in comparison with the HT1/HT2 ones. The obtained results shown in Figure 9.10 correspond to the registered increase in the surface macrohardness (Figure 9.11) and microhardness (Figure 9.12) of the

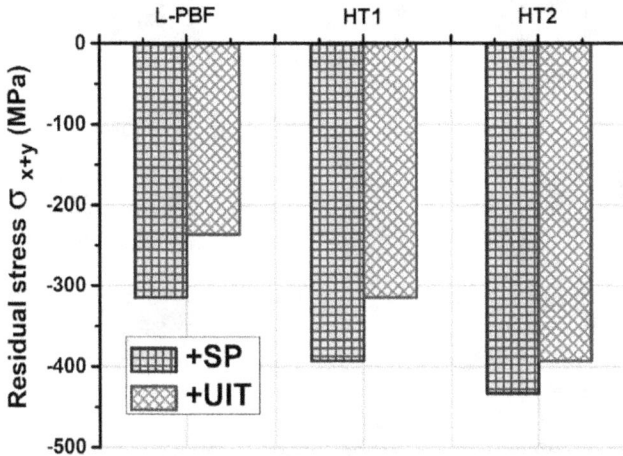

Figure 9.10 Residual macrostresses formed in the L-PBF-fabricated IN 718 alloy parts after mechanical post-processing by SP and UIT.

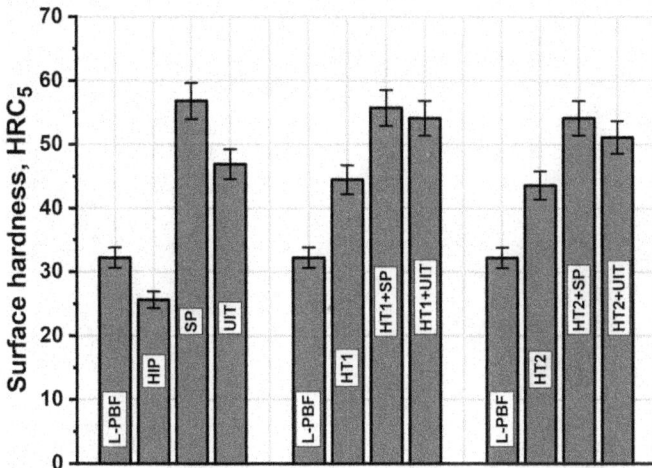

Figure 9.11 Surface macrohardness HRC$_5$ of the L-PBF-fabricated IN 718 alloy parts under different post-processing conditions.

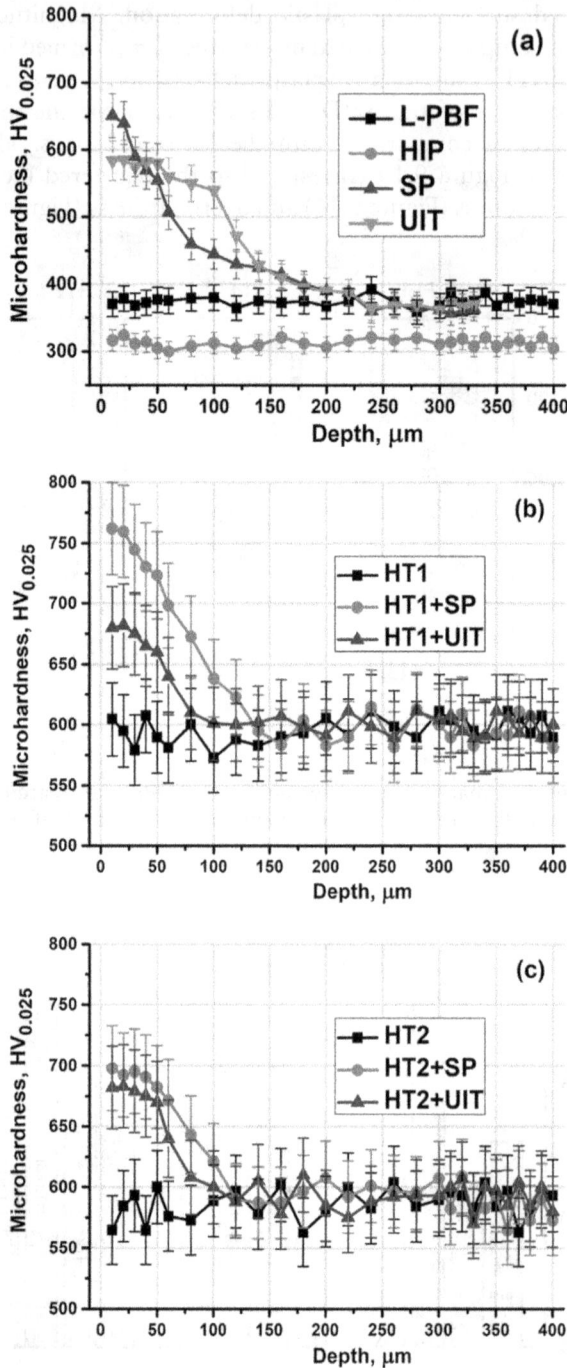

Figure 9.12 Microhardness depth profiles of the L-PBF-fabricated IN 718 alloy parts under different post-processing conditions.

near-surface layers after SP/UIT-induced deformation considered in the following section.

The magnitudes of HRC_5 are always higher for the specimens' surface-finished by SP than those obtained after the UIT process (Figure 9.11). Again, it is related to the energy applied at SP and UIT which was higher for the SP process in this study. This conclusion also correlates to the results of the comparative investigation of the surface finishing of the L-PBF-fabricated samples of IN 718 alloy using various mechanical post-processing methods that showed a direct dependence of the surface compressive residual macrostresses, hardness, and roughness on the mechanical energy accumulated in the modified near-surface layer [89].

The depth distribution of microhardness in the subsurface layers of the L-PBF-fabricated, heat-processed, and mechanically surface-finished are shown in Figure 9.12. Compared with the L-PBF condition, HIP led to some softening of the material while both UIT and SP processes resulted in the strain hardening of the near-surface layers of about 150–180 μm thick (Figure 9.12a).

Application of both HT1 and HT2 thermal post-processes led to similar precipitation strengthening of the IN718 alloy (Figure 9.12b, c) – the near-surface microhardness of the L-PBF-fabricated sample was enlarged by ~50%. Subsequent severe plastic deformation by SP or UIT further increases the surface macrohardness (see HRC_5 data in Figure 9.11) and near-surface microhardness (Figure 9.12b, c). The thickness of the strain-hardened near-surface layers is slightly larger in the case of the SP process than that after UIT, which is in good correlation with the aforementioned higher level of the mechanical energy accumulated at SP [89]. The thickness of the SP and UIT-produced layers registered for the HT1 processed samples, respectively are 150 and 100 μm, and for the HT2 processed samples, they respectively are 120 and 100 μm. The strain-induced hardening of the surface is associated with the sub-grain formation and an increase in the dislocation density and their rearrangement [44, 60, 63, 64, 71].

Slightly lower hardness magnitudes and slightly thinner hardened near-surface layers of the UIT/SP-peened samples underwent prior to the HP2 treatment. This is due to the smaller volume fractions and precipitated phases' size, as well as the larger grain size of the γ-phase matrix in the HP2-treated parts. As a consequence, these features can influence the rate of hardening during plastic deformation at UIT/SP mechanical post-processing.

9.3.3 Porosity

Along with the residual stresses, phase state, and microstructure, the residual surface and bulk defects, especially in the near-surface layers, are also of special importance for the corrosion/wear and fatigue resistance of the L-PBF-fabricated Ni-based alloys.

Cross-sectional LOM/SEM images of the subsurface residual porosity in the L-PBF-fabricated and post-treated IN 718 alloy samples are presented in Figures 9.13 and 9.14. The residual porosity near the surface of the samples was statically analyzed and compared in the build direction (Figure 9.15). The bulk and near-surface porosity in the L-PBF-fabricated specimens can be observed in Figures 9.13a–c and 9.14a, respectively. The aspherical pores and spherical pores are visible in Figure 9.13b. It is well-known that the gas/ spherical pores are caused by the incomplete release of the shielding gas from the melt during the laser powder bed fusion while the aspherical pores are located between the pre-fused layers and scanning passes owing to incomplete melting of the powder layer before the solidification stage [38, 90, 91]. The bulk defects varied in size from 1 to 60 μm in the L-PBF-fabricated parts (Figure 9.13c). The average residual porosity was ~0.3% in the near-surface layer of the L-PBF-fabricated part (Figure 9.15), indicating the risk of having multiple stress raisers that can initiate crack initiation sites.

The HIP treatment significantly reduced the bulk defects in the L-PBF-fabricated parts, providing a material density of ~99.987% (Figures 9.13d and 9.15). Several pores were detected in the HIP-processed specimen (Figure 9.13e). These observations correlate well with the literature data [43, 44, 62]. Tillmann et al. [92] reported that all bulk defects cannot completely be eliminated by HIT owing to the trapped shielding gas during the L-PBF manufacturing. The residual porosity is almost unchanged after both HT1 and HT2 heat treatments applied after HIP (Figures 9.14 and 9.15). It seems that the high-temperature heating during the homogenization led to a further increase in both grain and defects in the case of the HT2-processed specimen as compared to the HT1-processed specimen. Nevertheless, the averaged residual porosity varied in the range of 0.012–0.014% for the HIP, HT1, and HT2 (Figure 9.15).

The applied mechanical surface treatments lead to a further decrement of the near-surface porosity in both pre-heated and post-heated specimens (Figure 9.15). The subsurface porosity in the as-fabricated parts was decreased from ~0.3% to ~0.05% after surface plastic deformation by SP and UIT. It is also important to note that the UIT and SP processes slightly enlarged the density of the material (~99.99%) in the subsurface layer in comparison with the HT1 and HT2 post-treated specimens. On the other hand, the surface defects, such as balls, partially melted powder particles, and signs of the laser passes were eliminated from surface-modified specimens by SP and UIT (Figures 9.7, 9.14 and 9.16).

9.3.4 Surface texture

Figure 9.16 shows the surface textures of the L-PBF-fabricated IN 718 alloy parts in the as-fabricated, heat-treated, and thermo-mechanically post-processed conditions. The surface textures of the L-PBF-fabricated and

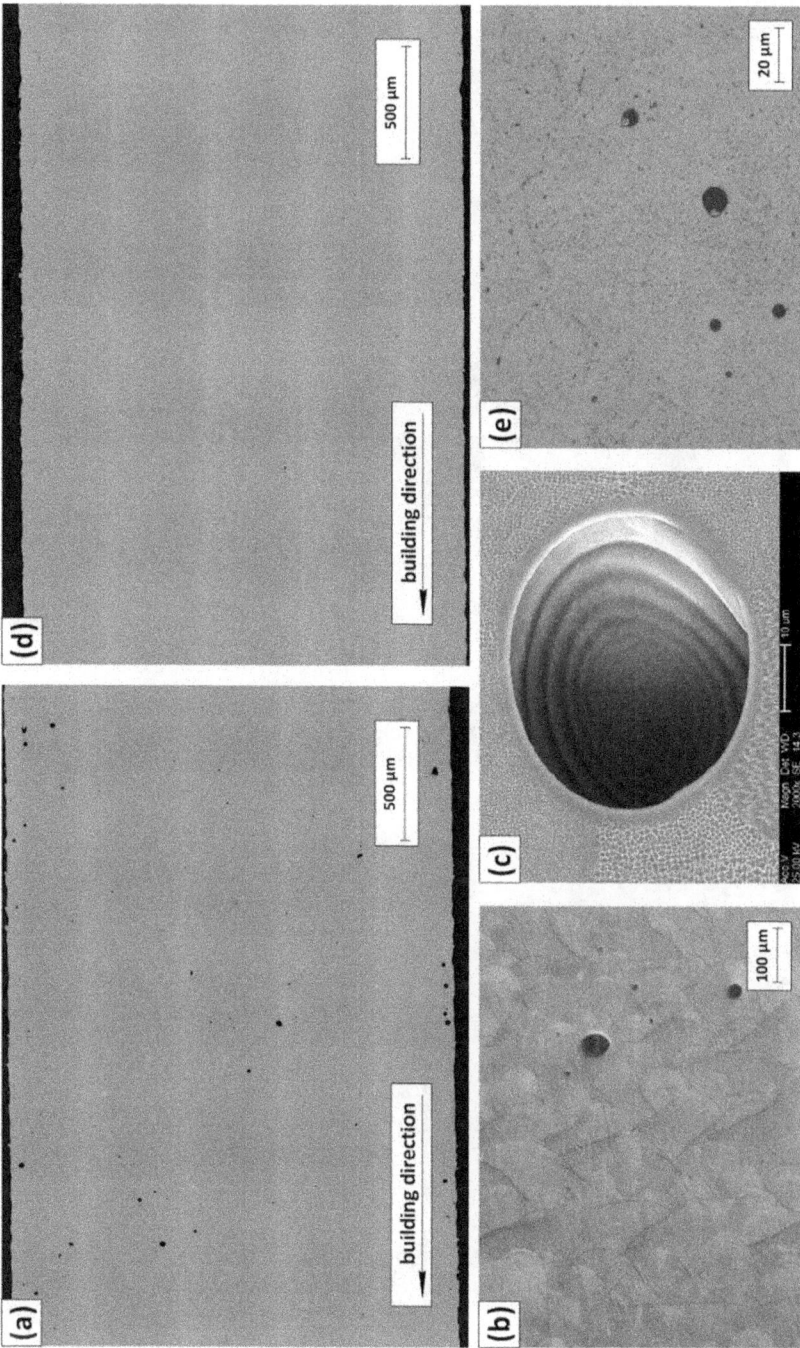

Figure 9.13 LOM micrographs of residual porosity in the L-PBF-fabricated (a–c) and HIP-processed (d, e) IN 718 alloy parts.

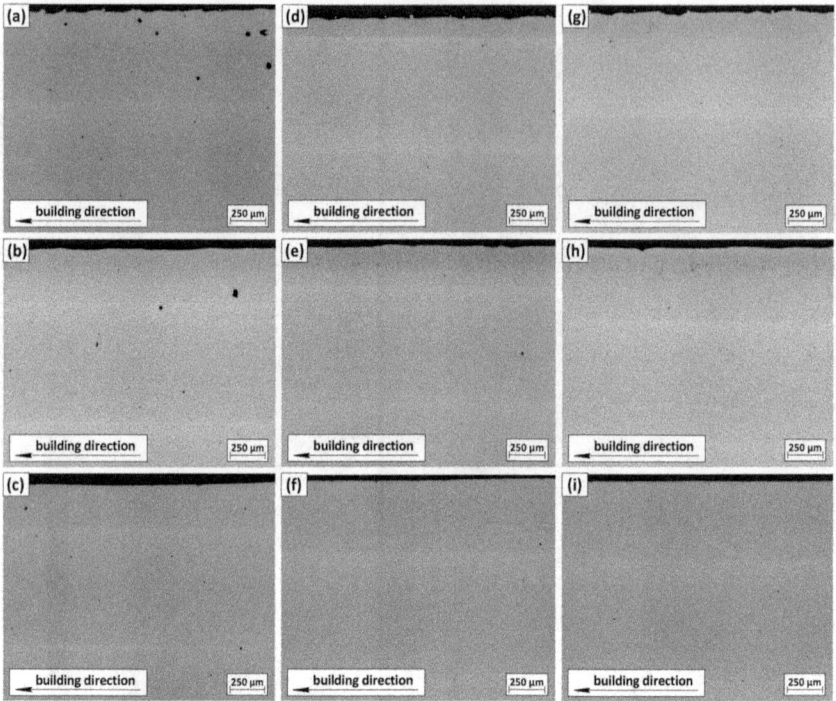

Figure 9.14 LOM micrographs of subsurface residual porosity in the L-PBF-fabricated (a) and SP (b), UIT (c), HT1 (d), HT1+SP (e), HT1+UIT (f), HT2 (g), HT2+SP (h), HT2+UIT-treated (i) IN 718 alloy parts.

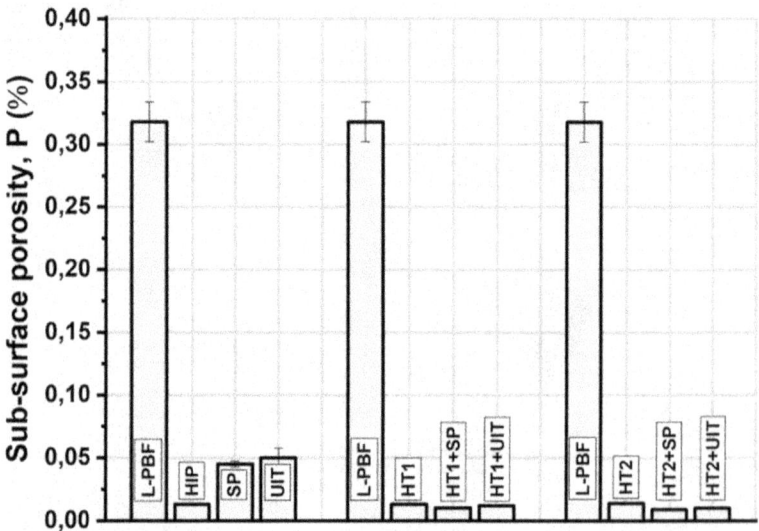

Figure 9.15 Subsurface residual porosity of the L-PBF-fabricated IN 718 alloy parts under different post-processing conditions.

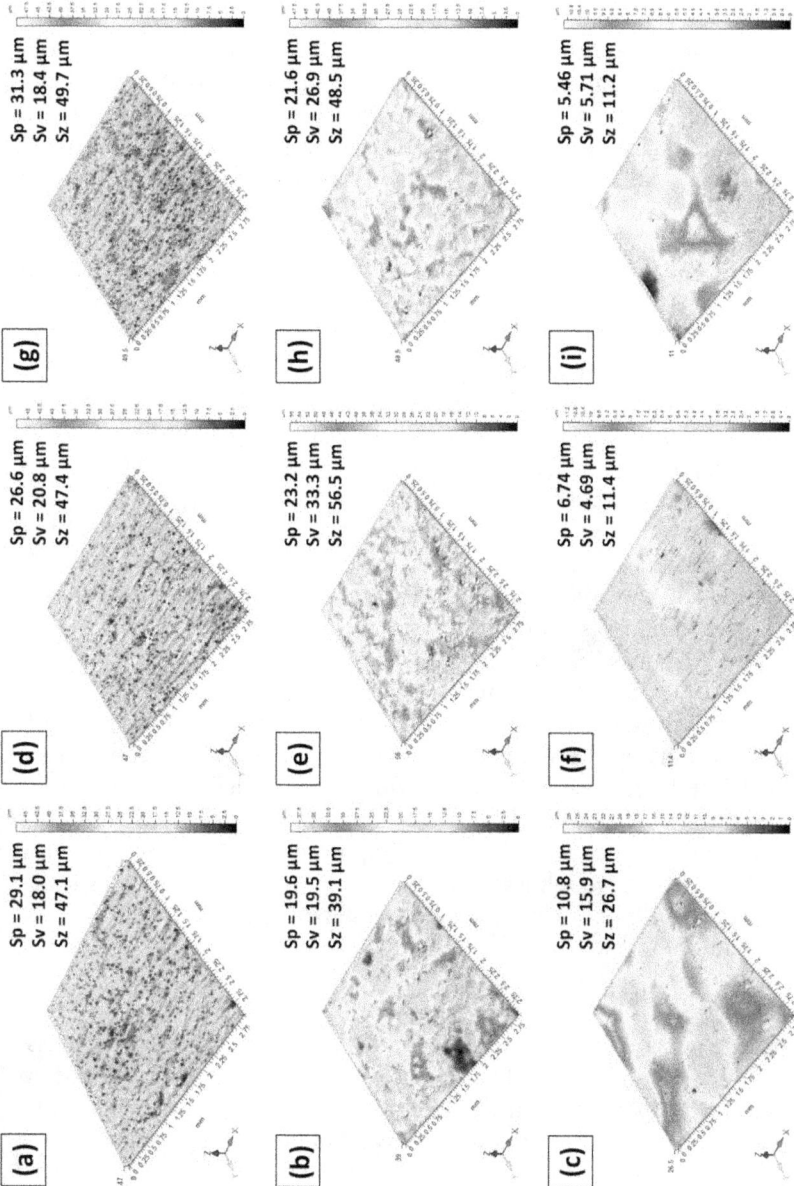

Sp = 31.3 μm
Sv = 18.4 μm
Sz = 49.7 μm

(g)

Sp = 21.6 μm
Sv = 26.9 μm
Sz = 48.5 μm

(h)

Sp = 5.46 μm
Sv = 5.71 μm
Sz = 11.2 μm

(i)

Sp = 26.6 μm
Sv = 20.8 μm
Sz = 47.4 μm

(d)

Sp = 23.2 μm
Sv = 33.3 μm
Sz = 56.5 μm

(e)

Sp = 6.74 μm
Sv = 4.69 μm
Sz = 11.4 μm

(f)

Sp = 29.1 μm
Sv = 18.0 μm
Sz = 47.1 μm

(a)

Sp = 19.6 μm
Sv = 19.5 μm
Sz = 39.1 μm

(b)

Sp = 10.8 μm
Sv = 15.9 μm
Sz = 26.7 μm

(c)

Figure 9.16 Surface topography in the L-PBF-fabricated (a) and SP (b), UIT (c), HT1 (d), HT1+SP (e), HT1+UIT (f), HT2 (g), HT2+SP (h), HT2+UIT-treated (i) IN 718 alloy parts.

heat-treated specimens are characterized by multidirectional inequalities with the area roughness Sp parameter of ~30 µm owing to various manufacturing defects (Figure 9.16a,d,g). The signs of the laser passes, shrinkage cavities, spattering powder particles or partially melted powder particles, and open pores are usually formed on the surface [43]. The HIP and HT1/HT2 heat treatments did no effect on the surface texture of the L PBF-fabricated parts (the maximum profile height Sz parameter ~ 48 µm) (Figure 9.16). The arithmetical mean height Sa parameter in the L-PBF-fabricated and heat-treated parts was varied in a range of 3.73–4.2 µm.

Application of SP and UIT treatments before and after heat treatments resulted in the formation of a new surface texture, regardless of the heat treatment used (Figure 9.16). The uniformly textured surface of the SP-peened specimens is characterized by irregular peaks and dimples formed due to the collision of hard shots. The surface texture is more regular and flattened after UIT (Figures 9.16i and 9.16c). As a consequence, the Sz parameter in the pre-heat processing specimen was slightly decreased by the SP technique (Figure 9.16b) in comparison with the UIT-peened specimen (Figure 9.16c). Compared to the HT1+SP and HT2+SP post-processed specimens (Figures 9.16h and 9.16i), the HT1/HT2 followed by UIT led to a reduction of the Sz parameter from ~50 µm to ~10 µm (Figures 9.16f and 9.16i). Moreover, the pre-treated surface by HT1 and HT2 promotes better UIT surface modification due to larger hardness. In contrast to the UIT-treated specimen, the area roughness Sz parameter was reduced by double after the heat treatments followed by the ultrasonic treatment. The lowest area surface roughness Sa parameter is observed after the combined HT1 followed by UIT (Sa = 0.723 µm). The figures point to the surface severe plastic strain with post-UIT using the rotated ultrasonic tool leads to a sliding impact of pins by the sample surface hardened by HT. In comparison with the L-PBF-built specimens, the maximum profile height Sz parameter was decreased by 75.80% and 76.85% after the HT1+UIT and HT2+UIP post-processing, respectively. It is also important to point out that the observed area surface roughness tendencies correlate well with the surface roughness profile Ra parameters presented in Figure 9.17.

9.3.5 Roughness and waviness

The surface roughness is another important parameter affecting the resistance of the L-PBF-fabricated IN 718 alloy parts to fatigue and corrosion behaviour. In comparison with the HIT treatment followed by heat treatments conducted under vacuum, which have a minor impact on the surface roughness profile parameters (Figure 9.17), the SP and UIT surface treatments were observed to markedly change it similar to other severe plastic deformation methods [40, 61–65]. This is in good agreement with the

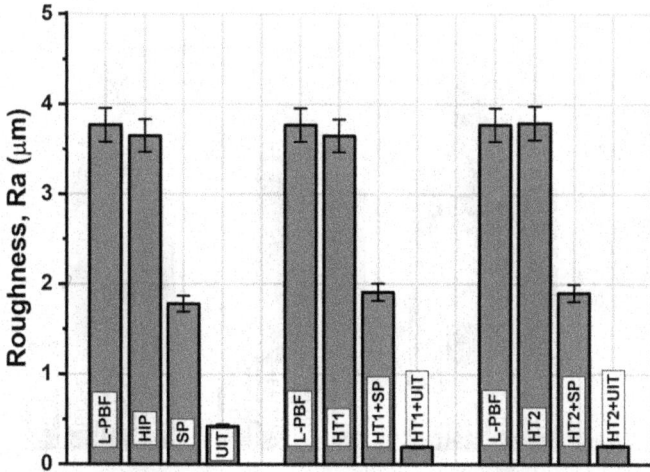

Figure 9.17 Surface roughness of the L-PBF-fabricated IN 718 alloy parts under different post-processing conditions.

findings of several studies [62, 63, 68] that estimated the effects of different peening post-treatments after heat treatment of the additive manufactured metal parts. The surface roughness of the L-PBF-built parts was decreased by double after SP, HT1+SP, and HT2+SP post-processing (Figure 9.17). This result also corresponds to the conclusions of literature data, where the SP treatment decreased the surface roughness Ra parameter in the L-PBF-fabricated superalloys by 40–50% [63].

The surface roughness was remarkably decreased after the application of the UIP treatment both pre-HT (Ra ~0.4 μm) and post-HT (Ra ~0.2 μm) (Figure 9.17). In comparison with the L-PBF-fabricated IN 718 alloy specimen, the surface roughness Ra parameter was decreased by about 53%, 49.5%, 49.5%, 89%, 95%, and 94.5% after the SP, HT1+SP, HT2+SP, UIT, HT1+UIT, and HT2+UIP post-processing, respectively.

The surface waviness was slightly increased after the ultrasonic treatment in comparison with the L-PBF-fabricated and HIP-treated specimens (Wa = ~1.65 μm) (Figure 9.18). The application of UIT after HT1/HT2 resulted in a significant reduction of the surface waviness Wa parameter (~0.6 μm) due to the treated higher surface hardness. In contrast to the HT1/HT2 followed by UIT treatment, the SP treatment caused an increment of the surface waviness Wa parameter in comparison with the as-fabricated and heat-treated specimens (Figure 9.18).

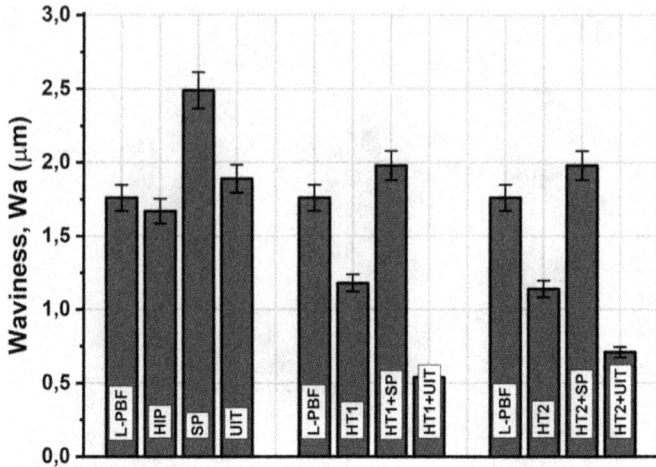

Figure 9.18 Surface waviness of the L-PBF-fabricated IN 718 alloy parts under different post-processing conditions.

9.4 CONCLUSIONS

The plane test IN 718 superalloy parts were fabricated by the L-PBF technology. This work evaluates the effect of the combination of heat treatment (hot isostatic pressing followed by homogenization and ageing) and mechanical surface treatment (shot peening and ultrasonic impact peening) on the microstructure, phase state, residual stress, hardness, porosity, surface texture, roughness, and waviness of the IN718 alloy parts fabricated by the optimized L-PBF parameters. The results of this work allow drawing the following conclusions:

- Application of the HT1 and HT2 thermal post-processing leads to the formation of homogeneous microstructure with the increased volume fractions of fine γ'/γ'' precipitates and NbC/TiC carbides providing significant precipitation strengthening and preventing grains from overgrowth. Homogenization used in HT2 results in a larger grain size. Considering very similar phase compositions and hardness magnitudes obtained and aiming to decrease the overall processing consumables the high enough strength and relatively low time/energy consumables can be supported by avoiding the homogenization treatment.
- Mechanical post-processing by the surface finishing using UIT or SP respectively leads to the crystallites refinement down to ~110 nm and ~80 nm, to the increased surface hardness, and formation of the hardened near-surface layer of 180–150 μm thick compressed with residual stresses. The magnitudes of the surface hardness, layer thickness and compressive stress level depend on the processed material state and mechanical energy accumulated in the processed surface layer during the SP/UIT treatments.

- After SP and UIT, the surface hardness HRC_5 magnitude, hardened layer thickness, and residual stresses of the L-PBF-fabricated surface were respectively reached 58 HRC_5 (81% increase) and 48 HRC_5 (50%), 200 μm and 180 μm, and −320 MPa and −230 MPa. These magnitudes become 55 HRC_5 (72%) and 54 HRC_5 (69%), 140 μm and 80 μm, and −395 MPa and −315 MPa for HT1-processed samples, and 54 HRC_5 (69%) and 51 HRC_5 (59%), 120 μm and 80 μm, and −430 MPa and −395 MPa for HT2-processed samples.
- Compared to the subsurface residual porosity of the L-PBF-fabricated parts (0.318%) in the build direction, the average porosity values were respectively increased by ~95.6%, ~85.0%, ~96.6%, ~97.0%, ~83.3%, ~96.0%, and ~96.6%, after the HIP, SP, HT1+SP, HT2+SP, UIT, HT1+UIT, and HT2+UIT post-processing techniques. The SP and UIT process eliminated the surface defects from both as-fabricated and HT-treated specimens.
- Application of the SP and UIT peening treatments after HT1/HT2 heat treatments leads to the formation of the uniformly textured surface, reducing the Sa parameter from ~4 μm to ~1 μm and ~3 μm, respectively. The UIT-textured surface is more regular and flattened (Sz = 11.4 μm after HT1+UIT, Sz = 11.2 μm after HT2+UIT) as compared to the SP (Sz = 56.5 μm after HT1+SP, Sz = 48.5 μm after HT2+SP).
- Surface roughness Ra parameter in the L-PBF-fabricated IN 718 alloy specimen was respectively reduced by ~53%, ~49.5%, ~49.5%, ~89%, ~95%, and ~94.5% after the SP, HT1+SP, HT2+SP, UIT, HT1+UIT, and HT2+UIP post-processing. Unlike the SP/UIT, HT1+SP, and HT2+SP post-processing techniques, the UIT combined with the HT1/HT2 techniques resulted in a significant reduction of the surface waviness (Wa = ~0.6 μm).

Fundings

This work was partially supported by the Polish Agency for Academic Exchange (NAWA) grant within the Stanislaw Ulam programme (Grant Number BPN/ULM/2021/1/00153) and National Academy of Sciences of Ukraine (programme 1230). The authors would like also to thank G.I. Prokopenko (G.V. Kurdyumov Institute for Metal Physics of the National Academy of Sciences of Ukraine, Kyiv, Ukraine) for inspiration the ultrasonic peening experiments assistance, and S. Faust and O. Stamann (Otto von Guericke University Magdeburg, Magdeburg, Germany) for the surface texture measurements.

REFERENCES

[1] Q.B. Nguyen, M.L. Sharon Nai, Z. Zhu, C.-N. Sun, J. Wei, W. Zhou, Characteristics of Inconel powders for powder-bed additive manufacturing, *Eng.* 3 (2017) 695–700. http://dx.doi.org/10.1016/J.ENG.2017.05.012

[2] S. Vock, B. Klöden, A. Kirchner, T. Weißgärber, and B. Kieback, Powders for powder bed fusion: A review, *Prog. Addit. Manuf.* 4 (2019) 383–397. https://doi.org/10.1007/s40964-019-00078-6

[3] C.-H. Su, K. Rodgers, P. Chen, E. Rabenberg, S. Gorti, Design, processing, and assessment of additive manufacturing by laser powder bed fusion: A case study on INCONEL 718 alloy, *J. Alloys Compd.* 902 (2022) 163735. https://doi.org/10.1016/j.jallcom.2022.163735

[4] D. Herzog, V. Seyd, E. Wycisk and C. Emmelmann, Additive manufacturing of metals, *Acta Mater.*, 117 (2016) 371–392. https://doi.org/10.1016/j.actamat.2016.07.019

[5] N. Kladovasilakis, P. Charalampous, K. Tsongas, I. Kostavelis, D. Tzovaras, D. Tzetzis, Influence of selective laser melting additive manufacturing parameters in Inconel 718 superalloy, *Mater.* 15 (2022) 1362. https://doi.org/10.3390/ma15041362

[6] Y. Zhang, L. Wu, X. Guo, S. Kane, Y. Deng, Y.-G. Jung, J.-H. Lee, J. Zhang, Additive manufacturing of metallic materials: A review, *J. Mater. Eng. Perform.*, 27 (2018) 1–13. https://doi.org/10.1007/s11665-017-2747-y

[7] W.J. Sames, F.A. List, S. Pannala, R.R. Dehoff, S.S. Babu, The metallurgy and processing science of metal additive manufacturing, *Int. Mater. Rev.*, 61 (2016) 315–360. https://doi.org/10.1080/09506608.2015.1116649

[8] A. Sinha, B. Swain, A. Behera, P. Mallick, S. Kumar Samal, H.M. Vishwanatha, A. Behera, A review on the processing of aero-turbine blade using 3D print techniques, *J. Manuf. Mater. Process.*, 6 (2022) 16. https://doi.org/10.3390/jmmp6010016

[9] S. Torres-Carrillo, H.R. Siller, C. Vila, C. Lopez, C.A. Rodríguez, Environmental analysis of selective laser melting in the manufacturing of aeronautical turbine blades, *J. Clean. Prod.*, 246 (2020) 119068. https://doi.org/10.1016/j.jclepro.2019.119068

[10] B. Blakey-Milner, P. Gradl, G. Snedden, M. Brooks, J. Pitot, E. Lopez, M. Leary, F. Berto, A. Plessis, Metal additive manufacturing in aerospace: A review, *Mater. Des.*, 209 (2021) 110008. https://doi.org/10.1016/j.matdes.2021.110008

[11] Lou, X., Gandy, D. Advanced manufacturing for nuclear energy, *JOM* 71 (2019) 2834–2836. https://doi.org/10.1007/s11837-019-03607-4

[12] K.A. Small, M.L. Taheri, Role of processing in microstructural evolution in Inconel 625: A comparison of three additive manufacturing techniques, *Integr. Mater. Manuf. Innov.* 52 (2021) 2811–2820. https://doi.org/10.1007/s11661-021-06273-x

[13] S. Sanchez, P. Smith, Z. Xu, G. Gaspard, C.J. Hyde, W.W. Wits, I.A. Ashcroft, H. Chen, A.T. Clare, Powder bed fusion of nickel-based superalloys: A review, *Int. J. Mach. Tools Manuf.*, 165 (2021) 103729. https://doi.org/10.1016/j.ijmachtools.2021.103729

[14] J.A. Gonzalez, J. Mireles, S.W. Stafford, M.A. Perez, C.A. Terrazas, R.B. Wicker, Characterization of Inconel 625 fabricated using powder-bed-based additive manufacturing technologies, *J. Mater. Process. Technol.*, 264 (2019) 200–210. https://doi.org/10.1016/j.jmatprotec.2018.08.031

[15] G. Vastola, W.J. Sin, C.-N. Sun, N. Sridhar, Design guidelines for suppressing distortion and buckling in metallic thin-wall structures built by powder-bed fusion additive manufacturing, *Mater. Des.* 215 (2022) 110489. https://doi.org/10.1016/j.matdes.2022.110489

[16] M. Renderos, A. Torregaray, M. Esther Gutierrez-Orrantia, A. Lamikiz, N. Saintier, F. Girot, Microstructure characterization of recycled IN718 powder and resulting laser clad material, *Mater. Charact.* 134 (2017) 103–113. https://doi.org/10.1016/j.matchar.2017.09.029

[17] M. Scendo, K. Staszewska-Samson, H. Danielewski, Corrosion behavior of Inconel 625 coating produced by laser cladding, *Coatings* 11 (2021) 759. https://doi.org/10.3390/coatings11070759

[18] M. Bambach, I. Sizova, F. Kies, C. Haase, Directed energy deposition of Inconel 718 powder, cold and hot wire using a six-beam direct diode laser set-up, *Addit. Manuf.* 47 (2021) 102269. https://doi.org/10.1016/j.addma.2021.102269

[19] P. Nandwana, A.M. Elliott, D. Siddel, A. Merriman, W.H. Peter, S.S. Babu, Powder bed binder jet 3D printing of Inconel 718: Densification, microstructural evolution and challenges, *Curr. Opin. Solid State Mater. Sci.*, 21 (2017) 207–218. https://doi.org/10.1016/j.cossms.2016.12.002

[20] P.D. Enrique, E. Marzbanrad, Y. Mahmoodkhani, Z. Jiao, E. Toyserkani, N.Y. Zhou, Surface modification of binder-jet additive manufactured Inconel 625 via electrospark deposition, *Surf. Coat. Technol.*, 362 (2019) 141–149. https://doi.org/10.1016/j.surfcoat.2019.01.108

[21] R. Jiang, L. Monteil, K. Kimes, A. Mostafaei, M. Chmielus, Influence of powder type and binder saturation on binder jet 3D–printed and sintered Inconel 625 samples, *Int. J. Adv. Manuf. Technol.* 116 (2021) 3827–3838. https://doi.org/10.1007/s00170-021-07496-3

[22] T.G. Spears and S.A. Gold, In-process sensing in selective laser melting (SLM) additive manufacturing, *Integr. Mater. Manuf. Innov.* 5 (2016) 2. https://doi.org/10.1186/s40192-016-0045-4

[23] B.B. Ravichander, K. Mamidi, V. Rajendran, B. Farhang, A. Ganesh-Ram, M. Hanumantha, N.S. Moghaddam, A. Amerinatanzi, Experimental investigation of laser scan strategy on the microstructure and properties of Inconel 718 parts fabricated by laser powder bed fusion, *Mater. Charact.* 186 (2022) 111765. https://doi.org/10.1016/j.matchar.2022.111765

[24] D.A. Lesyk, O.S. Lymar, V.V. Dzhemelinskyi, Surface characterization of the cobalt-based alloy stents fabricated by 3D laser metal fusion technology, *Lect. Notes Netw. Syst.* 233 (2021) 357–364. https://doi.org/10.1007/978-3-030-75275-0_40

[25] S. Goel, M. Ahlfors, F. Bahbou, S. Joshi, Effect of different post-treatments on the microstructure of EBM-built alloy 718, *J. Mater. Eng. Perform.* 28 (2019) 673–680. https://doi.org/10.1007/s11665-018-3712-0

[26] B. Lim, H. Chen, Z. Chen, N. Haghdadi, X. Liao, S. Primig, S.S. Babu, A.J. Breen, S.P. Ringer, Microstructure-property gradients in Ni-based superalloy (Inconel 738) additively manufactured via electron beam powder bed fusion, *Addit. Manuf.* 46 (2021) 102121. https://doi.org/10.1016/j.addma.2021.102121

[27] D.C. Kong, C.F. Dong, X.Q. Ni, L. Zhang, R.X. Li, X. He, C. Man, X.G. Li, Microstructure and mechanical properties of nickel-based superalloy fabricated by laser powder-bed fusion using recycled powders, *Int. J. Miner. Metall. Mater.*, 28 (2021) 266–278. https://doi.org/10.1007/s12613-020-2147-4

[28] G.H. Cao, T.Y. Sun, C.H. Wang, Xing Li, M. Liu, Z.X. Zhang, P.F. Hu, A.M. Russell, R. Schneider, D. Gerthsen, Z.J. Zhou, C.P. Li, G.F. Chen, Investigations of γ', γ'' and δ precipitates in heat-treated Inconel 718 alloy fabricated by

selective laser melting, *Mater. Charact.*, 136 (2018) 398–406. https://doi.org/10.1016/j.matchar.2018.01.006

[29] D.A. Lesyk, S. Martinez, V.V. Dzhemelinskyi, A. Lamikiz, Additive manufacturing of the superalloy turbine blades by selective laser melting: Surface quality, microstructure and porosity, *Lect. Notes Netw. Syst.* 128 (2020) 267–275. https://doi.org/10.1007/978-3-030-46817-0_30

[30] T. Huynh, A. Mehta, K. Graydon, J. Woo, S. Park, H. Hyer, L. Zhou, D.D. Imholte, N.E. Woolstenhulme, D.M. Wachs, Y. Sohn, Microstructural development in Inconel 718 nickel-based superalloy additively manufactured by laser powder bed fusion, *Metallogr. Microstruct. Anal.* 11 (2022) 88–107. https://doi.org/10.1007/s13632-021-00811-0

[31] L. Zhang, X. Shi, N. Li, L. Zhao, W. Chen, Heterogeneities of microstructure and mechanical properties for inconel 718 strut tensile sample fabricated by selective laser melting, *J. Mater. Res. Technol.* 12 (2021) 2396–2406. https://doi.org/10.1016/j.jmrt.2021.04.029

[32] E.R. Lewis, M.P. Taylor, B. Attard, N. Cruchley, A.P.C. Morrison, M.M. Attallah, S. Cruchley, Microstructural characterisation and high-temperature oxidation of laser powder bed fusion processed Inconel 625, *Mater. Lett.* 311 (2022) 131582. https://doi.org/10.1016/j.matlet.2021.131582

[33] I. Koutiri, E. Pessard, P. Peyre, O. Amlou, T. de Terris, Influence of SLM process parameters on the surface finish, porosity rate and fatigue behavior of as-built Inconel 625 parts, *J. Mater. Process. Technol.* 255 (2018) 536–546. https://hal.archives-ouvertes.fr/hal-01826611

[34] E.E. Covarrubias and M. Eshraghi, Effect of build angle on surface properties of nickel superalloys processed by selective laser melting, *J. Mater. Process. Technol.* 70 (2018) 336–342. https://doi.org/10.1007/s11837-017-2706-y

[35] D.A. Lesyk, S. Martinez, B.N. Mordyuk, V.V. Dzhemelinskyi, A. Lamikiz, G.I. Prokopenko, Post-processing of the Inconel 718 alloy parts fabricated by selective laser melting: Effects of mechanical surface treatments on surface topography, porosity, hardness and residual stress, *Surf. Coat. Technol.* 381 (2020) 125136. https://doi.org/10.1016/j.surfcoat.2019.125136

[36] C. Ye, C. Zhang, J. Zhao, Y. Dong, Effects of post-processing on the surface finish, porosity, residual stresses, and fatigue performance of additive manufactured metals: A review, *J. Mater. Eng. Perform.* 30 (2021) 6407–6425. https://doi.org/10.1007/s11665-021-06021-7

[37] A. Sola and A. Nouri, Microstructural porosity in additive manufacturing: The formation and detection of pores in metal parts fabricated by powder bed fusion, *J. Adv. Manuf. Process.* 1 (2019) e10021. https://doi.org/10.1002/amp2.10021

[38] M.C. Brennan, J.S. Keist, T.A. Palmer, Defects in metal additive manufacturing processes, *J. Mater. Eng. Perform.* 30 (2021) 4808–4818. https://doi.org/10.1007/s11665-021-05919-6

[39] C. Pauzon, T. Mishurova, M. Fischer, J. Ahlström, T. Fritsch, G. Bruno, E. Hryha, Impact of contour scanning and helium-rich process gas on performances of Alloy 718 lattices produced by laser powder bed fusion, *Mater. Des.* 215 (2022) 110501. https://doi.org/10.1016/j.matdes.2022.110501

[40] D.A. Lesyk, S. Martinez, O.O. Pedash, V.V. Dzhemelinskyi, A. Lamikiz, Porosity and surface defects characterization of hot isostatically pressed Inconel 718 alloy turbine blades printed by 3D laser metal fusion technology, *MRS Adv.* 7 (2022) 197–201. https://doi.org/10.1557/s43580-021-00187-x

[41] R. Barros, F.J.G. Silva, R.M. Gouveia, A. Saboori, G. Marchese, S. Biamino, A. Salmi, E. Atzeni, Laser powder bed fusion of Inconel 718: Residual stress analysis before and after heat treatment, *Metals* 4 (2019) 97–107. https://doi.org/10.3390/met9121290

[42] I. Serrano-Munoz, A. Ulbricht, T. Fritsch, T. Mishurova, A. Kromm, M. Hofmann, R.C. Wimpory, A. Evans, G. Bruno, Scanning manufacturing parameters determining the residual stress state in LPBF IN718 small parts, *Adv. Eng. Mater.* 23 (2021) 2100158. https://doi.org/10.1002/adem.202100158

[43] A. Mostafa, I. Picazo Rubio, V. Brailovski, M. Jahazi, M. Medraj, Structure, texture and phases in 3D printed IN718 alloy subjected to homogenization and HIP treatments, *Metals* 7 (2017) 196. https://doi.org/10.3390/met7060196

[44] D.A. Lesyk, S. Martinez, O.O. Pedash, B.N. Mordyuk, V.V. Dzhemelinskyi, A. Lamikiz, Nickel superalloy turbine blade parts printed by laser powder bed fusion: Thermo-mechanical post-processing for enhanced surface integrity and precipitation strengthening, *J. Mater. Eng. Perform.* (2022). https://doi.org/10.1007/s11665-022-06710-x

[45] R. Seede, A. Mostafa, V. Brailovski, M. Jahazi and M. Medraj, Microstructural and microhardness evolution from homogenization and hot isostatic pressing on selective laser melted Inconel 718: Structure, texture, and phases, *Manuf. Mater. Process*, 2 (2018) 30–51. https://doi.org/10.3390/jmmp2020030

[46] W.M. Tucho, P. Cuvillier, A. Sjolyst-Kverneland, V. Hansen, Microstructure and hardness studies of Inconel 718 manufactured by selective laser melting before and after solution heat treatment, *Mater. Sci. Eng. A* 689 (2017), 220–232. https://doi.org/10.1016/j.msea.2017.02.062

[47] Y.L. Kuo, T. Nagahari, and K. Kakehi, The effect of post-processes on the microstructure and creep properties of alloy718 built up by selective laser melting, *Mater.* 11 (2018) 996–1009. https://doi.org/10.3390/ma11060996

[48] E.M. Fayed, M. Saadati, D. Shahriari, V. Brailovski, M. Jahazi and M. Medraj, Optimization of the post-process heat treatment of inconel 718 superalloy fabricated by laser powder bed fusion process, *Metals*, 11 (2021) 144. https://doi.org/10.3390/met11010144

[49] L.M. Roncery, I. Lopez-Galilea, B. Ruttert, D. Burger, P. Wollgramm, G. Eggeler, and W. Theisen, On the effect of hot isostatic pressing on the creep life of a single crystal superalloys, *Adv. Eng. Mater.*, 18 (2016) 1383–1387.

[50] G.E. Bean, T.D. McLouth, D.B. Witkin, S.D. Sitzman, P.M. Adams and R.J. Zaldivar, Build orientation effects on texture and mechanical properties of selective laser melting Inconel 718, *J. Mater. Eng. Perform.*, 28 (2019) 1942–1949. https://doi.org/10.1007/s11665-019-03980-w

[51] J. Xu, H. Brodin, R.L. Peng, V. Luzin, J. Moverare, Effect of heat treatment temperature on the microstructural evolution of CM247LC superalloy by laser powder bed fusion, *Mater. Charact.* 185 (2022) 111742. https://doi.org/10.1016/j.mtcomm.2022.103139

[52] W. Wang, S. Wang, X. Zhang, F. Chen, Y. Xu, Y. Tian, Process parameter optimization for selective laser melting of Inconel 718 superalloy and the effects of subsequent heat treatment on the microstructural evolution and mechanical properties, *J. Manuf. Process.* 64 (2021) 530–543. https://doi.org/10.1016/j.jmapro.2021.02.004

[53] K. Gruber, W. Stopyra, K. Kobiela, B. Madejski, M. Malicki, T. Kurzynowski, Mechanical properties of Inconel 718 additively manufactured by laser powder

bed fusion after industrial high-temperature heat treatment, *J. Manuf. Process.* 73 (2022) 642–659. https://doi.org/10.1016/j.jmapro.2021.11.053

[54] J. Xu, F. Schulz, R.L. Peng, E. Hryha, J. Moverare, Effect of heat treatment on the microstructure characteristics and microhardness of a novel γ′ nickel-based superalloy by laser powder bed fusion, *Result. Mater.* 12 (2021) 100232. https://doi.org/10.1016/j.mtcomm.2022.103139

[55] L.S.B. Ling, Z. Yin, Z. Hu, J.H. Liang, Z.-Y. Wang, J. Wang, B.D. Sun, Effects of the γ″-Ni3Nb phase on mechanical properties of Inconel 718 superalloys with different heat treatments, *Mater.* 13 (2020) 151. https://doi.org/10.3390/ma13010151

[56] R.J. Vikram, A. Singh, S. Suwas, Effect of heat treatment on the modification of microstructure of selective laser melted (SLM) IN718 and its consequences on mechanical behavior, *J. Mater. Res.* 35 (2020) 1949–1962. https://doi.org/10.1557/jmr.2020.129

[57] A. Kreitcberg, V. Brailovski and S. Turenne, Effect of heat treatment and hot isostatic pressing on the microstructure and mechanical properties of Inconel 625 alloy processed by laser powder bed fusion, *Mater. Sci. Eng. A*, 689 (2017) 1–10. https://doi.org/10.1016/j.msea.2017.02.038

[58] Y. Kaynak and E. Tascioglu, Post-processing effects on the surface character-istics of Inconel 718 alloy fabricated by selective laser melting additive man-ufacturing, *Prog. Addit. Manuf.* 5 (2020) 221–234. https://doi.org/10.1007/s40964-019-00099-1

[59] A. Boschetto, L. Bottini, L. Macera, F. Veniali, Post-processing of complex SLM parts by barrel finishing, *Appl. Sci.* 10 (2020) 1382. https://doi.org/10.3390/app10041382

[60] D.A. Lesyk, V.V. Dzhemelinskyi, S. Martinez, B.N. Mordyuk, A. Lamikiz, Surface shot peening post-processing of Inconel 718 alloy parts printed by laser powder bed fusion additive manufacturing, *J. Mater. Eng. Perform.* 30 (2021) 6982–6995. https://doi.org/10.1007/s11665-021-06103-6

[61] D.A. Lesyk, S. Martinez, O.O. Pedash, V.V. Dzhemelinskyi, and B.N. Mordyuk, Combined thermo-mechanical techniques for post-processing of the SLM-printed Ni-Cr-Fe alloy parts, *Lect. Notes Mech. Eng.* (2020) 295–304. https://doi.org/10.1007/978-3-030-50794-7_29

[62] C. Yu, Z. Huang, Z. Zhang, J. Wang, J. Shen, Z. Xu, Effects of sandblasting and HIP on very high cycle fatigue performance of SLM-fabricated IN718 superalloy, *J. Manuf. Process.* 73 (2022) 642–659. https://doi.org/10.1016/j.jmapro.2021.11.053

[63] O.V. Mythreyi, A. Raja, B.K. Nagesha, R. Jayaganthan, Corrosion study of selective laser melted IN718 alloy upon post heat treatment and shot peening, *Metals* 10 (2020) 1562. https://doi.org/10.3390/met10121562

[64] D.T. Ardi, L. Guowei, N. Maharjan, B. Mutiargo, S.H. Leng, R. Srinivasan, Effects of post-processing route on fatigue performance of laser powder bed fusion Inconel 718, *Addit. Manuf.* 36 (2020) 101442. https://doi.org/10.1016/j.addma.2020.101442

[65] H. Soyama, F. Takeo, Effect of various peening methods on the fatigue proper-ties of titanium alloy Ti6Al4V manufactured by direct metal laser sintering and electron beam melting, *Materials* 13 (2020) 2216. https://doi:10.3390/ma13102216

[66] Z. Baicheng, L. Xiaohua, B. Jiaming, G. Junfeng, W. Pan, S. Chen-Nan, N. Muiling, Q. Guojun and W. Jun, Study of selective laser melting (SLM) Inconel 718 part surface improvement by electrochemical polishing, *Mater. Des.*, 116 (2017) 531–537.

[67] D.A. Lesyk, S. Martinez, V.V. Dzhemelinskyi, O. Stamann, B.N. Mordyuk, A. Lamikiz, Surface polishing of laser powder bed fused superalloy components by magnetic post-treatment, *2020 IEEE 10th Int. Conf. Nanomater.: Appl. Propert. NAP-2020* (2020) 02SAMA17-1–02SAMA17-4. https://doi.org/10.1109/NAP51477.2020.9309600

[68] C. Yu, Z. Huang, Z. Zhang, J. Wang, J. Shen, Z. Xu, Effect of laser shock peening on microstructure and mechanical properties of TiC strengthened inconel 625 alloy processed by selective laser melting, *Mater. Sci. Eng. A.* 835 (2022) 142610. https://doi.org/10.1016/j.msea.2022.142610

[69] E. Maleki, O. Unal, M. Gugliano, S. Bagherifard, The effects of shot peening, laser shock peening and ultrasonic nanocrystal surface modification on the fatigue strength of Inconel 718, *Mater. Sci. Eng. A* 810 (2021) 141029. https://doi.org/10.1016/j.msea.2021.141029

[70] M. Munther, A. Tajyar, N. Holtham, L. Hackel, A. Beheshti, K. Davami, An investigation into the mechanistic origin of thermal stability in thermal-microstructural-engineered additively manufactured Inconel 718, *Vacuum* 119 (2022) 110971. https://doi.org/10.1016/j.vacuum.2022.110971

[71] D.A. Lesyk, S. Martinez, B.N. Mordyuk, O.O. Pedash, V.V. Dzhemelinskyi, A. Lamikiz, Ultrasonic surface post-processing of hot isostatic pressed and heat treated superalloy parts manufactured by laser powder bed fusion, *Addit. Manuf. Lett*, 3 (2022) 100063, https://doi.org/10.1016/j.addlet.2022.100063

[72] L. Lemarquis, P.F. Girouxa, H. Maskrot, B. Barki, O. Hercher, P. Castany, Cold-rolling effects on the microstructure properties of 316L stainless steel parts produced by Laser Powder Bed Fusion (LPBF), *J. Mater. Res. Technol.* 21 (2021) 4725–4736. https://doi.org/10.1016/j.matdes.2022.110405

[73] M. Sadeghi, A. Diaz, P. McFadden, E. Sadeghi, Chemical and mechanical post-processing of Alloy 718 built via electron beam-powder bed fusion: Surface texture and corrosion behavior, *Mater. Des.* 214 (2022) 110405. https://doi.org/10.1016/j.matdes.2022.110405

[74] M. Krawczyk, B. Powalka, M. Matuszak, S. Fryska, D. Grzesiak, P. Figiel, The role of the additive manufacturing process parameters in the shaping of the surface geometric structure during micro-milling, *J. Mach. Eng.* 20 (2020) 86–93. https://doi.org/10.36897/jme/119673

[75] D. Grzesiak, A. Terelak-Tymczyna, E. Bachtiak-Radka, K. Filipowicz, Technical and economic implications of the combination of machining and additive manufacturing in the production of metal parts on the example of a disc type element, *Lect. Notes Mech. Eng.* (2020) 128–137. https://doi.org/10.1007/978-3-030-49910-5_12

[76] S.K. Mishra, G. Gomez-Escudero, H. Gonzalez-Barrio, A. Calleja-Ochoa, S. Martinez, M. Barton, L.N. Lopez de Lacalle, Machining-induced characteristics of microstructure-supported LPBF-IN718 curved thin walls, *Procedia CIRP* 108 (2022) 176–181. https://doi.org/10.1016/j.procir.2022.03.031

[77] M.B. Krawczyk, M.A. Krolikowski, D. Grochala, B. Powalka, P. Figiel, S. Wojciechowski, Evaluation of surface topography after face turning of CoCr

alloys fabricated by casting and selective laser melting, *Mater.* 13 (2020) 2448. https://doi.org/10.3390/ma13112448

[78] S. Sendino, M. Gardon, F. Lartategui, S. Martinez, A. Lamikiz, The effect of the laser incidence angle in the surface of L-PBF processed parts, *Coatings* 10 (2020) 1024. https://doi.org/10.3390/coatings10111024

[79] S. Martinez, N. Ortega, D. Celentano, A.J.S. Egea, E. Ukar, A. Lamikiz, Analysis of the part distortions for Inconel 718 SLM: A case study on the NIST test artifact, *Mater.*, 13 (2020) 5087. https://doi.org/10.3390/ma13225087

[80] D.A. Lesyk, S. Martinez, B.N. Mordyuk, V.V. Dzhemelinskyi, A. Lamikiz, Surface finishing of complexly shaped parts fabricated by selective laser melting, *Lect. Notes Mech. Eng.* (2020) 186–195. https://doi.org/10.1007/978-3-030-40724-7_19

[81] D.A. Lesyk, S. Martinez, V.V. Dzhemelinskyi, A. Lamikiz, B.N. Mordyuk, G.I. Prokopenko, Surface microrelief and hardness of laser hardened and ultrasonically peened AISI D2 tool steel, *Surf. Coat. Technol.*, 278 (2015) 108–120. https://doi.org/10.1016/j.surfcoat.2015.07.049

[82] Y. Tian, D. McAllister, H. Colijn, M. Mills, D. Farson, M. Nordin, S. Babu, Rationalization of microstructure heterogeneity in Inconel 718 builds made by the direct laser additive manufacturing process, *Met. Mat. Trans. A*, 45 (2014) 4470–4483. https://doi.org/10.1007/s11661-014-2370-6

[83] J. Schroder, T. Mishurova, T. Fritsch, I. Serrano-Munoz, A. Evans, M. Sprengel, M. Klaus, C. Genzel, J. Schneider, G. Bruno, On the influence of heat treatment on microstructure and mechanical behavior of laser powder bed fused Inconel 718, *Mater. Sci. Eng. A*, 805 (2021) 140555. https://doi.org/10.1016/j.msea.2020.140555

[84] L. Zhou, A. Mehta, B. McWilliams, K. Cho, Y. Sohn, Microstructure, precipitates and mechanical properties of powder bed fused Inconel 718 before and after heat treatment, *J. Mater. Sci. Technol.*, 35 (2019) 1153–1164. https://doi.org/10.1016/j.jmst.2018.12.006

[85] N.C. Ferreri, S.C. Vogel, M. Knezevic, Determining Volume Fractions of γ, γ', γ'', δ, and MC-carbide phases in Inconel 718 as a function of its processing history using an advanced neutron diffraction procedure, *Mater. Sci. Eng. A*, 781 (2020) 139228. https://doi.org/10.1016/j.msea.2020.139228

[86] Y. Zhao, K. Guan, Z. Yang, Z.P. Hu, Z. Qian, H. Wang, Z.Q. Ma, The effect of subsequent heat treatment on the evolution behavior of second phase particles and mechanical properties of the Inconel 718 superalloy manufactured by selective laser melting, *Mater. Sci. Eng. A*, 794 (2020) 139931. https://doi.org/10.1016/j.msea.2020.139931

[87] D. Cai, P. Nie, J. Shan, W. Liu, Y. Gao and M. Yao, Precipitation and residual stress relaxation kinetics in shot-peened Inconel 718, *J. Mater. Eng. Perform.*, 15 (2006) 614–617. https://doi.org/10.1361/105994906X124613

[88] A. Bunsch, J. Kowalska, M. Witkowska, Influence of die forging parameters on the microstructure and phase composition of Inconel 718 alloy, *Arch. Met. Mater.*, 57 (2012) 929–935, https://doi.org/10.2478/v10172-012-0102-8

[89] D.A. Lesyk, B.N. Mordyuk, S. Martinez, M.O. Iefimov, V.V. Dzhemelinskyi, A. Lamikiz, Influence of combined laser heat treatment and ultrasonic impact treatment on microstructure and corrosion behavior of AISI 1045 steel, *Surf. Coat. Technol.*, 401 (2020) 126275, https://doi.org/10.2478/v10172-012-0102-8

[90] B. Zhang, Y. Li, Q. Bai, Defect formation mechanisms in selective laser melting: A review, *Chin. J. Mech. Eng.*, 30 (2017) 515–527, https://doi.org/10.1007/s10033-017-0121-5

[91] G.M. Volpato, U. Tetzlaff, M.C. Fredel, A comprehensive literature review on laser powder bed fusion of Inconel superalloys, *Addit. Manuf.* 55 (2022) 102871. https://doi.org/10.1016/j.addma.2022.102871

[92] W. Tillmann, C. Schaak, J. Nellesen, M. Schaper, M.E. Aydinoz, K.-P. Hoyer, Hot isostatic pressing of IN718 components manufactured by selective laser melting, *Addit. Manuf.*, 13 (2017) 93–102. http://dx.doi.org/10.1016/j.addma.2016.11.006

Chapter 10

Post-processing techniques of additively manufactured Ti-6Al-4V alloy: A complete review on property enhancement

Pankaj Kumar Singh, Santosh Kumar and Govind Kumar Verma

IIT (BHU), Varanasi, India

Pramod Kumar Jain

IIT Roorkee, Roorkee, India

CONTENTS

DOI: 10.1201/9781003288619-10

10.1 INTRODUCTION

Additive manufacturing (AM) is a collection of modern manufacturing techniques that can be used to create highly customized parts for low-volume applications [1]. The term "additive manufacturing" refers to technologies that manufacture physical items by layering materials based on geometrical representations. AM, as opposed to subtractive and formative manufacturing methods, is the process of joining materials to build items from 3D model data, usually layer by layer [2]. AM can create complex geometries without the use of moulds, potentially lowering part counts [3, 4]. Metals including stainless steel [5, 6], titanium [7, 8], cobalt chrome [9], and nickel alloys [10, 11] can be used to benefit the aerospace and healthcare industries [12–15]. While AM may accomplish complex geometry, the manufacturing process is extremely complicated, with a variety of factors influencing the part's qualities in which metal's microstructure is a significant property. There might be a range of microstructural characteristics that affect the mechanical properties of a metal component. Tensile strength and ductility are related to grain size, micro segregation of alloying elements, and phases within the metal [16]. Because the microstructure is generated in-situ during the AM process, it is mainly dependent on the process settings and material employed [9, 16]. The parameters of the process are determined by the metal AM technique employed.

Titanium and its alloys are frequently used as biomedical implants in comparison to other metal alloys due to attractive qualities, considering enhanced mechanical capabilities, good resistance to corrosion, and excellent biocompatibility [14, 15, 17]. The mechanical behaviour of AM parts might differ dramatically from that of conventionally made parts because of their layered microstructure. Furthermore, due to incorrect process parameter selection or process disruptions, AM components might develop a range of flaws. Incomplete wetting and balling effects, for example, are affected by inadequate input energy and result in pores or cavities in SLM sections. When the input energy is insufficient, consecutive scan paths do not blend together effectively, and faults occur through the scan paths. When the input process parameters aren't carefully selected, EBM products often have enormous holes or cavities extending across multiple layers. Gases entrapment during the formation of powders in gas atomization process can also cause smaller spherical pores to form in EBM parts.

During AM process, defects like porosity, cracks, residual stress are commonly produced [18]. These defects have a significant effect on the metallurgical and mechanical properties of the manufactured product. As a result, post-processing techniques are frequently utilized just after fabricating the component by the AM process, to enhance the tensile properties and surface roughness. There are numerous post-processing methods available, such as heat treatment to reduce the stresses induced during fabrication due to high solidification rate of the AM process. Shot peening is commonly used to reduce micro-sized faults, enhance the surface quality, and increase fatigue resistance in AM components [19].

This review article examines the consequences of hot isostatic pressing, shot peening, electropolishing, and abrasive blasting methods on the Additively Manufactured Ti-6Al-4V alloys. Section 10.2 compiles the different metal AM methods that are accessible. Section 10.3 explains the important defects in additively manufactured components, such as porosity formation, crack formation, build orientation effect, and surface roughness issues. The reason for these problems has also been mentioned in Section 10.3 based on the literature review. Different post-processing techniques including heat treatment (HT), shot peening (SP), abrasive flow machining, hot isostatic pressing (HIP), and other surface roughness improvement techniques and their effects on the additively manufactured components are discussed in Section 10.4.

10.2 CHRONOLOGICAL EVOLUTION OF AM TECHNOLOGIES FOR METALS

The metal AM systems may be classified into four groups on the basis of the feedstock of material, energy supply, and build volume, among other factors: (i) Sheet lamination or laminated object manufacturing (LOM); (ii) Powder bed systems; (iii) Powder feed systems or direct energy deposition (DED); and (iv) Wire feed systems [20]. Electron beam melting (EBM) and selective laser melting (SLM) are covered in powder bed systems, while powder feed systems include laser-engineered net shaping (LENS), direct metal deposition (DMD), and wire feed systems include wire arc additive manufacturing (WAAM) [21].

10.2.1 Sheet lamination or laminated object manufacturing (LOM)

The LOM method relies on cutting and laminating sheets in a layered fashion. Consecutive layers are precisely cut with either a laser beam (as shown in Figure 10.1) or a mechanical cutter and then bonded together

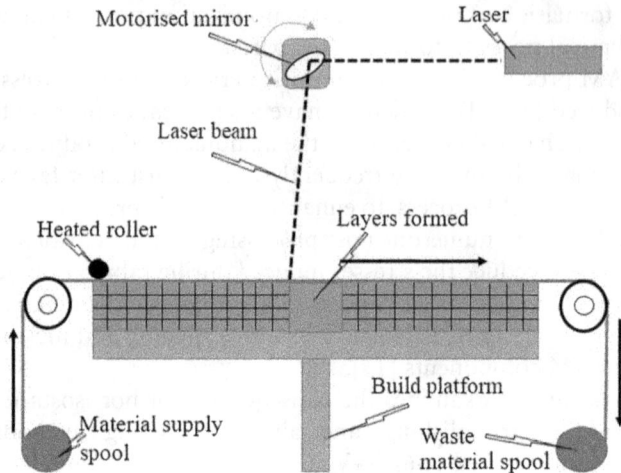

Figure 10.1 Schematic representation of laminated object manufacturing (LOM) process [28].

(a technique known as form-then-bond) or vice versa (i.e. called the bond-then-form technique) [22]. Thermal bonding of metallic and ceramic materials is accomplished using the form-then-bond technique. This technique can also be used to create internal features by removing extra materials before bonding. The extra materials can be used as a support structure, and once completed, the support can be removed [23–25]. Various materials such as paper sheets, ceramics, metallic tapes, and polymers could be used in this process, and post-processing techniques are also required depending on the different materials and required properties. This process has been used in a variety of industries, including foundry, paper, sensor, and processor manufacturing. Based on the LOM principle, ultrasonic AM is also on the rise, combining CNC milling with metal seam welding for the lamination process [26]. LOM has the distinct advantage of producing large structures while having low dimensional accuracy when compared to other AM processes. As a result, this process is not used for complex shapes or where precision is required [27].

10.2.2 Powder bed fusion systems

Powder bed fusion is employed in two common AM methods: selective laser melting (SLM) [29] and electron beam melting (EBM) [30]. A metallic powder is dispersed into a thin layer and melted selectively by a precisely focused laser, fusing it to the previous layer. Part fabrication of customized metals and their alloys, such as maraging steel, Ti-6Al-4V, and Inconel alloy, is done in an inert gas-filled environment. Before proceeding with the actual process, various process parameters should be optimized. The laser process parameters are power, spot size, wavelength, and laser type. Scanning process parameters include hatch spacing, scan strategy, and scan speed. Powder-related parameters include size, bed density, layer thickness, and powder morphology [16]. Unlike the laser beam in the SLM technique, the EBM method uses an electron beam to melt the metallic powder selectively. EBM, unlike SLM, fabricates parts inside a vacuum chamber with the powder surrounding the part kept at a high temperature. Currently, EBM is mostly utilized to manufacture titanium alloy parts [31]. The schematic representation of the selective laser melting method can be seen in Figure 10.2.

The EBM process is an AM method that involves melting metallic powders with an electron beam to obtain the geometry of a specific part. The use of an electron beam in the field of AM is relatively new. The schematic representation of the electron beam melting process can be seen in Figure 10.3.

10.2.3 Powder feed systems or direct energy depositions (DED)

The DED process (also known as LENS) is used to create high-performance alloys. This process is also called as direct metal deposition (DMD) and the electron beam melting (EBM) process [6, 34–37]. As shown in Figure 10.4, a laser or electron beam is concentrated on a small area of the build plate and is utilized to melt feedstock powder material coming from the

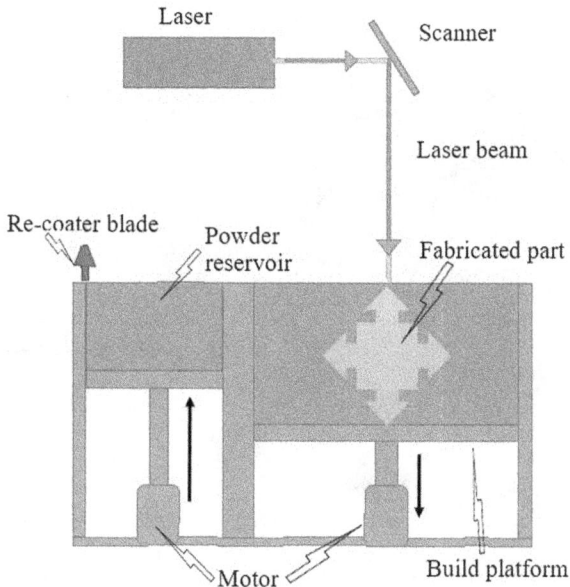

Figure 10.2 Schematic representation of powder bed system [32].

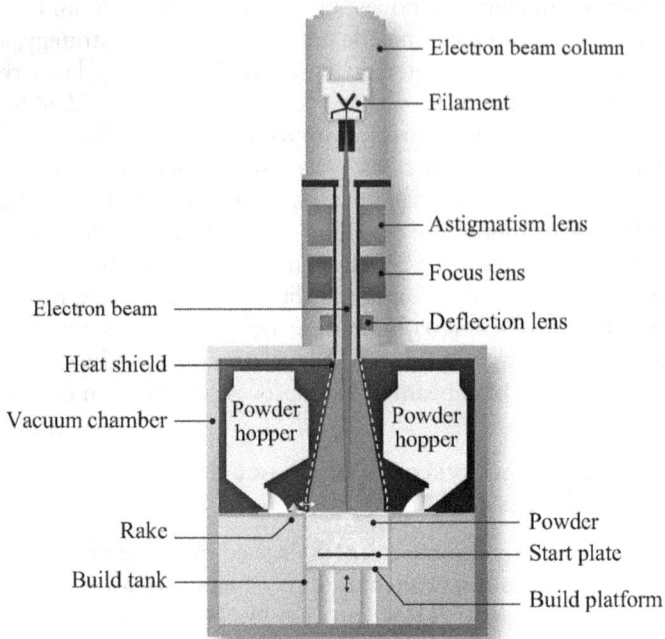

Figure 10.3 Schematic representation of Arcam electron beam melting process [33]. (Reproduced with permission from Elsevier.)

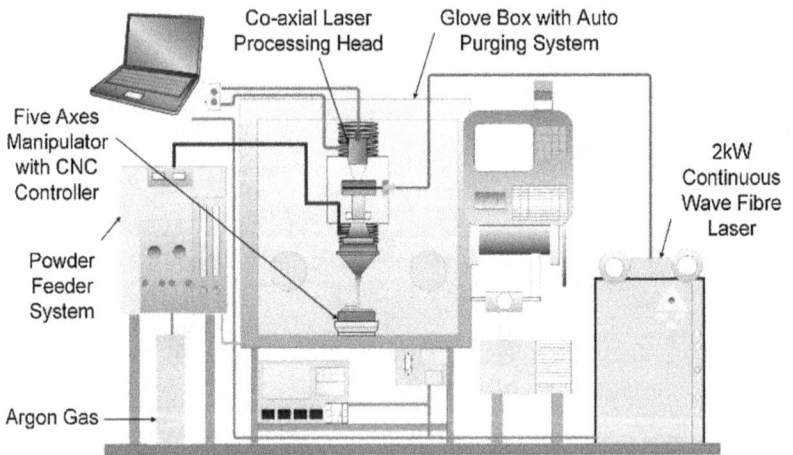

Figure 10.4 Schematic representation of direct energy deposition process [43]. (Reproduced with permission from Elsevier.)

deposition head at the same time in a powder feed system. After the laser beam is moved, the molten material is placed and fused on the melting substrate, and the part solidifies [38, 39]. There is no powder bed in this process, and similar to the fused deposition modelling process, the powder particles are melted with a lot of energy before deposition in a layered fashion. As a result, it is used for repairing, crack filling, and retrofitting manufactured components where SLM cannot be used. This method is extremely useful for creating functionally graded materials (FGMs) by combining multiple materials, and it also allows for multiple axis deposition [40–42]. The main advantages of this method are its large work envelope and its high speed. However, when compared to SLM, this process has poor surface quality and dimensional accuracy and cannot produce high-complexity components [37]. As a result, this procedure is appropriate for large components, less complex components, and large-scale repair turbine engines. [36]. The direct metal deposition method can be seen in Figure 10.4.

10.2.4 Wire arc additive manufacturing (WAAM)

In many manufacturing industries, wire arc additive manufacturing (WAAM) has emerged as another technique for fabricating large metal components with high value. By considering the different types of heat source, there are three primary WAAM techniques: Gas Tungsten Arc Welding (GTAW), Gas Metal Arc Welding (GMAW), and Plasma Arc Welding (PAW) [44]. WAAM (Figure 10.5) employs existing welding attachments, feedstock of welding

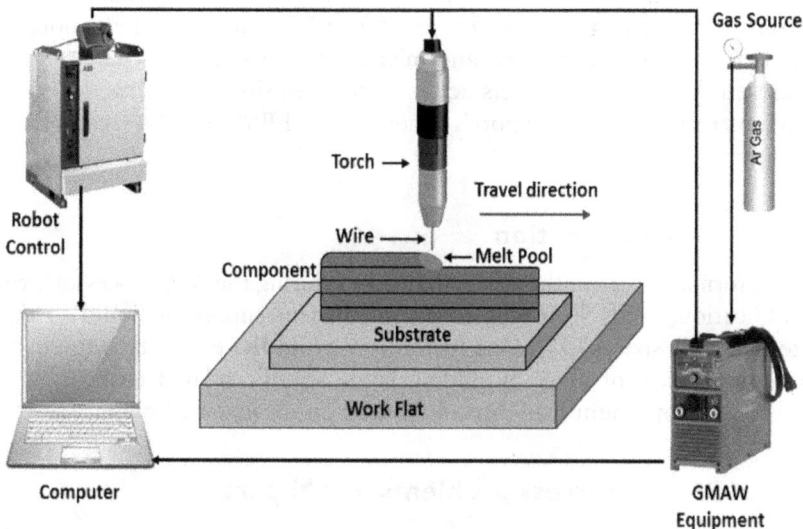

Figure 10.5 Schematic representation of wire arc additive manufacturing process [50]. (Reproduced with permission from Elsevier.)

wire, and electric arc as heat sources. When compared to other AM technologies, this process has a higher material deposition rate [45]. The energy and wire are fed simultaneously through consumable wire in the GMAW process [46]. The orientation of the welding wire feeding affects the accuracy and structure of the fabricated components [47, 48]. The GMAW offers some automation and productivity. The cold metal transfer technique based on the GMAW process has recently gained popularity due to its low heat input and efficient distribution of energy [49].

10.3 DEFECTS IN ADDITIVELY MANUFACTURED METAL COMPONENTS

The defects produced in AM processed samples are pores, anisotropy, residual stress, thermal stress, poor surface finish, cracks, and distortion in shape [51, 52]. The residual stresses were generated during the fabrication and are preserved in a component even after release of applied load [53]. During the AM process, there are numerous reasons for defects' generation, and we have collected a list of flaws and their causes also.

10.3.1 Formation of porosity

In theory, AM processes can produce a fully dense component, but non-optimal process parameters will result in defects. Components fabricated by the AM technique produce uncontrolled pores that will lower the tensile properties. Vilaro et al. [54] explained that the orientation and shape of pore defects have a strong influence on the elongation behaviour of components. The macroscopic experiments and microstructure analysis of Li et al. [55] disclosed that the pore defects act as nucleation sites for microcracks. The SEM microstructure of the pore's generation in EBM Ti-6Al-4V can be seen in Figure 10.6.

10.3.2 Crack formation

Crack formation is another issue that arises during the AM process. During solidification, voids form due to an insufficient supply of liquid to their interdendritic spaces [57]. One method to avoid these defects is to change the composition of alloy by adding some specific related element or by selecting the optimum scan direction based on the given alloy.

10.3.3 Residual stress problems in AM parts

The residual stress is another important defect that is generated due to the large temperature difference during the AM process. According to outcomes in many research papers, the induced residual stress can be eliminated by

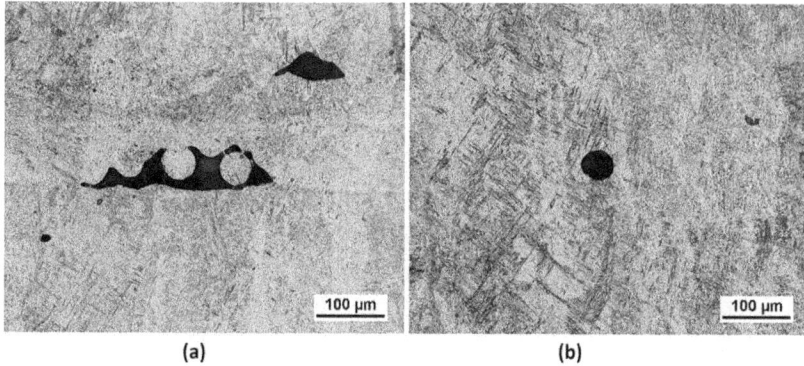

Figure 10.6 Porosity present in electron beam melting of Ti6Al4V: (a) pore due to lack-of-fusion; and (b) gas pore. (Reproduced with permission from Elsevier [56].)

heat treatment at approx. 800° C for 2 hours [54, 58]. Various parameters influence the generation of residual stresses, such as height of the component, build orientation, and conditions of heat treatment [58, 59]. The main reason for the generation of residual stress in as-built samples is non-uniform phase equilibrium and plastic deformation during melting and solidification. When metal powder is melted by the action of a laser, the nearby region of the melting zone generates sudden thermal distortion. When the laser beam is displaced from the region of molten zone, the molten zone solidifies and contracts. Due to the difference in temperature between the melting region and the nearby region, compressive stress is induced in the solidified region, therefore, residual stress is induced in the as-built AM components [53, 60].

10.3.4 Build orientation effect on AM parts

The build orientation effect on the microstructure is another type of defect in AM process. Thijs et al. [61] discovered that the anisotropy of the component could be restrained by selecting the best scanning strategy of the laser. Ren et al. [58] demonstrated that certain combination of process parameters could produce anisotropic properties in as-built AM parts. Consequently, the build orientation and the support structure utilized during the AM process play an important role in the anisotropy of the part.

10.3.5 Surface roughness problem in AM parts

The surface roughness of the majority of components produced by the AM process is poor. This issue could be solved by establishing a link between input process parameters and surface roughness [62]. But enhancing the surface properties could lead to other defects in the manufactured components. The used process parameters related to laser beam have a direct influence on the stability of molten pool and, as a result, the final bead's homogeneity [16]. AM is also prone to balling, which results in a coarse melt pool. The surface roughness of the final component increases due to the low laser scan speed [12]. Section 10.4 of this paper will go with different post-processing techniques in depth.

10.4 DEFECTS AND SIGNIFICANT CHALLENGES OCCURRING SPECIFICALLY IN ADDITIVELY MANUFACTURED TI-6AL-4V COMPONENTS

Due to process parameters (laser power, beam width, layer height, build rate etc.) involved in additive manufacturing of Ti6Al4V, defects like un-melted powder, pores, anisotropy in microstructure, residual stresses, and surface roughness are produced [56, 63]. Defects like un-melted powder and pores are created due to lack of fusion caused by insufficient power source, high built rate, and trapped gas, which usually occur between layers and may be diminished by increasing laser power [64]. Spherical or round pores are developed by trapped gas inside the atomized powder, while irregular pores are produced by un-melted areas as shown in Figure 10.6 [56].

Residual stresses are induced in additive manufactured Ti6Al4V due to high local temperature, low build rate, fast and uneven cooling may be eliminated by thermal treatment [65]. As high temperatures involved to melt metal powder, lead to hotter and softer of upper layer, while lower layer is colder and softer, resulting uneven solidification. Significant thermal gradients start to form around the molten metal pool caused by faster cooling and shrinking of materials on the top layer [66]. Rough surfaces are instigated by staircase effect, positioning errors in laser or beam spots at the counter, inclination angle, balling, and partially melted particles producing surface as well as subsurface roughness caused by improper sintering, and as a result of powder particles adhering to the molten surface. This roughness has an adverse influence on its performance and may lead to additional notch effects and fracture initiation [67].

The primary causes of these defects include variations of feedstock degradation in addition to processing parameters like laser/beam power, slice height, or scan speed etc. Attempts were made to improve additive manufacturing process parameters in order to get rid of cracks, pores, and the absence of un-melted powders. The holes and lack of un-melted particles, especially the surface holes, may barely be completely eradicated, although the cracks can be efficiently repressed. By causing plastic deformation like annealing, hot isotopic pressing etc. during subsequent processing, pores in metallic materials may be effectively closed [68]. To improve the internal pores and to eliminate residual stress, heat treatment is conducted; however, residual stresses are not necessarily bad. For instance, materials are frequently rapidly chilled to impart compressive stress into the plate's surface area, enhancing the overall loading resistance and halting the development of surface cracks [69]. Therefore, AM has a challenge to perform heat treatment for stress. At high laser power, metal powders are melted immediately, along with the occurrence of liquid splashing. This splashing accumulates layer by layer at different locations and creates an uneven layering of powder, resulting in a rough surface and minor spheroidization defects [64].

Poor surface finish is inherent feature in additive manufacturing which may be minimized by finer layer thickness, smaller laser spot, and by optimizing process parameters, but cannot eliminated completely, resulting machining, blasting and polishing are required to improve the surface quality [63, 66]. Major challenges in AM include the staircase effect and subsurface defects, which are almost impossible to eliminate.

10.5 POST-PROCESSING TECHNIQUES FOR ADDITIVELY MANUFACTURED TI-6AL-4V COMPONENTS

The components fabricated by AM processes have poor surface quality due to the use of large-sized powder particles (30–120 μm). However, ideally it will be beneficial to use the small-sized powder particles (equal to 20 μm or less) in order to have good surface characteristics after fabricating the components from AM processes [29]. But small-sized powder particles influence the flowability of the powder and fail to spread the powder uniformly on the substrate, hence it is not desirable [70]. In addition to this, the production of small-sized powder particles will rise the fabrication cost of the components. The quality of the surface in the case of the EBM method is also critical [71]. This reflects that the requirement for the quality of the surface plays a crucial role in considering the different post-processing techniques.

10.5.1 Heat treatment process

The microstructure of Ti–6Al–4V components manufactured using an AM process is typically composed of an acicular α' martensitic microstructure with low ductility. The microstructure produced by the AM process is neither equivalent to wrought nor cast part microstructures. Thus, post-heat treatment can help to increase ductility by reducing the strength of components. The heat treatment at 650°C for 1 h resulted in no microstructural changes, but heat treatment at the same temperature for 2 h resulted in fine β phase precipitation along the boundaries of the needle type α' phase [72, 73]. Heat-treated samples at 800°C for 2 hours underwent complete martensitic decomposition while retaining their refined microstructure, increasing ductility with a minor decrease in tensile strength, and microhardness [54, 74]. The influence of HT on the tensile properties and hardness of AM Ti-6Al-4V components has been tabulated in Table 10.1.

Table 10.1 Effect of heat treatment on tensile properties and hardness of AM Ti-6Al-4V components

Sample	Hardness (HV)	Yield strength (MPa)	Ultimate tensile strength (MPa)	Elongation (%)	References
DMLS as-built	–	1070	1155	4.1	[75]
DMLS after HT		948	1008	8.4	[75]
DED as-built	–	1105±19	1163±22	4±1	[76, 77]
DED after HT	–	907	956	10.8	[78]
SLM as-built	430±10	1002.15	1023.9	11.5±1.5	[17, 58, 79, 80]
SLM after HT	340±15	743±19	964±56	24.3±0.7	[17, 74, 80]
SLM according to ASTM F2924	–	825	895	6–10	[81]
EBM as-built		982.9±5.7	1029.7±7	12.2±0.8	[62]
EBM after HT		778–943	885–1015	3–9	[82]
Wrought	370±30	849±1	934±1	16±1	[83]

From Table 10.1, it can be seen the results of the mechanical properties of AM components before heat treatment, which showed high strength but low elongation at fracture. IT can be seen from this that these components should not be used in high-loaded conditions. After HT, elongation at fracture of the components increased significantly, while strength decreased. Phase composition changes from α' to α and β phases, i.e. transformation of martensitic α' phase to α and β phases, which increases ductility. Components obtained after HT could be used in high-loaded parts because of their high ductility with a decent level of strength.

10.5.2 Hot isostatic pressing (HIP) process

The tensile strength of AM Ti–6Al–4V parts is ranging from 900 to 1100 MPa and elongation at fracture is up to 22% [84, 85]. The large variation of process parameters generated by the various energy sources produces flaws, such as residual stress, pores, and micro-sized cracks, that would generate in between the successive layers during the manufacturing process. These defects have a huge influence on the final product's microstructure and mechanical properties [86, 87]. Therefore, post-processing techniques are mandatory just after components are made to improve mechanical properties and the quality of the surface [88].

Hot Isotopic Processing (HIP) is a typical thermomechanical process which uses extreme temperatures, ranging from 900°C to 2,000°C with an executing pressure of up to 200 MPa. With high pressure and temperature, the created components are equally pressed from all sides. As a result, the components are produced with a high density, exceptional homogeneity, and outstanding performance, as shown in Figure 10.7. The HIP has the ability to eliminate internal defects like pores in powder bed fusion parts [89, 90]. HIP has been proven in several studies to significantly improve the fatigue properties of AM Ti-6Al-4V produced by the EBM process. This advancement is attributable to the material's reduced crack initiation locations. HIP procedures have improved the mechanical properties of EBM products. By eliminating the pores and un-melted powders from the as-built component during the HIP process, the AM can attain good mechanical properties as well [91].

The HIP process of AM Ti-6Al-4V alloys is usually carried out below the β-transus at temperatures between 900°C and 950°C at 100 MPa within an inert gas atmosphere, where the $\alpha + \beta$ phase develops, leading to an increase in the α-lath thickness [92, 93].

10.5.3 Shot peening process

Shot peening is a mechanical surface treatment post-processing method for metallic components. In this process, compressive residual stresses (i.e. the generation of plastic deformation) have been induced by striking the surface of a metallic component with small shots (round metallic or ceramic particles) with sufficient force. Several indentations form on the surfaces of components when a series of shots collide, enveloping the part in a compressive stress layer on the surface of the metal without any material removal [94]. Repeated shots produce uniform compressive residual stress, which

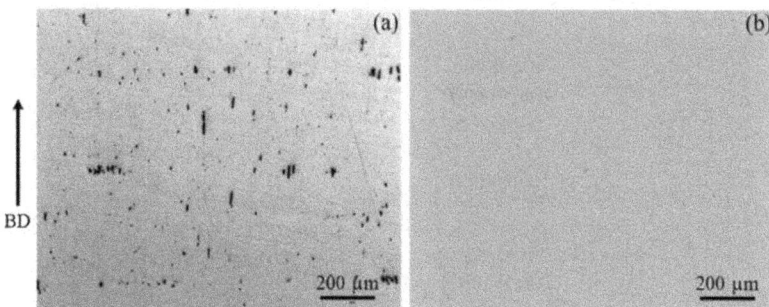

Figure 10.7 Optical micrographs of (a) the as-built material and (b) the Hot Isostatic Processed material displaying defects. The arrow points in the direction of build [89]. (Reproduced with permission from Elsevier.)

improves the surface hardness, mechanical properties, and fatigue life of the sample. Figure 10.8 shows a schematic representation of an ultrasonic shot peening process instrument.

The main advantage of shot peening is that it improves the fatigue life by increasing the fatigue resistance of a part by creating an induced compressive stress layer while also assisting in the prevention of crack initiation and propagation. Stress corrosion cracking is decreased by the plastic deformation produced by different types of shot peening processes. Within a material, the tensile stresses are less problematic because cracks on the surface are lese expected to begin from inside the material [65, 95].

The ultrasonic shot peening process has also improved the microhardness and corrosion resistance of additively manufactured Ti-6Al-4V alloy. The sonotrode as shown in Figure 10.8 transmits energy to shots via continuous vibration; the shots then move through the chamber in multiple directions and strike the sample, causing microstructure evolution such as surface layer plastic deformation, grain refining, and work hardening. The shot diameter, number of balls, vibration amplitude, and processing time are thus the main characteristics of ultrasonic shot peening. The shots were chosen from 400 to 500 zirconia or stainless-steel balls with a diameter of 2 to 2.5 mm [96, 97]. Generally, the ultrasonic shot peening durations for the AM Ti-6Al-4V samples were kept between 480 and 1,920 seconds. The microhardness was increased by 25% as compared to the microhardness obtained before ultrasonic shot peeing of the sample [98, 99]. But, the surface roughness of ultrasonic shot peened AM Ti-6Al-4V samples increased

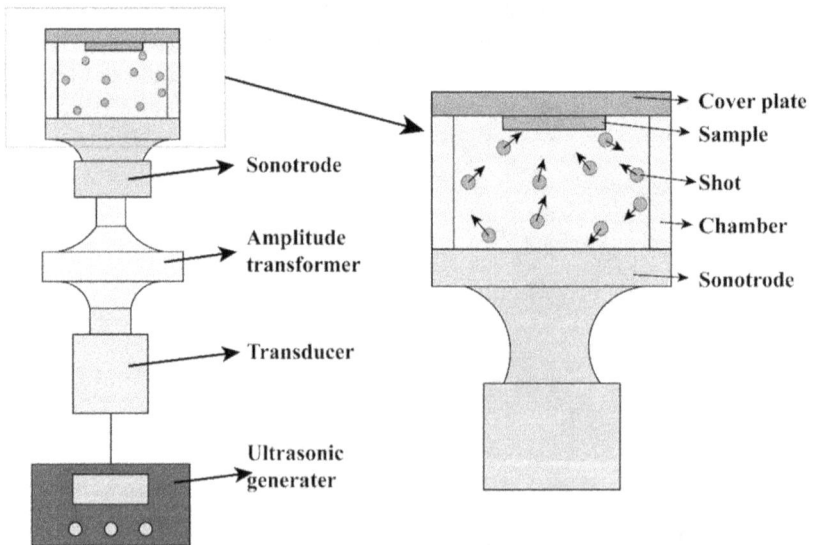

Figure 10.8 Schematic representation of ultrasonic shot peening process instrument [94]. (Reproduced with permission from Elsevier.)

by increasing the duration of ultrasonic shot peening. Proper selection of ultrasonic shot peening parameters can reduce the corrosion current density and increase the corrosion potential, therefore showing improved corrosion resistance [94, 100].

10.5.4 Surface finish methods

10.5.4.1 Polishing

Mechanical polishing is a type of mechanical surface processing that works by removing material from the top surface of the samples. It involves the use of fine abrasive particles to achieve a certain desired surface texture by removing the micron-sized chips from the surface [101]. Abrasive particles are stacked on polishing pads that are rubbed over the workpiece gradually. With the help of different grit sizes of polishing pads, mechanical polishing is used in stages to decrease the surface roughness gradually. After polishing by pads, a cloth pad is used to finish the surface until it is mirror finished [18, 102]. A comparative optical microscopic image and colour map can be seen in Figure 10.9. It can be seen from Figure 10.10 that the surface finish is improved from 15.22 µm to 9.12 µm after mechanical polishing [103]. Again, a comparative 3D surface profiles of as-built and mechanical polished Ti-6Al-4V samples can be seen in Figure 10.10 (c) and (d).

10.5.4.2 Electropolishing

As a part of the post-processing method, electropolishing is used to provide a superior surface finish profile of the as-built Ti-6Al-4V alloy. When compared to other post-processing techniques, electropolishing has the advantage of being able to polish complex designed shapes quickly. Furthermore, despite the chemical and electrical combination in the process, the mass loss on the samples can be controlled [103]. Titanium alloys are recognized for their high oxygen affinity, which results in the formation of a durable oxide coating that is resistant to disintegration. Despite utilizing an electrolyte solution with low water content, uncontrolled oxide film formation is formed during the electropolishing of Ti-6Al-4V alloy. A laboratory power supply is used for the electropolishing of Ti-6Al–4V. The anode is built from a Ti–6Al–4V specimen, and the cathode is made of a stainless-steel plate. The purpose of electropolishing is to improve the surface finish of as-built additively manufactured samples [103, 104]. Generally, the electrolytic polishing condition is determined after a significant number of trials. Using a roughness measuring tool, the surface roughness is assessed before and after electropolishing [105, 106]. The lowest surface roughness is measured as 9.52 µm [103]. The final SEM micrographs and 3D surface profile of electropolished AM Ti-6Al-4V alloy in electrolytes 1, 2, and 3 can be seen in Figure 10.11.

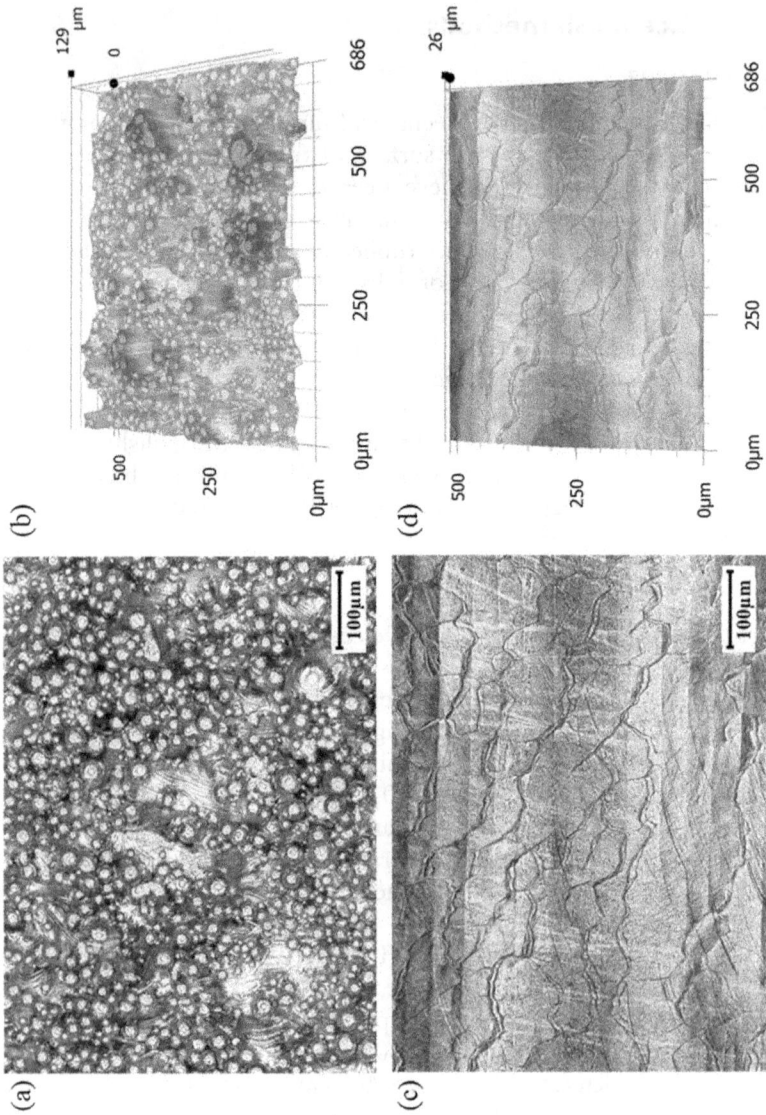

Figure 10.9 (a, b) Optical image and colour map before polishing, and (c, d) optical image and colour map after polishing [102]. (Reproduced with permission from Elsevier.)

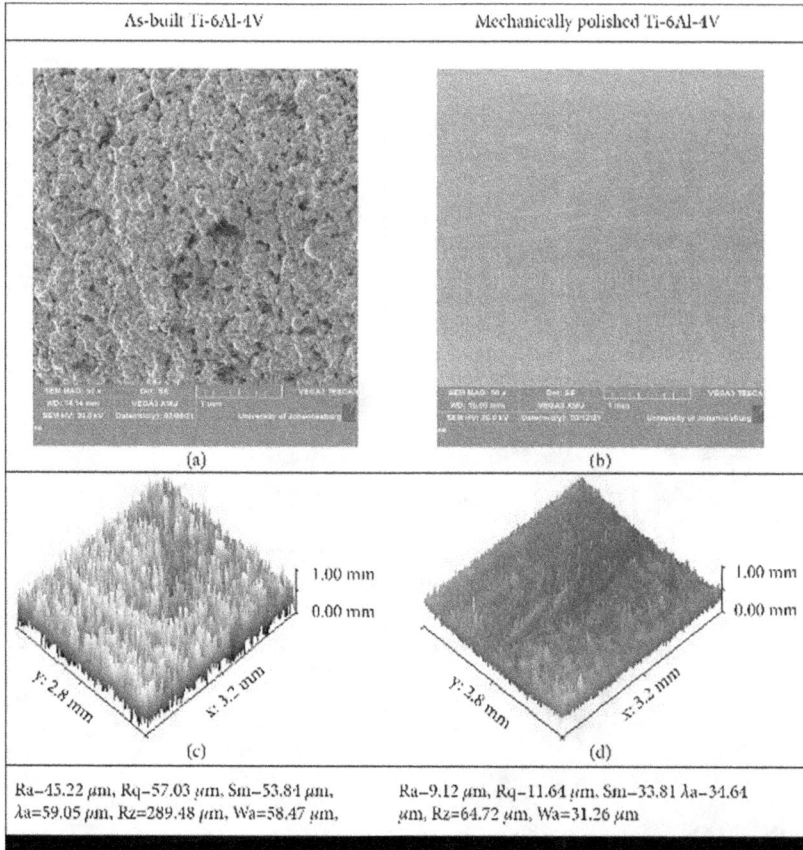

As-built Ti-6Al-4V	Mechanically polished Ti-6Al-4V
(a)	(b)
(c)	(d)
Ra–45.22 μm, Rq–57.03 μm, Sm–53.84 μm, λa=59.05 μm, Rz=289.48 μm, Wa=58.47 μm,	Ra–9.12 μm, Rq–11.64 μm, Sm–33.81 λa–34.64 μm, Rz=64.72 μm, Wa=31.26 μm

Figure 10.10 The SEM images of (a) as-built; (b) mechanically polished; (c) as-built 3D surface profile; and (d) polished 3D surface profile of AM Ti-6Al-4V samples [103]. (Reproduced under CC BY 4.0.)

10.5.4.3 Abrasive flow machining

Abrasive flow machining polishes surfaces and edges by driving a flowable abrasive medium through or across the samples. Abrasion occurs where the flow of media is blocked and other regions are unaffected [107, 108]. It may process many passages on a sample at the same time, reaching locations that are normally inaccessible. This process is advantageous for finishing of difficult to access components [109, 110]. A single fixture may handle dozens of pieces at a time, resulting in output speeds of hundreds of parts per hour [111]. With the help of this process, the surface roughness can be reduced from 53 μm to 2 μm [112, 113]. This process is applied to remove the surface defects produced due to balling defects and the sticking of powder. The comparative analysis of the as-built and final polished surface texture can be seen in Figure 10.12.

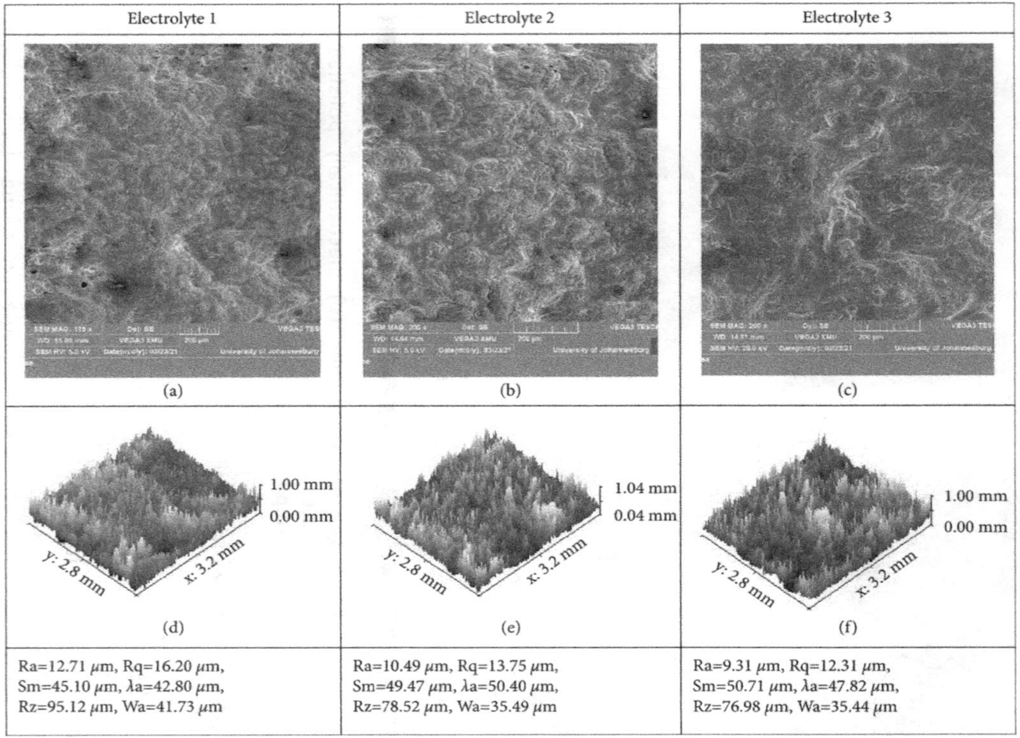

Figure 10.11 The SEM images of electropolished AM Ti-6Al-4V samples in (a) electrolyte 1, (b) electrolyte 2, (c) electrolyte 3, and (d), (e), and (f) 3D surface profiles of same samples in electrolyte 1, 2 and 3 respectively [103]. (Reproduced under CC BY 4.0.)

Figure 10.12 Comparison of the as-built and the finished surface after AFM process [109]. (Reproduced with permission from Elsevier.)

10.5.4.4 Abrasion finishing/grinding

Grinding is an abrasive finishing technique that can provide smooth surfaces and very accurate dimensions on machined components [114]. This process involves grinding wheels bound with abrasive grits as a cutting tool. Grinding is a shear deformation process in which microchips are formed. The grinding process might quickly accomplish the desired workpiece shape and finishing in a single pass [115, 116].

10.5.4.5 Abrasive blasting/bead blasting

Abrasive blasting is a type of mechanical abrasive machining process that involves the use of fine abrasive particles that are accelerated in a gas stream towards the surface that is treated to remove the material and change the texture of the surface of the as-built AM components [117, 118]. The particles slowly remove surface material when they hit the upper surface of the samples. The desired surface texture is achieved by repeating the process until the desired surface texture is reached, as shown in Figure 10.13. When abrasive particles collide on the surface of workpiece they may scatter or collide and produce indentations in ductile and craters in brittle materials. For this process, the impact velocity of abrasives is the important parameter. If this velocity is lower than the critical value, then abrasive particles will rotate only and material removal cannot occur because of insufficient energy [119]. In Figure 10.13 (a) and (b), SEM micrographs of as-built and hot isostatic pressed AM samples have been shown, where unmelted and partially melted powder particles are clearly visible. In Figure 10.13 (c), SEM micrograph of sandblasted samples has been shown, where sandblasting process is used to increase the surface finish of top surface of AM samples. If the impact velocity is high enough, then the abrasives rotate, scatter, and create a crater on the surface of the samples [119]. Other than impact velocity, the feeding medium, geometry, and type of abrasive are also important factors to improve the surface texture and tribological behaviour of the samples. The final micrographs of sandblasted, polished Ti-6Al-4V components can be seen in Figure 10.13 (c) and (d), respectively.

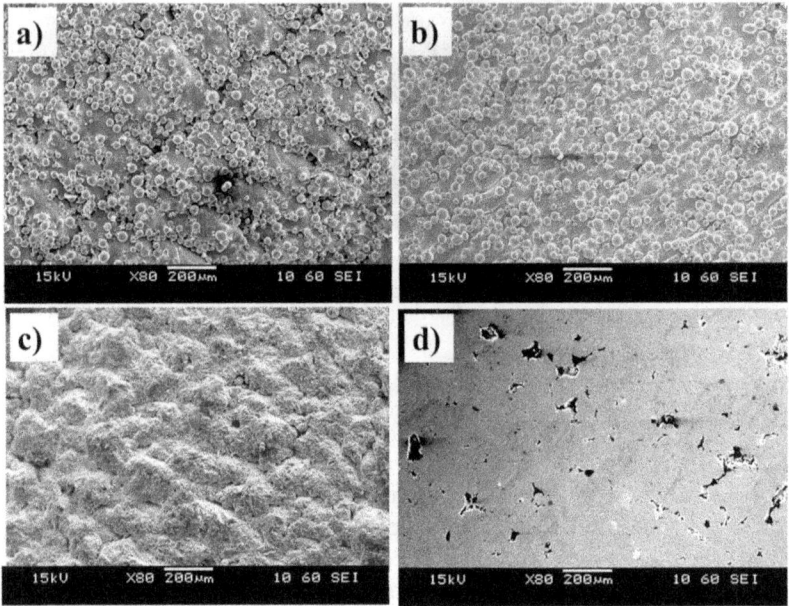

Figure 10.13 SEM micrographs of top surfaces of (a) as-built with samples after post-processing; (b) HIP; (c) sandblasted; and (d) polished AM Ti-6Al-4V components [118]. (Reproduced under MDPI open-access licence.)

In Table 10.2, the values of average surface roughness (R_a) and peak to valley height (R_z) are reported for evaluating the surface quality of the samples. The average surface roughness (R_a) parameter is the most important parameter according to applications [122, 123]. The values of surface roughness vary between different AM processes. In addition to this, it also depends on the metal powder particle size used in different AM processes. Finally, the surface quality of the components depends on the type of

Table 10.2 Summary of surface roughness before surface finishing and after surface finishing roughness parameters in the literature

| Sample | Before surface finishing | | After surface finishing | | |
	R_a (μm)	R_z (μm)	R_a (μm)	R_z (μm)	References
Manual polishing	19.3±2.4	133.9±11.6	1.4±0.5	7.0±2.6	[104, 120]
Abrasive grinding	19.3±2.4	133.9±11.6	0.7–0.9	–	[116, 121]
Abrasive blasting/ bead blasting	6.2	29.5	5.1	27.0	[52, 116]

application they are required for. Small powder particle size is beneficial for mirror finished samples so that the as-built surfaces become less rough before going for applications and achieve the required surface without going for post-processing [116].

10.6 CONCLUSIONS

In this paper, an in-depth exploration of numerous metal AM technologies, various defects, and different post-processing techniques has been carried out. The AM system can be categorized on the basis of the feedstock of powder particles, source of energy, and volume of the build platform. Metal and its alloys are printed utilizing the powder bed fusion process. The DED process melts metal powder by concentrating a high energy beam (laser or electron beam) on a small area of the substrate and using it to melt feedstock powder material coming from the deposition head at the same time. There is no powder bed in this process, and the powder is melted before deposition in a layered fashion, similar to the FDM process, except the powder particles are melted with a lot of energy. The LOM method relies on cutting and laminating sheets in a layered fashion. Consecutive layers are precisely cut with either a laser beam or a mechanical cutter and then bonded together. Finally, the WAAM process is based on an arc-based AM technique that feeds both the energy input and the wire filament at the same time.

There are various defects, such as pores, cracks, effects of build orientation, residual stress, and poor surface roughness, generated in AM components.

Despite the fact that AM processes have numerous advantages over conventional manufacturing processes, they have several significant drawbacks, including poor surface roughness and low mechanical properties. To achieve the required mechanical properties, good surface finish, and dimensional accuracy, nearly all AM components require post-processing. This post-processing includes heat treatment, HIP, support structure removal, component removal from the build plate, surface finishing procedures, and so on. In the AM processes, residual stress is created during processing and is relieved by heating the parts before removing them from the build. The HIP process is used to produce fully dense parts by consolidating metals, as well as to provide microstructure homogeneity by removing porosity, voids, and internal cracks. Surface treatment methods include bead blasting, abrasive grinding, polishing, and abrasive flow machining (used for internal surfaces). Surface roughness is one of the primary drawbacks of the AM process. The following significant conclusions have been drawn from this study:

- Different methods of thermal post-processing are frequently utilized to remove porosity, improve resistance to corrosion, and different tensile properties. These methods include heat treatment (HT) and hot isostatic pressing (HIP). These methods involve refinement of grains and increasing the compactness of deposited layers at high temperature, which reduce the porosity of AM Ti-6Al-4V samples.

- Shot peening is a mechanical surface treatment post-processing method for metallic components. In this process, compressive residual stresses (i.e. the generation of plastic deformation) have been induced by striking the surface of a metallic component with small shots (round metallic or ceramic particles) with enough force. The ultrasonic shot peening process has also improved the resistance to corrosion and microhardness of AM Ti-6Al-4V alloy.
- Mechanical and electropolishing play an important role in controlling surface roughness and microhardness of AM components. Polishing acts on the top of a given surface, increasing its reflectivity. Furthermore, polishing with different grades of emery paper determines the surface quality of a specimen. Polishing has been shown to reduce surface roughness by up to 95% when compared to the as-built specimen. The microhardness value of the polished specimen is maximum at the top surface of the AM samples.

REFERENCES

1. Zhang, L.C. and L.Y. Chen, A review on biomedical titanium alloys: recent progress and prospect. *Advanced Engineering Materials*, 2019. **21**(4): p. 1801215.
2. Sun, C., et al., Additive manufacturing for energy: A review. *Applied Energy*, 2021. **282**: p. 116041.
3. Roopavath, U.K., et al., Optimization of extrusion based ceramic 3D printing process for complex bony designs. *Materials & Design*, 2019. **162**: p. 263–270.
4. Herzog, D., et al., Additive manufacturing of metals. *Acta Materialia*, 2016. **117**: p. 371–392.
5. Tey, C.F., et al., Additive manufacturing of multiple materials by selective laser melting: Ti-alloy to stainless steel via a Cu-alloy interlayer. *Additive Manufacturing*, 2020. **31**: p. 100970.
6. Kim, D.-K., et al., Microstructure and mechanical characteristics of multi-layered materials composed of 316L stainless steel and ferritic steel produced by direct energy deposition. *Journal of Alloys and Compounds*, 2019. **774**: p. 896–907.
7. Baghi, A.D., et al., Experimental realisation of build orientation effects on the mechanical properties of truly as-built Ti-6Al-4V SLM parts. *Journal of Manufacturing Processes*, 2021. **64**: p. 140–152.
8. Barba, D., et al., On the size and orientation effect in additive manufactured Ti-6Al-4V. *Materials & Design*, 2020. **186**: p. 108235.
9. Wang, J.-H., et al., Effect of selective laser melting process parameters on microstructure and properties of co-cr alloy. *Materials*, 2018. **11**(9): p. 1546.

10. Wen, S., et al., Research status and prospect of additive manufactured nickel-titanium shape memory alloys. *Materials*, 2021. 14(16): p. 4496.

11. Dhinakaran, V., et al., Wire Arc Additive Manufacturing (WAAM) process of nickel-based superalloys–A review. *Materials Today: Proceedings*, 2020. 21: p. 920–925.

12. Najmon, J.C., S. Raeisi, and A. Tovar, Review of additive manufacturing technologies and applications in the aerospace industry. *Additive Manufacturing for the Aerospace Industry*, 20191: p. 7–31.

13. Liu, R., et al., Aerospace applications of laser additive manufacturing, in *Laser additive manufacturing*. 2017, Elsevier. p. 351–371.

14. Soro, N., et al., Investigation of the structure and mechanical properties of additively manufactured Ti-6Al-4V biomedical scaffolds designed with a Schwartz primitive unit-cell. *Materials Science and Engineering: A*, 2019. 745: p. 195–202.

15. Wally, Z.J., et al., Selective laser melting processed Ti6Al4V lattices with graded porosities for dental applications. *Journal of the Mechanical Behavior of Biomedical Materials*, 2019. 90: p. 20–29.

16. Khorasani, A., et al., The effect of SLM process parameters on density, hardness, tensile strength and surface quality of Ti-6Al-4V. *Additive Manufacturing*, 2019. 25: p. 176–186.

17. Gupta, S.K., et al., Enhanced biomechanical performance of additively manufactured Ti-6Al-4V bone plates. *Journal of the Mechanical Behavior of Biomedical Materials*, 2021. 119: p. 104552.

18. Mahmood, M.A., et al., Post-processing techniques to enhance the quality of metallic parts produced by additive manufacturing. *Metals*, 2022. 12(1): p. 77.

19. Hung, W., Postprocessing of additively manufactured metal parts. *Journal of Materials Engineering and Performance*, 2021. 30(9): p. 6439–6460.

20. Li, H., et al., The effect of process parameters in fused deposition modelling on bonding degree and mechanical properties. *Rapid Prototyping Journal*, 2018. 24(1): p. 80–92.

21. Duda, T. and L.V. Raghavan, 3D metal printing technology. *IFAC-PapersOnLine*, 2016. 49(29): p. 103–110.

22. Himmer, T., T. Nakagawa, and M. Anzai, Lamination of metal sheets. *Computers in Industry*, 1999. 39(1): p. 27–33.

23. Parandoush, P. and D. Lin, A review on additive manufacturing of polymer-fiber composites. *Composite Structures*, 2017. 182: p. 36–53.

24. Dermeik, B. and N. Travitzky, Laminated object manufacturing of ceramic-based materials. *Advanced Engineering Materials*, 2020. 22(9): p. 2000256.

25. Himmer, T., et al. Metal laminated tooling-A quick and flexible tooling concept. in *International congress on applications of lasers & electro-Optics*. 2004. Laser Institute of America.

26. Yamasaki, H., Applying laminated die to manufacture automobile part in large size. *Die Mould Technol*, 2000. 15(7): p. 36–45.

27. Mekonnen, B.G., G. Bright, and A. Walker, A study on state-of-the-art technology of laminated object manufacturing (LOM), in *CAD/CAM, robotics and factories of the future*, 2016, Springer. p. 207–216.

28. Group, A.M.R Sheet lamination. Available from: https://www.lboro.ac.uk/research/amrg/about/the7categoriesofadditivemanufacturing/sheetlamination/

29. Singh, D.D., T. Mahender, and A.R. Reddy, Powder bed fusion process: A brief review. *Materials Today: Proceedings*, 2021. **46**: p. 350–355.

30. Wang, X. and K. Chou, Effect of support structures on Ti-6Al-4V overhang parts fabricated by powder bed fusion electron beam additive manufacturing. *Journal of Materials Processing Technology*, 2018. **257**: p. 65–78.

31. Zhao, X., et al., Comparison of the microstructures and mechanical properties of Ti–6Al–4V fabricated by selective laser melting and electron beam melting. *Materials & Design*, 2016. **95**: p. 21–31.

32. Group, A.M.R Powder Bed Fusion. Available from: https://www.lboro. ac.uk/research/amrg/about/the7categoriesofadditivemanufacturing/powder bedfusion/

33. Galati, M. and L. Iuliano, A literature review of powder-based electron beam melting focusing on numerical simulations. *Additive Manufacturing*, 2018. **19**: p. 1–20.

34. Froend, M., et al., Thermal analysis of wire-based direct energy deposition of Al-Mg using different laser irradiances. *Additive Manufacturing*, 2019. **29**: p. 100800.

35. Anderson, R., et al., Characteristics of bi-metallic interfaces formed during direct energy deposition additive manufacturing processing. *Metallurgical and Materials Transactions B*, 2019. **50**(4): p. 1921–1930.

36. Oh, W.J., et al., Repairing additive-manufactured 316L stainless steel using direct energy deposition. *Optics & Laser Technology*, 2019. **117**: p. 6–17.

37. Koike, R., et al., Evaluation for mechanical characteristics of Inconel625–SUS316L joint produced with direct energy deposition. *Procedia Manufacturing*, 2017. **14**: p. 105–110.

38. Ian Gibson, I.G., *Additive manufacturing technologies 3D printing, rapid prototyping, and direct digital manufacturing*, 2015, Springer.

39. Edgar, J. and S. Tint, Additive manufacturing technologies: 3D printing, rapid prototyping, and direct digital manufacturing. *Johnson Matthey Technology Review*, 2015. **59**(3): p. 193–198.

40. Dev Singh, D., S. Arjula, and A. Raji Reddy, Functionally graded materials manufactured by direct energy deposition: A review. *Materials Today: Proceedings*, 2021. **47**: p. 2450–2456.

41. Meng, W., et al., Additive manufacturing of a functionally graded material from Inconel625 to Ti6Al4V by laser synchronous preheating. *Journal of Materials Processing Technology*, 2020. **275**: p. 116368.

42. Bobbio, L.D., et al., Additive manufacturing of a functionally graded material from Ti-6Al-4V to Invar: Experimental characterization and thermodynamic calculations. *Acta Materialia*, 2017. **127**: p. 133–142.

43. Jinoop, A., C. Paul, and K. Bindra, Laser assisted direct energy deposition of Hastelloy-X. *Optics & Laser Technology*, 2019. **109**: p. 14–19.

44. Wu, Q., et al., Obtaining uniform deposition with variable wire feeding direction during wire-feed additive manufacturing. *Materials and Manufacturing Processes*, 2017. **32**(16): p. 1881–1886.

45. Derekar, K., A review of wire arc additive manufacturing and advances in wire arc additive manufacturing of aluminium. *Materials science and technology*, 2018. **34**(8): p. 895–916.

46. Ding, J., et al., Thermo-mechanical analysis of Wire and Arc Additive Layer Manufacturing process on large multi-layer parts. *Computational Materials Science*, 2011. **50**(12): p. 3315–3322.

47. Spencer, J., P. Dickens, and C. Wykes, Rapid prototyping of metal parts by three-dimensional welding. *Proceedings of the Institution of Mechanical Engineers, Part B: Journal of Engineering Manufacture*, 1998. **212**(3): p. 175–182.
48. Dickens, P., et al. Rapid prototyping using 3-D welding. in *1992 International Solid Freeform Fabrication Symposium*. 1992.
49. Fang, X., et al., Microstructure evolution and mechanical behavior of 2219 aluminum alloys additively fabricated by the cold metal transfer process. *Materials*, 2018. **11**(5): p. 812.
50. Xia, C., et al., A review on wire arc additive manufacturing: Monitoring, control and a framework of automated system. *Journal of Manufacturing Systems*, 2020. **57**: p. 31–45.
51. Sola, A. and A. Nouri, Microstructural porosity in additive manufacturing: The formation and detection of pores in metal parts fabricated by powder bed fusion. *Journal of Advanced Manufacturing and Processing*, 2019. **1**(3): p. e10021.
52. Rifat, M., et al., Effect of prior surface textures on the resulting roughness and residual stress during bead-blasting of electron beam melted Ti-6Al-4V. *Crystals*, 2022. **12**(3): p. 374.
53. Mukherjee, T., W. Zhang, and T. DebRoy, An improved prediction of residual stresses and distortion in additive manufacturing. *Computational Materials Science*, 2017. **126**: p. 360–372.
54. Vilaro, T., C. Colin, and J.-D. Bartout, As-fabricated and heat-treated microstructures of the Ti-6Al-4V alloy processed by selective laser melting. *Metallurgical and Materials Transactions A*, 2011. **42**(10): p. 3190–3199.
55. Li, S., et al., Additive manufacturing-driven design optimization: Building direction and structural topology. *Additive Manufacturing*, 2020. **36**: p. 101406.
56. Galarraga, H., et al., Effects of the microstructure and porosity on properties of Ti-6Al-4V ELI alloy fabricated by electron beam melting (EBM). *Additive Manufacturing*, 2016. **10**: p. 47–57.
57. Qiu, Y.-D., et al., Balling phenomenon and cracks in alumina ceramics prepared by direct selective laser melting assisted with pressure treatment. *Ceramics International*, 2020. **46**(9): p. 13854–13861.
58. Ren, S., et al., Effect of build orientation on mechanical properties and microstructure of Ti-6Al-4V manufactured by selective laser melting. *Metallurgical and Materials Transactions A*, 2019. **50**(9): p. 4388–4409.
59. Li, Y., et al., Modeling temperature and residual stress fields in selective laser melting. *International Journal of Mechanical Sciences*, 2018. **136**: p. 24–35.
60. Ahmad, B., et al., Residual stress evaluation in selective-laser-melting additively manufactured titanium (Ti-6Al-4V) and inconel 718 using the contour method and numerical simulation. *Additive Manufacturing*, 2018. **22**: p. 571–582.
61. Thijs, L., et al., A study of the microstructural evolution during selective laser melting of Ti–6Al–4V. *Acta Materialia*, 2010. **58**(9): p. 3303–3312.
62. Hrabe, N. and T. Quinn, Effects of processing on microstructure and mechanical properties of a titanium alloy (Ti–6Al–4V) fabricated using electron beam melting (EBM), Part 2: Energy input, orientation, and location. *Materials Science and Engineering: A*, 2013. **573**: p. 271–277.
63. Chan, K.S., Characterization and analysis of surface notches on Ti-alloy plates fabricated by additive manufacturing techniques. Surface *Topography: Metrology and Properties*, 2015. **3**(4): p. 044006.

64. Jin, N., et al., Effects of heat treatment on microstructure and mechanical properties of selective laser melted Ti-6Al-4V lattice materials. *International Journal of Mechanical Sciences*, 2021. **190**: p. 106042.

65. Aguado-Montero, S., et al., Fatigue behaviour of PBF additive manufactured TI6AL4V alloy after shot and laser peening. *International Journal of Fatigue*, 2022. **154**: p. 106536.

66. Vastola, G., et al., Controlling of residual stress in additive manufacturing of Ti6Al4V by finite element modeling. *Additive Manufacturing*, 2016. **12**: p. 231–239.

67. Strano, G., et al., Surface roughness analysis, modelling and prediction in selective laser melting. *Journal of Materials Processing Technology*, 2013. **213**(4): p. 589–597.

68. Yu, H., et al., Fatigue performances of selective laser melted Ti-6Al-4V alloy: Influence of surface finishing, hot isostatic pressing and heat treatments. *International Journal of Fatigue*, 2019. **120**: p. 175–183.

69. Mercelis, P. and J.P. Kruth, Residual stresses in selective laser sintering and selective laser melting. *Rapid Prototyping Journal*, 2006. **12**(5): p. 254–265.

70. Wilkinson, S., et al., A parametric evaluation of powder flowability using a Freeman rheometer through statistical and sensitivity analysis: A discrete element method (DEM) study. *Computers & Chemical Engineering*, 2017. **97**: p. 161–174.

71. Gokuldoss, P.K., S. Kolla, and J. Eckert, Additive manufacturing processes: Selective laser melting, electron beam melting and binder jetting—Selection guidelines. *Materials*, 2017. **10**(6): p. 672.

72. Sallica-Leva, E., et al., Ductility improvement due to martensite α′ decomposition in porous Ti–6Al–4V parts produced by selective laser melting for orthopedic implants. *Journal of the Mechanical Behavior of Biomedical Materials*, 2016. **54**: p. 149–158.

73. Yan, X., et al., Effect of heat treatment on the phase transformation and mechanical properties of Ti6Al4V fabricated by selective laser melting. *Journal of Alloys and Compounds*, 2018. **764**: p. 1056–1071.

74. Wauthle, R., et al., Effects of build orientation and heat treatment on the microstructure and mechanical properties of selective laser melted Ti6Al4V lattice structures. Additive Manufacturing, 2015. 5: p. 77–84.

75. Becker, T.H., M. Beck, and C. Scheffer, Microstructure and mechanical properties of direct metal laser sintered Ti-6Al-4V. *South African Journal of Industrial Engineering*, 2015. **26**(1): p. 1–10.

76. Dinda, G.P., L. Song, and J. Mazumder, Fabrication of Ti-6Al-4V Scaffolds by Direct Metal Deposition. *Metallurgical and Materials Transactions A*, 2008. **39**(12): p. 2914–2922.

77. Zhai, Y., H. Galarraga, and D.A. Lados, Microstructure, static properties, and fatigue crack growth mechanisms in Ti-6Al-4V fabricated by additive manufacturing: LENS and EBM. *Engineering Failure Analysis*, 2016. **69**: p. 3–14.

78. Wang, F., et al., Laser fabrication of Ti6Al4V/TiC composites using simultaneous powder and wire feed. *Materials Science and Engineering: A*, 2007. **445–446**: p. 461–466.

79. Hartunian, P. and M. Eshraghi, Effect of build orientation on the microstructure and mechanical properties of selective laser-melted Ti-6Al-4V alloy. *Journal of Manufacturing and Materials Processing*, 2018. **2**(4): p. 69.

80. Leuders, S., et al., On the mechanical behaviour of titanium alloy TiAl6V4 manufactured by selective laser melting: Fatigue resistance and crack growth performance. *International Journal of Fatigue*, 2013. **48**: p. 300–307.
81. Popovich, A., et al. Microstructure and mechanical properties of Ti-6Al-4V manufactured by SLM. in *Key engineering materials*, 2015. Trans Tech Publ.
82. Hayes, B.J., et al., Predicting tensile properties of Ti-6Al-4V produced via directed energy deposition. *Acta Materialia*, 2017. **133**: p. 120–133.
83. Okazaki, Y. and A. Ishino, Microstructures and mechanical properties of laser-sintered commercially pure Ti and Ti-6Al-4V alloy for dental applications. *Materials*, 2020. **13**(3): p. 609.
84. Baufeld, B., E. Brandl, and O. Van der Biest, Wire based additive layer manufacturing: Comparison of microstructure and mechanical properties of Ti–6Al–4V components fabricated by laser-beam deposition and shaped metal deposition. *Journal of Materials Processing Technology*, 2011. **211**(6): p. 1146–1158.
85. Keist, J.S. and T.A. Palmer, Role of geometry on properties of additively manufactured Ti-6Al-4V structures fabricated using laser based directed energy deposition. *Materials & Design*, 2016. **106**: p. 482–494.
86. Leung, C.L.A., et al., The effect of powder oxidation on defect formation in laser additive manufacturing. *Acta Materialia*, 2019. **166**: p. 294–305.
87. DebRoy, T., HL Wei, JS Zuback, T. Mukherjee, JW Elmer, JO Milewski, AM Beese, A. Wilson-Heid, A. De, and W. Zhang, "Additive manufacturing of metallic components—Process, structure and properties". *Progress in Materials Science*, 2018. **92**: p. 112–224.
88. Varela, J., et al., Performance characterization of laser powder bed fusion fabricated Inconel 718 treated with experimental hot isostatic processing cycles. *Journal of Manufacturing and Materials Processing*, 2020. **4**(3): p. 73.
89. Goel, S., et al., Effect of post-treatments under hot isostatic pressure on microstructural characteristics of EBM-built Alloy 718. *Additive Manufacturing*, 2019. **28**: p. 727–737.
90. Tillmann, W., et al., Hot isostatic pressing of IN718 components manufactured by selective laser melting. *Additive Manufacturing*, 2017. **13**: p. 93–102.
91. Leon, A., et al., The effect of hot isostatic pressure on the corrosion performance of Ti-6Al-4 V produced by an electron-beam melting additive manufacturing process. *Additive Manufacturing*, 2020. **33**: p. 101039.
92. Keist, J., S. Nayir, and T. Palmer, Impact of hot isostatic pressing on the mechanical and microstructural properties of additively manufactured Ti–6Al–4V fabricated using directed energy deposition. *Materials Science and Engineering: A*, 2020. **787**: p. 139454.
93. Atkinson, H. and S. Davies, Fundamental aspects of hot isostatic pressing: an overview. *Metallurgical and Materials Transactions A*, 2000. **31**(12): p. 2981–3000.
94. Zhang, Q., et al., Effect of ultrasonic shot peening on microstructure evolution and corrosion resistance of selective laser melted Tie6Ale4V alloy. *Journal of Materials Research and Technology*, 2021. **11**(1090): p. e1099.
95. Guo, W., et al., Effect of laser shock processing on oxidation resistance of laser additive manufactured Ti6Al4V titanium alloy. *Corrosion Science*, 2020. **170**: p. 108655.
96. Agrawal, R.K., et al., Effect of ultrasonic shot peening duration on microstructure, corrosion behavior and cell response of cp-Ti. *Ultrasonics*, 2020. **104**: p. 106110.

97. Gou, J., et al., Effects of ultrasonic peening treatment in three directions on grain refinement and anisotropy of cold metal transfer additive manufactured Ti-6Al-4V thin wall structure. *Journal of Manufacturing Processes*, 2020. **54**: p. 148–157.

98. Kahlin, M., et al., Improved fatigue strength of additively manufactured Ti6Al4V by surface post processing. *International Journal of Fatigue*, 2020. **134**: p. 105497.

99. Kumar, C.S., et al., Role of ultrasonic shot peening on microstructure, hardness and corrosion resistance of nitrogen stabilised stainless steel without nickel. *Materials Research Express*, 2019. **6**(9): p. 096578.

100. Hu, Y., et al., Effect of multiple laser peening on surface integrity and microstructure of laser additive manufactured Ti6Al4V titanium alloy. *Rapid Prototyping Journal*, 2019. **25**(8): p. 1379–1387.

101. Bagehorn, S., J. Wehr, and H. Maier, Application of mechanical surface finishing processes for roughness reduction and fatigue improvement of additively manufactured Ti-6Al-4V parts. *International Journal of Fatigue*, 2017. **102**: p. 135–142.

102. Lee, S., et al., Laser polishing for improving fatigue performance of additive manufactured Ti-6Al-4V parts. *Optics & Laser Technology*, 2021. **134**: p. 106639.

103. Tsoeunyane, G., et al., Electropolishing of additively manufactured Ti-6Al-4V surfaces in nontoxic electrolyte solution. *Advances in Materials Science and Engineering*, 2022. **2022**: p. 1–12.

104. Bagehorn, S., et al. Surface finishing of additive manufactured Ti-6Al-4V-a comparison of electrochemical and mechanical treatments. in *6th Eur conf aerosp sci*. 2015.

105. Urlea, V. and V. Brailovski, Electropolishing and electropolishing-related allowances for powder bed selectively laser-melted Ti-6Al-4V alloy components. *Journal of Materials Processing Technology*, 2017. **242**: p. 1–11.

106. Fayazfar, H., I. Rishmawi, and M. Vlasea, Electrochemical-based surface enhancement of additively manufactured Ti-6Al-4V complex structures. *Journal of Materials Engineering and Performance*, 2021. **30**(3): p. 2245–2255.

107. Jalui, S., et al. Abrasive Flow Machining of Additively Manufactured Titanium: Thin Walls and Internal Channels. in *2021 International Solid Freeform Fabrication Symposium*. 2021. University of Texas at Austin.

108. Shiyas, K. and R. Ramanujam, A review on post processing techniques of additively manufactured metal parts for improving the material properties. *Materials Today: Proceedings*, 2021. **46**: p. 1429–1436.

109. Peng, C., et al., Study on Improvement of Surface Roughness and Induced Residual Stress for Additively Manufactured Metal Parts by Abrasive Flow Machining. *Procedia CIRP*, 2018. **71**: p. 386–389.

110. Kum, C.W., et al., Prediction and compensation of material removal for abrasive flow machining of additively manufactured metal components. *Journal of Materials Processing Technology*, 2020. **282**: p. 116704.

111. Pahuja, R. and M. Ramulu, Abrasive water jet machining of Titanium (Ti6Al4V)–CFRP stacks–A semi-analytical modeling approach in the prediction of kerf geometry. *Journal of Manufacturing Processes*, 2019. **39**: p. 327–337.

112. Bergmann, C., A. Schmiedel, and E. Uhlmann. Postprocessing of selective laser melting components using abrasive flow machining and cleaning. in *International Additive Manufacturing Symposium* 2013.

113. Uhlmann, E., C. Schmiedel, and J. Wendler, CFD simulation of the abrasive flow machining process. *Procedia Cirp*, 2015. **31**: p. 209–214.

114. Wang, S., *Investigation into the grinding of Titanium Alloys*, 2000, Cranfield University.

115. de Mello, A., et al., Surface grinding of Ti-6Al-4V alloy with SiC abrasive wheel at various cutting conditions. *Procedia Manufacturing*, 2017. **10**: p. 590–600.

116. Fashanu, F.F., D.J. Marcellin-Little, and B.S. Linke. Review of surface finishing of additively manufactured metal implants. in *International manufacturing science and engineering conference*, 2020. American Society of Mechanical Engineers.

117. Sagbas, B., Surface texture properties and tribological behavior of additive manufactured parts. *Tribology and Surface Engineering for Industrial Applications*, 2021. **3**: p. 85.

118. Jamshidi, P., et al., Selective laser melting of Ti-6Al-4V: the impact of post-processing on the tensile, fatigue and biological properties for medical implant applications. *Materials*, 2020. **13**(12): p. 2813.

119. Karoluk, M. and M. Madeja, The Influence of Abrasive Blasting Parameters on Surface Quality of Titanium Alloy Ti-6Al-4V Parts Produced by Electron Beam Melting. *Proceedings Book*, 2019: p. 44.

120. Zaitsev, D., et al., Effect of mechanical polishing of aluminum alloy surfaces on wetting and droplet evaporation at constant and cyclically varying pressure in the chamber. *Journal of Materials Science*, 2021. **56**(36): p. 20154–20168.

121. Anusavice, K. and S. Antonson, Materials and processes for cutting, grinding, finishing and polishing. Phillips' Science of *Dental Materials*. 2014.

122. Ramos, J., et al. Surface roughness enhancement of indirect-SLS metal parts by laser surface polishing. in *2001 International Solid Freeform Fabrication Symposium*. 2001.

123. Balaji, V., D. Ravi, and D.P.N. Chandran, Evaluation of cutting forces, cutting temperature and surface roughness in Cryogenic CO_2 cooling in conventional milling machine using AISI-D2 steel. *Journal of Engineering Technology*, 2018. **6**(2): p. 55–69.

Index

Page numbers in *italics* refer to figures, and those in **bold** to tables.

For Product Safety Concerns and Information please contact our EU
representative GPSR@taylorandfrancis.com
Taylor & Francis Verlag GmbH, Kaufingerstraße 24, 80331 München, Germany

9 781032 265117